Human Perspectives in Health Sciences and Technology

Volume 5

The Human Perspectives in Health Sciences and Technology series publishes volumes that delve into the coevolution between technology, life sciences, and health sciences. The distinctive mark of the series is a focus on the human, as a subject and object of research. The series provides an editorial forum to present both scientists' cutting-edge proposals in health sciences that are able to deeply impact our human biological, emotional and social lives and environments, and thought-provoking theoretical reflections by philosophers and scientists alike on how those scientific achievements affect not only our lives, but also the way we understand and conceptualize how we produce knowledge and advance science, so contributing to refine the image of ourselves as human knowing subjects and active participants in a constantly evolving environment. The series addresses ethical issues in a unique way, i.e. an ethics seen not as an external limitation on science, but as internal to scientific practice itself; as well as an ethics characterized by a positive attitude towards science, trusting the history of science and the resources that, in science, may be promoted in order to orient science itself towards the common good for the future. This is a unique series suitable for an interdisciplinary audience, ranging from philosophers to ethicists, from bio-technologists to epidemiologists as well to public health policy makers.

More information about this series at https://link.springer.com/bookseries/16128

Anne Bremer • Roger Strand
Editors

Precision Oncology and Cancer Biomarkers

Issues at Stake and Matters of Concern

Springer

Editors
Anne Bremer
Centre for Cancer Biomarkers, Centre for the Study of the Sciences and the Humanities
University of Bergen
Bergen, Norway

Roger Strand
Centre for Cancer Biomarkers, Centre for the Study of the Sciences and the Humanities
University of Bergen
Bergen, Norway

ISSN 2661-8915 ISSN 2661-8923 (electronic)
Human Perspectives in Health Sciences and Technology
ISBN 978-3-030-92614-4 ISBN 978-3-030-92612-0 (eBook)
https://doi.org/10.1007/978-3-030-92612-0

This Springer imprint is published by the registered company Springer Nature Switzerland AG
The registered company address is: Gewerbestrasse 11, 6330 Cham, Switzerland

Preface: What Is Responsible Cancer Research?

New possibilities in cancer treatment mean better health but also involve a risk of increased costs and more dilemmas in setting priorities. A critical discussion of the future projections made by the cancer research community may encourage more responsible research and health policies. Responsible cancer research should combine biomedical research activities with critical, scholarly analysis of the same research. At the Centre for Cancer Biomarkers CCBIO, a cancer research centre affiliated to the University of Bergen, we have chosen to integrate analyses based on the social sciences and humanities (SSH) into the scientific activities (Blanchard 2016). The present volume is a collection of these analyses, performed by our SSH scholars and some of their international collaborators, but also by our own biomedical researchers. Indeed, a unique feature of CCBIO is the long-standing and fertile interdisciplinary collaboration between biomedical researchers and SSH scholars.

When CCBIO came into existence in 2013, the buzzword acronym for such activities was 'ELSI/ELSA' – the study of ethical, legal and societal issues/aspects of biomedical research. At its worst, ELSA studies were external exercises with little interaction with the research that it set out to analyse. In our initial vision, ELSA should be integrated with the core activities of the biomedical research itself, and that vision proved to be timely and well conceptualised by what came to be known in Europe as Responsible Research and Innovation (RRI). In this preface we will present the core ideas of RRI and explain how they have supplied both our work and this volume with a general frame of understanding.

The framework for Responsible Research and Innovation (RRI) gained prominence in Europe when it became a cross-cutting principle for the EU's Horizon 2020 research programme. Our main funder, the Research Council of Norway, has published the first version of a Norwegian RRI framework. The Research Council defines responsibility as meaning that 'the processes in the research and innovation system shall increasingly be characterised as anticipatory, reflexive, inclusive and dynamic/flexible' (Norges Forskningsråd 2021, Engineering and Physical Sciences

A previous version of this text was published in Norwegian in Tidsskrift for Den Norske Legeforening 2017; 137: 292–294, https://tidsskriftet.no/2017/02/kronikk/hva-er-ansvarlig-kreftforskning. The journal has kindly given their permission to reuse and modify the piece.

Research Council 2016). It might be tempting to dismiss such characteristics as empty buzzwords. We believe, however, that the RRI framework provides useful concepts for understanding how good decision-making processes can be established in future cancer treatment.

Anticipatory and Reflexive – The Importance of Sociotechnical Imaginaries

RRI thinking links responsibility to the willingness and ability to imagine and reflect critically on possible social consequences of one's own research results. This is based on an insight derived from philosophy of science that the American Society of Clinical Oncology (ASCO) (2012) affirmed in its vision statement *Shaping the Future of Oncology: Envisioning Cancer Care in 2030*: 'By anticipating the future, we can shape it'. In order to make choices, we need notions of the future. In sectors where the forefront of research is rapidly moving, sociotechnical imaginaries play a key role, because the existing evidence base can be expected to become quickly outdated (Jasanoff and Kim 2009).

Sociotechnical imaginaries are defined as collective visions about the good technology and well-functioning society of the future. They postulate society's future needs and challenges as they appear to those who promote these visions, how these can be addressed technologically and the kind of scientific development that will be required to produce the necessary technology. Such ideas may influence not only political decisions, but also big and small choices in the research areas themselves. Research is not a blind walk towards truth. The choices of research foci also influence the direction in which knowledge and technology develop. Those who have the power to formulate sociotechnical imaginaries therefore wield power in society. We will return to this issue below.

Moreover, sociotechnical imaginaries are partly descriptive and partly normative, and always uncertain. When researchers and the institutions that fund research formulate their visions, they engage in what is essentially a creative exercise that not so much predicts the future as helps shape it. Such visions are often characterised by optimism. Those who are directly engaged in the field are at risk of an overly optimistic bias that causes them to overestimate the usefulness of their own research. When the RRI literature calls for reflexivity, it alludes not least to the need for self-critical reflection on our own optimism.

ASCO's vision statement is an example of exaggerated optimism. In their future scenario for personalised cancer therapy in 2030, the linkage between research and treatment is closer and more immediate, and this will change a number of roles in the health services. Specialists in oncology will increasingly act as supervisors that provide quality assurance for treatment interventions that can be delegated to other health personnel. The patients will be better informed and more involved in their own diagnostics and treatment. The requirements for quality will rise even further.

This American vision statement identifies some challenges and problems, primarily a rising cost level combined with increasing uncertainty associated with the business models of the pharmaceutical industry. The document also envisages that a system based on a more active and well-informed patient role may give rise to new forms of inequality, since not all patients are motivated or have the personal resources required to assume such a role. However, these challenges are not discussed in any detail. Instead, it is presumed that increased resources will become available for cancer treatment. It is also presumed that cancer treatment will become precision medicine to an extent that adverse effects will diminish, and that the costs of treatment will not necessarily increase significantly.

This is a familiar motif in sociotechnical imaginaries created by the researchers and innovators themselves. They identify possible advances and problems, ending up with a generally positive vision by anticipating the advances while failing to anticipate the problems. This imparts a bias towards technological optimism and diverts attention from the increasing complexity and cost level involved in technological systems.

Responsible cancer research in the RRI sense involves taking this optimistic bias seriously and teaching the researchers how to exercise self-criticism. In practical terms, this can be implemented in a number of ways (Strand 2014). At the Centre for Cancer Biomarkers, CCBIO, we have chosen to introduce critical perspectives from science and technology studies, philosophy of science and ethics in our research seminars, in addition to training younger researchers through dedicated PhD courses.

Inclusive – The Relationship to the Public

The choice of sociotechnical imaginaries may have considerable political and scientific influence. Helping ensure that these notions are realistic, fair and sustainable is therefore an important social responsibility.

Cancer is associated with especially strong cultural and political notions (Mukherjee 2011). The media have a number of standard narratives about cancer, but very few of them appear to be useful when it comes to understanding the complex associations between disease, science and economics. One such narrative is the technologically optimistic one, which portrays research findings as revolutionary and as the advance that will solve the cancer puzzle. Another narrative is the scandalmongering one, which describes individual patients who have been denied costly treatment and portrays the authorities in a negative light. An interview with representatives of the pharmaceutical industry is frequently included, presenting them as the adversary of the authorities and the patients' friend.

Brekke and Sirnes (2011) describes the emergence of a new type of identity in the Western world, referring to it as 'the hypersomatic individual' – the human whose identity fully and completely consists in its existence as a mortal body, but who refuses to accept this fate. The hypersomatic individual believes that science is potentially omnipotent, and that disease and death are avoidable. These people therefore hold the authorities publicly accountable for their disease and death, since they have failed to provide medical science with sufficient latitude and resources. Brekke and Sirnes highlight an increasing alienation from disease and death. The medical communities need to ask themselves whether they inadvertently contribute to this trend, and the question of how we can establish an informed public debate about the cancer therapies of the future remains unanswered. Several of the chapters in this volume dive into the many aspects of how to deal with disease and death in an informed debate about cancer research and not the least priority-setting of limited resources for cancer treatment.

Responsive – Are There Any Roads from Criticism to Action?

What is described as 'responsive' in the English original is referred to by the Research Council of Norway as 'dynamic and flexible' (Engineering and Physical Sciences Research Council 2016; Norges forskningsråd 2021). The efforts involved in imagining the future, practising self-criticism and including other stakeholders in the discussions should be more than just an intellectual exercise; it should give rise to more reflective choices in terms of health policy, in the implementation of technologies and even in the research process itself.

Translating this requirement into practice is an experiment in its own right. We can provide an example from our own experience. A vital topic for our centre is the relationship between academic research and industrial innovation. Our research focus is biomarkers. In simple terms, biomarkers are molecules or other measurable biological parameters that provide diagnostic, prognostic or predictive information, for example about therapeutic response. In drug-based cancer therapy, many patients draw little benefit, and sometimes even considerable risk or harm, from therapies that nevertheless provide benefits at the group level. Potentially, biomarkers can help pave the way for a future when therapies can be better targeted. Moreover, if a biomarker does not involve major diagnostic costs, it can help improve the therapy without raising costs to the same extent as new drugs.

This, so to speak, is the optimistic side of the coin. On the other hand, it is less easy to draw up a complete sociotechnical imaginary about biomarkers that also promises a viable business model for the pharmaceutical industry. It is a boon for patients to avoid taking drugs that are of no benefit to them, but from the industry's point of view, the sales of each drug will decline. Biomarkers in combination with personalised medicine will result in small patient groups that will undergo the same treatment. The price per patient will therefore rise and conflict with the limits that the authorities will fund through the public purse. In informal conversations with some industry stakeholders, we have therefore seen a lukewarm attitude to biomarkers.

Moreover, there are major knowledge gaps, for example with regard to what makes a biomarker a good biomarker. Kern (2012) points out that only one per cent of all biomarkers that are launched from biomedical research end up being applied clinically. As conscious and reflexive cancer researchers, we therefore need to combine our belief in working for a future where there are more precise therapies that do not accelerate costs, with doubts regarding the realism and economic viability of this vision. This is a challenge, including in terms of motivational psychology. However, we believe that removal of the false security provided by exaggerated optimism about the potential in basic research will spur creative thinking.

Could other principles for payment of drug-based cancer therapies be envisaged (Dutch 2012)? Could we envisage a future when patents and profits play a lesser role, thus making biomarkers less of a concern for the industry's business models? Such questions are relevant for choices in clinical trials – for example whether the researchers primarily seek to test new drugs or whether they would rather attempt new combinations of known drugs. These are difficult considerations not only for individual researchers and research groups, but also for institutions that fund research when designing their programmes.

'Responsible cancer research' as defined by the RRI framework does not make a researcher's life easier, nor more productive, perhaps, when measured in the short term and according to conventional criteria. Nevertheless, given that the complexity in the relationship between science, technology and society has been recognised, the alternative appears problematic, both ethically an intellectually. Indeed, it is our hope that such broader reflections can inspire scientific creativity, for instance, by rethinking the design and choice of endpoints as well as diagnostic and therapeutic models. The Norwegian government recently launched its 2021–2025 Action Plan for clinical trials, in the acknowledgement that Norway in fact has been lagging behind in this area, at least what quantity is concerned (Helse og omsorgsdepartementet 2021). Both the biomedical science and the scholarly developments within our cancer biomarker research centre seem to point towards the need for more personalised and adaptive therapies based on a deeper understanding of cancer as system diseases, as phenomena that play out in real time in the biologically and existentially complex systems that human beings always are.

Bergen, Norway Roger Strand

Bergen, Norway Lars A. Akslen

References

American Society of Clinical Oncology. 2012. *Shaping the Future of Oncology: Envisioning Cancer Care in 2030*. Alexandria, VA.

Blanchard, A. 2016. Mapping ethical and social aspects of cancer biomarkers. *New Biotechnology* 33: 763–772.

Brekke, O.A., and T. Sirnes. 2011. Biosociality, biocitizenship and the new regime of hope and despair: Interpreting «Portraits of Hope» and the «Mehmet Case». *New Genetics and Society* 30: 347–374.

Dutch, T.L. 2012. «no cure no pay» scheme for some new drugs. *PharmaTimes Online*. http://www.pharmatimes.com/news/dutch_no_cure_no_pay_scheme_for_some_new_drugs_976705. Accessed 30 Mar 2021.

Engineering and Physical Sciences Research Council. 2016. *Framework for Responsible Innovation*. London. https://www.epsrc.ac.uk/research/framework/. Accessed 30 Mar 2021.

Helse- og omsorgsdepartementet. 2021. *Nasjonal handlingsplan for kliniske studier 2021–2025*. https://www.regjeringen.no/contentassets/59ffc7b38a4f46fbb062aecae50e272d/207035_kliniske_studier_k6_b.pdf. Accessed 30 Mar 2021.

Jasanoff, S., and S.-H. Kim. 2009. Containing the atom: Sociotechnical imaginaries and nuclear power in the United States and South Korea. *Minerva* 47: 119–146.

Kern, S.E. 2012. Why your new cancer biomarker may never work: Recurrent patterns and remarkable diversity in biomarker failures. *Cancer Research* 72: 6097–6101.

Mukherjee, S. 2011. *The Emperor of All Maladies*. New York.

Norges Forskningsråd. 2021. *A Framework for Responsible Innovation*. https://www.forskningsradet.no/contentassets/558d5b1a9f53421f81371ecf96cf1692/framework-responsible-innovation.pdf. Accessed 30 Mar 2021.

Regulation (EU) No 1291/2013 of the European Parliament and of the Council of 11 December 2013 establishing Horizon 2020. *Official Journal of the European Union Law* 347: 104, 20.12.2013. http://eur-lex.europa.eu/LexUriServ/LexUriServ.do?uri=OJ:L:2013:347:0104:0173:EN:PDF.

Strand, R. 2014. Indicators for promoting and monitoring RRI. *Report from the Expert Group on Policy Indicators for RRI*. Brussels: European Commission Directorate-General for Research and Innovation.

Acknowledgements

We first wish to thank the Centre for Cancer Biomarker (CCBIO), and particularly its director, Lars A. Akslen, for the invaluable opportunity and trust that was given to us to work within CCBIO, and for the enthusiasm and support we received throughout the writing of this anthology.

We are also extremely grateful for and humbled by the participation of the brilliant authors in this anthology: Lars A. Akslen, John Cairns, Dominique Chu, Hanna Dillekås, Stacey D'mello, Caroline Engen, Leonard M. Fleck, Karen Gissum, Line Hillersdal, Jiyeon Kang, Maria Lie Lotsberg, Irmelin Nilsen, Mille Stenmarck, Mette Svendsen, Eirik Tranvåg and Elisabeth Wik. We want to thank them for their patience, as the writing process was lengthened due to the pandemic, for having shared their valuable thought-provoking reflections that have materialised into important chapters, and for their engagement throughout the process – it was a collaborative effort.

Our most sincere thanks furthermore go to the research assistance we have received by Irmelin Nilsen and Maria Skjelbred Meyer, without which we simply would not have been able to complete the book preparation and submission with its myriad of small and larger tasks.

Finally, we thank the Centre for the Study of the Sciences and the Humanities, Bergen, our academic home that always provides us with collegial support and constructive critical feedback.

The work was supported by the University of Bergen and the Research Council of Norway through its Centres of Excellence funding scheme (CCBIO, project number 223250).

Anne Bremer

Roger Strand

Contents

Introduction

Anne Bremer and Roger Strand

With the exception of the COVID-19 pandemic, which has triggered an unprecedented mobilisation of resources and political will, no disease (or rather group of diseases) attracts more attention than cancer. This holds true for many different public spheres, and most certainly in the world of scientific research and technology. Indeed, as the panorama of diseases change with human development, cancer has become increasingly prominent as a cause of death and suffering. For this reason, cancer research, its agendas and trajectories, is an important site for understanding modern societies. What cancer researchers, patients and healthcare workers do, think, fear and desire is not only interesting in its own right but an important part of how our future science, technology and society are conceived, imagined and produced.

This book is the result of close collaborations between researchers and members of the extended network of the Centre for Cancer Biomarkers (CCBIO). CCBIO is a Norwegian centre of excellence located at the University of Bergen, funded for a ten-year period over 2013–2023, which does research on "new cancer biomarkers and targeted therapy, […] how cancer cells are affected by the microenvironment in the tumours, and what significance this has for cancer proliferation and poor prognosis".[1] More precisely, the research at CCBIO is articulated around four overlapping research programmes, that respectively look at: (i) the mechanisms of tumour-microenvironment interactions, looking at how tumour cells interact with the surrounding and supporting microenvironment with different types of cells; (ii) the discovery of cancer biomarkers, aiming at validating different types of

[1] On the website of CCBIO: https://www.uib.no/en/ccbio#

A. Bremer (✉) · R. Strand
Centre for Cancer Biomarkers, Centre for the Study of the Sciences and the Humanities, University of Bergen, Bergen, Norway
e-mail: anne.bremer@uib.no; roger.strand@uib.no

A. Bremer, R. Strand (eds.), *Precision Oncology and Cancer Biomarkers*, Human Perspectives in Health Sciences and Technology 5,
https://doi.org/10.1007/978-3-030-92612-0_1

biomarkers in tissue samples from patients; (iii) the clinical applications and trial studies, through performing clinical trials with associated biomarker studies; and (iv) the questions of health ethics, prioritisation of care, economics and other societal issues pertaining to cancer biomarkers and precision oncology. The CCBIO was funded both on the basis of its potential for excellent research in these fields, and the innovative set up of these research teams across seven departments at the University of Bergen. While most of the CCBIO activity is located at the Department of Clinical Medicine, the Department of Clinical Science, and the Department of Biomedicine, there are also ongoing collaborations with the Department of Informatics, the Department of Economics, the Department of Global Public Health and Primary Care, and last but not least, the Centre for the Study of the Sciences and the Humanities.

Of the 18 co-authors in this book, 14 are affiliated with CCBIO and spread across these various departments, as follows: the editors, Anne Bremer and Roger Strand, are researchers in the fields of Science and Technology Studies and philosophy of science at the Centre for the Study of the Sciences and the Humanities. Four more collaborators and authors are affiliated to the same centre: Irmelin W. Nilsen, Caroline Engen, Mille S. Stenmarck and Karen Gissum. With the exception of Nilsen, who is a media scholar, all three are health professionals and early career researchers who have combined biomedical research with building their own research expertise in STS/philosophy, a demanding combination. Several co-authors are biomedical researchers who, in the course of development of CCBIO, have developed if not an additional research track in STS, philosophy etc, then definitely a strong interest and affinity towards such work, including the CCBIO director Lars A. Akslen, Elisabeth Wik, Hanna Dillekås, Maria Lie Lotsberg and Stacey D'mello Peters. In addition, our interdisciplinary team has included the medical ethicist Eirik Tranvåg at the Centre for Ethics and Priority Setting, Department of Global Public Health and Primary Care and the health economists John Cairns and Jiyeon Kang, health economists respectively working economic evaluation in the field of cancer, who are long-distance affiliates of CCBIO from their home institution at the London School of Hygiene and Tropical Medicine. Beyond CCBIO, the network of co-authors extends to the Center for Ethics, College of Human Medicine, Michigan State University (USA), where Len Fleck, philosopher of medicine and medical ethicist, focuses on just health care rationing and democratic deliberation processes supporting those debates; as well as to the Department of Anthropology, University of Copenhagen, and Centre for Medical Science and Technology Studies, Department of Public Health, University of Copenhagen, where Line Hillersdal and Mette N. Svendsen, anthropologists, share research interests on cancer patients in experimental treatment with personalised medicine. Finally, we have Dominique Chu, computer scientist, complex systems theoretician and philosopher at the School of Computing, University of Kent, which has been collaborating with Roger Strand on critically looking at the limits of models in the life sciences.

To further understand the interdisciplinary collaborations at play between the co-authors of this book, we think it is important to specifically look at the role of the editors within CCBIO. We have been part of the CCBIO research team 'Health ethics, prioritisation and economics' from the beginning. This team, composed of philosophers of science, Science and Technology Studies scholars, health ethicists and health economists, is charged with linking the research on cancer biomarkers that is being done at the centre to the ethical, legal and social aspects and implication of this research; in other words, we have a role as *critical* social science and humanities scholars within CCBIO. Worthy of note, it is through this team 'Health ethics, prioritisation and economics' that we have met and kept ongoing interdisciplinary collaborations with John Cairns, Jiyeon Kang and Eirik Tranvåg, co-authors of this book and part of the team. Particularly, our research interests have converged into exploring how the social, political and economic debates around prioritisation of health care and the medicalisation of society (unfair cut-offs and 'ragged edges', the constitution of new 'bio-communities' of patients, the emerging side-effects of pursuing precision oncology, etc.) are deeply anchored in the complexity of and uncertainties around cancer biomarker research. For the most part, our collaborations were concretely articulated around teaching a common CCBIO PhD course (see below) and the co-supervision by Roger Strand of Eirik Tranvåg's PhD project.

But what is our explicit role and mandate within CCBIO, as formalised in the project proposal? What is our less explicit agenda, that has developed through our experience with working with CCBIO? What are some of the activities we do in CCBIO, and how is all of that received? These considerations form a background from which this anthology was elaborated, and therefore contribute to the reader's apprehension of the book.

The role we take on the CCBIO 'Health ethics, prioritisation and economics' team can be said to be twofold. First, we have the explicit mandate to call attention to the concrete and visible Ethical, Legal and Social Aspects (ELSA) of cancer biomarkers, and look at how these are linked to what happens in the laboratory. For instance, we discuss the challenges of reproducibility and validation in the lab, the complexity of cancer biology and tumour heterogeneity that cannot be grasped even by sophisticated models, and the ethical, legal and social aspects these lead to: questions of how to justly and fairly prioritise health care, nationally and globally, in a context of expensive drugs and limited efficacy, or the complicated alignment between academia, pharmaceutical companies and regulatory agencies, when it comes to getting a scientific discovery to the clinical setting (Blanchard 2016). This part of our role is therefore about making the social and political context of cancer biomarker research more explicit, and integrating awareness of these ELSA-type issues into everyday research practices. This is mainly done through regular informal interaction (even friendship) with cancer researchers at CCBIO over the years, but also laboratory visits, participating in CCBIO meetings and events, ranging from junior researcher meetings, PI meetings and our annual symposia, co-authoring of papers and opinion pieces (for instance between Lars A. Akslen and Roger

Strand[2]), co-supervising students (for instance by Elisabeth Wik and Anne Bremer[3]), as well as a yearly course that we organise for CCBIO PhD candidates (we return to this below).

By way of example, collaborations between CCBIO cancer researchers and the 'health ethics, prioritisation and economics' team led us to a first anthology titled 'Cancer Biomarkers: Ethics, Economics and Society' (Blanchard and Strand 2017). Based on the ongoing collaborations we were having within the 'health ethics' team, and with CCBIO more generally, several co-authors of this book already participated in the first anthology: John Cairns, on the evaluation of targeted cancer therapies, Eirik Tranvåg on the influence of cancer biomarkers on priority settings, Elisabeth Wik, on what is a good biomarker, Caroline Engen, on concepts of good life and health in a context of cancer, and Len Fleck, on ethical ambiguity around cancer biomarkers. That first anthology was rooted in the interdisciplinary collaborations and reflections ongoing at CCBIO, and aimed to provide a map of different ethical, social, political, institutional, economic and existential issues around cancer biomarker research. It began by questioning what a 'good' biomarker might look like in a context of hypes, high hopes and substantial biological complexity, to then explore how the complex terrain of cancer biomarker research is structurally entangled with questions of what a 'good' (just, fair and caring) society is, and what the 'good' life is (with or without cancer). In that sense, this first anthology aimed to *map* the different aspects of this terrain to each other, and to the high levels of complexity and uncertainty that characterise cancer biology; whereas this second anthology is more concerned with *critically scrutinising* the ideal of precision oncology, through actor-centred perspectives – what it really means to pursue ideals of precision oncology for patients, for society at large, for oncology research or for priority-setting institutions, for instance. We come back to the essence and key themes of this anthology below.

The second aspect of our role on the ELSA team in CCBIO is somewhat less explicit, and developed through our experience with working in CCBIO. We quickly realised, by discussing and reflecting with CCBIO researchers, that there was a need to go beyond the immediate issues faced by cancer researchers, to reflect on the underlying endeavour that these researchers are part of. We chose to approach that by a deeper analysis of the sociotechnical imaginaries surrounding cancer biomarker research, notably the imaginary of precision oncology.[4] A sociotechnical imaginary being defined in brief as the "collectively held and performed visions of desirable futures, animated by shared understandings of forms of social life and social order attainable through, and supportive of, advances in science and technol-

[2] See for instance: Strand, R., and Akslen, L. A. 2017. What is responsible cancer research? *Tidsskr Nor Legeforen* 137(4): 292–294.

[3] In 2019, Elisabeth Wik and Anne Bremer co-supervised the research assignment of two students on the topic of uncertainties in the use of biomarkers in breast cancer and monogenic diabetes.

[4] In this book, see the chapter by Bremer, Wik and Akslen: "HER2 revisited - Reflections on the future of cancer biomarker research", and the chapter by Stenmarck and Nilsen: "Precision oncology in the news".

ogy" (Jasanoff 2015), we critically call into question and discuss with CCBIO researchers why the sociotechnical imaginary of precision oncology has been deemed a 'desirable' and 'feasible' future in the first place, and explore what is 'co-produced' – what things mutually emerge – in pursuing this imaginary.

A particularly important way we manage to convey these reflections is via a yearly PhD course that we organise, primarily targeted at CCBIO PhD candidates, and titled: 'Cancer research: Ethical, economic and social aspects'. This course introduces the various 'ELSA'-type issues mentioned above, but also allows for what Åm (2019) calls 'moments of dislocation'. These dislocatory moments occur when one realises that there are differences and discrepancies between the practices one claims to follow ('espoused theories'); and the practices one actually adopts and implements (or 'theories-in-use'), that can be made explicit by studying the individual's actions, views, identities or organisational policies (Argyris et al. 1990). Hesjedal et al. (2020) argue that such dislocatory moments "may trigger learning processes that encompass the revision of mental maps, that is, double-loop learning" (p. 6). Double-loop learning (Argyris and Schön 1974, Schön 1983) distinguishes itself from single-loop learning insofar as reflection on the discrepancies between espoused theories and theories-in-use results in a learning process which entails a revision of one's mental maps and models. It is therefore not about 'simply' learning about ways to incrementally adjust our practices around challenges or problems, like introducing new policies to hire more women in research positions for example. Rather, it is about deeply reflecting on institutionalised practices, values and ontologies, so that everyday practices and theories-in-use can be questioned and potentially revised (Hesjedal et al. 2020); rethinking the gendered aspects of oncology, and rationales and approaches for incorporating gender perspectives to use this example.

This is what we aim for in our CCBIO PhD course, to provide opportunity and support for participants who want to, to experience dislocatory moments as a first step to a double-loop learning. We observed that double-loop learning was triggered by discussions around broad themes, such as the lack of ambivalence and the power of goodness (Loga 2004) that characterise discourses and practices around precision medicine, the resulting framing and overflowing dynamics (Callon 1998), or the importance of sustaining an economy of hope (Rose and Novas 2004) in fuelling the imaginary of precision oncology. These themes are central in this anthology as well, and we come back to them later in the introduction.

Unsurprisingly, we have witnessed tensions between espoused theory and theory-in-use among the course participants. Our course runs over two weeks, with one month in-between where participants proceed with their research work, including their duties in the lab. It was frequent that at the beginning of the second week of our course, participants would raise the discomfort they had experienced when trying to apply their new reflections or insights into their everyday practices. Either they felt 'locked-up' in a tightly-designed project with very little room to manoeuvre, or overwhelmed by the duties in the lab that leave little space for reflection, or again met with resistance from the disciplinary or hierarchical structures of their field.

Reflexivity is part of the researcher's practice, and we all engage in some kind of reflections in the course of our work. But as Schön (1983) argues, "[scientists] seldom reflect on their reflection-in-action" (p. 243), or in other words, they do not often engage in double-loop learning (Hesjedal et al. 2020). The lack of reflexive discourse and practices around the context, status and inherently complex nature of cancer biology is a source of naivety in the field (Strand 2000), and arguably contributes to developing blind spots around important concerns that lie to the side of the main trajectory of precision oncology. However, as we saw above, our invitation to a double-loop learning and a reflexive critique of precision oncology within CCBIO was sometimes justly met by resistance and unease: 'are you against cancer biomarkers?' Our answer is a profound "no" but whenever that remained unclear it was evidence of immature reflection or communication on our side. Indeed, we soon came to realise that we were asking (mostly) early career scientists to carry responsibility for the trajectory of current cancer research – which was unfair from our side. This responsibility is too heavy to be carried by single individuals, or even research groups; Åm et al. (2020) have rightly questioned the way in which scientists are being imagined in certain imaginaries of RRI.

We therefore had to readjust the way we wanted to convey double-loop learning, and as such, we became very explicit in our PhD courses that our invitation to critically reflect on cancer biomarkers and precision oncology aimed at mutual learning and the uncovering of blind spots in those fields: Which other research areas receive less attention because of the focus on biomarkers? What are the scientific, structural, organisational limits of biomarker research? What is the political economy of precision oncology? We think that double-loop learning is crucially important when working within the field of biomarkers, as it highlights the fusion of hope and reality around precision oncology, and helps us realise that the current efforts and resources placed in this endeavour are to a large extent justified by optimistic future imaginaries of precision oncology. It is important to note that this course was key in further consolidating collaborations between several of the co-authors of this book: Anne Bremer, Roger Strand and John Cairns being the main instructors, Elisabeth Wik being a recurrent guest lecturer in the course, and Caroline Engen, Mille Stenmarck, Irmelin Nilsen, Hanna Dillekås, Karen Gissum, Maria Lie Lotsberg being first participants in the course, and presenting their work and reflections in subsequent editions of the course. Holding a course together was an important way to meaningfully discuss each other's visions, assumptions and overlapping research interests. Mutual learning and an interdisciplinary approach have been key to our efforts, as they draw on a multitude of knowledge fields, professions and disciplines. Thus, the authors of this book are medical doctors, pathologists, philosophers, nurses, media researchers, molecular biologists, STS scholars, sociologists, computer scientists, economists and ethicists – individuals frequently belonging to more than one of these categories. Long-term collaborations built on mutual trust developed in real time is another key component. Indeed, as noted above, nine of the authors in our first anthology contribute also to the present volume. In our view, we have enjoyed and sustained a high degree of mutual reflexivity and openness between the biomedical perspective on one hand, and the various SSH (social sciences and humanities) perspectives on the other hand, including the STS tradition that was created with the explicit purpose of providing social critique of science and technology. To us, this is an indication of a growing distance from the polarized past of the "science wars" in which STS scholars and sociologists of science – rightly or unfairly – were accused of relativising scientific knowledge and undermining public trust in science. At least in the Norwegian context, with decades of SSH-STEM collaboration in and around biotechnology and the life sciences, this mostly feels as a distant past while the tensions and conflicts may still be strong in other parts of the world. Our SSH scholars and STEM scientists could all agree with Andrew Pickering's famous claim that high quality scientific knowledge is both objective and relative: It is objective in the sense of being the outcome of well-organized intersubjective practices and processes of experimental work, observation, analysis, peer review and so on. But it is also relative to the problem context where it emerged, in the sense that certain research questions were asked and certain model systems were employed, rather than others. With the science wars well behind us, this insight should not threaten anyone. As explained in the preface, co-authored by CCBIO director Lars Akslen, a pathologist and cancer researcher, and Roger Strand, a professor of philosophy of science, our vision is to employ the critical resources from STS and other SSH disciplines to *improve* cancer research, make it stronger, more relevant and more aligned to the needs and concerns of society. In this way, a conceptual basis can be developed to rigorously identify, describe and discuss the difficult social and *socio-technical* issues that exist within cancer research and cancer care itself, problems for which biomedicine by itself does not provide theories or concepts. SSH, such as the STS, philosophy, ethics, economics and media studies traditions represented in this volume, provide such theories and concepts as well as methods to identify, observe and analyse these issues within and around biomedicine as *phenomena*. This is the essence of the collaboration between the "two cultures": We are all researchers who create knowledge.

In sum, interdisciplinary exchange in an atmosphere of trust gives the opportunity to enter fearlessly into rigorous critique. As mentioned, revealing and critically discussing "blind spots" is central to our approach within CCBIO. It is also central to this anthology, and we have articulated this attempt around three overarching themes: (i) uncomfortable knowledge and lack of ambivalence in the discourses and practices around precision oncology; (ii) dynamics of framing and overflowing, when trying to control biological, social and ethical complexity; and (iii) the role of the economy of hope in legitimising and sustaining the imaginary of precision oncology, and the starch dichotomy between illness and disease it leads to. We will now go through these themes, and present how the various chapters broadly relate to them.

(i) Uncomfortable knowledge and lack of ambivalence

The first overarching theme in this anthology is the all-encompassing vagueness and lack of ambivalence found in discourses and practices around precision medicine. Is precision oncology already here? Is it working? What is it supposed to achieve? We know, as of today, that less than 1% of published cancer biomarkers actually enter clinical practice (see Kern 2021; but the trends mapped almost ten years ago are seen to largely hold true). And we know that 'we have done an about face', from a period where molecular and genetic research gave hope that cancer could be understood through simple and reductionist thinking, to now where we struggle to interpret and make sense of the complex data that is being accumulated by sophisticated imaging and sequencing techniques (Weinberg 2014). Kern and Weinberg's observations are in the domain of 'uncomfortable knowledge': they undermine the legitimacy of the imaginary of precision oncology by demonstrating that it faces huge failure rates, and that it is deeply limited by biological complexity. Rayner (2012) defines uncomfortable knowledge as knowledge that contradicts the simplified, predictable and closed models that we use for making sense of our complex world. For those simplified models to 'work', uncomfortable knowledge needs to be excluded, either by denial, dismissal or diversion. In that sense, to exist and survive as an imaginary worthy of interest despite the uncomfortable knowledge conveyed by Kern, Weinberg and others, precision oncology needs to dismiss these claims, notably by constantly being surrounded by vagueness and a lack of ambivalence.

In chapter "Precision Oncology in the News", Stenmarck and Nilsen look at the lack of diversity in how the news media frame issues related to cancer treatment and research. They show how new cancer drugs are framed as future revolutions, and how their efficacy and high cost are left unquestioned. Similarly, precision oncology is depicted as a way to achieve "the right therapeutic strategy for the right person at the right time, and/or to determine the predisposition to disease, and/or to deliver timely and targeted prevention" (EC 2015, p. 3). Uncomfortable knowledge about the significant opportunity costs of precision oncology, the problems relative to prolongment of life for cancer patients, and the reality for already fragile healthcare systems with limited healthcare resources, are being diverted from by rather pointing at the tragic stories of suffering cancer patients. The 'truth' about cancer seems to be owned by the patients and their doctors, and other, outside perspectives are seen as unwelcome and irrelevant, and dismissed as being pessimistic views. This reflects the 'power of goodness' (Loga 2004) that is at play here. According to Loga, an argument that openly represents goodness gains a superior stance that makes it difficult for other arguments to get a foothold as legitimate. It is therefore less controversial for news media and policy debates to speculate on the potential for win-win solutions where there is going to be better health for everyone, and to drive forward these developments as urgently needed by cancer patients.

Critical claims about the feasibility and desirability of precision oncology are also invalidated as having no solid scientific foundation. As Lakatos (1970) argues, we have indeed no means to evaluate whether a 'research programme' is 'degenerating' or 'progressing' before reaching some historical hindsight. But Kern's uncomfortable knowledge about the 99% failure rate in cancer biomarker research is telling. Chapter "HER2 Revisited: Reflections on the Future of Cancer Biomarker Research", by Bremer, Wik and Akslen, relies both on oncology research and perspectives from Science and Technology Studies to show how even successful cancer biomarkers face important limitations, and cannot be seen as the solution to solving ethical, social and clinical dilemmas. The authors revisit the story of the most successful cancer biomarker: HER2 for breast cancer, and discuss how HER2 has become the standard reference for showcasing the success of precision oncology. However, despite its important applications in the clinic, is not a perfect biomarker. Notably, there are challenges related to inter- and intra-tumoral heterogeneity, which question the reliability and quality of biopsies taken from patients. The determination of HER2 positivity is not straightforward either, which means that treatment options are chosen on the basis of best, but uncertain, knowledge, and that questions of where to place the cut-off between subgroup of patients remain. This uncomfortable knowledge, however, is often overshadowed by the extraordinary success consisting in HER2 finding important applications in the clinic, and therefore being one of the key arguments in validating precision oncology.

The lack of nuances and ambivalence in discourses and practices exacerbates, perhaps ironically so, the ambiguity with regard to whether precision oncology is a reality now, a soon-to-be realised miracle, or a 'mirage of health' (Callahan 2003). It fuels confusion about the temporalities of precision medicine, and results in a fusion of hope and reality. According to Callahan (2003, p. 261): "Medical miracles are expected by those who will be patients, predicted by those seeking research funds, and profitably marketed by those who manufacture them. [...] The "mirage of health" – a perfection that never comes – is no longer taken to be a mirage, but solidly out there on the horizon." Not only have hope and reality fused, but the current predicament is justified by the future imaginary of precision oncology. The legitimacy of the current efforts put into precision oncology lies precisely in the future: "targeted drugs will work"; or "every patient will get his/her targeted cancer treatment". In chapter "Introduction to the Imaginary of Precision Oncology", Engen notes that more than two decades have passed since precision medicine was projected to bring significant advances to cancer research, treatment and care. The author reviews several studies that display uncomfortable knowledge by showing how the overwhelming majority of novel oncological agents are approved without clear evidence of clinical benefit and utility. This further contributes to illustrate the increasing gap between hope and reality around precision oncology.

(ii) Framing and overflowing

The second overarching theme in this anthology Callon's notion of *framing and overflowing* (1998). Callon defines framing as "the identification, measuring and containment of [...] overflows" (p. 244), and overflows as positive or negative externalities, or in other words emergent products or practices that result, expectedly or not, from the scientific work of framing. Callon further explains that in some cases "framing is either impossible to achieve or is deliberately transgressed by the actors: this produces overflows which cause the barriers to become permeable" (p. 251). Precision oncology is, to a great extent, a project about removing ambivalence and reducing uncertainties by providing an illusion of molecular certainty that would allow us to solve any kind of social, ethical or clinical dilemma. However, the harder we try to domesticate or frame the highly uncertain and complex biology of cancer and associated dilemmas, the more there is a risk of an 'overflow'. Framing and overflowing dynamics are indeed particularly relevant when discussing the limits of biological and mathematical models in addressing the complexity of cancer biology. Every time a model tries to capture or frame tumour heterogeneity, it overflows in the shape of a reproducibility crisis, as this specific aspect of cancer biology we thought we had control of, dissolves into hundreds of different complexities.

In chapter "The Dynamics of the Labelling Game: An Essay On FLT3 Mutated Acute Myeloid Leukaemia", Engen looks at how the 20 years of trying to 'label' and frame the FLT3 mutated acute myeloid leukaemia through temporal, spatial, multidimensional and high-resolution analyses, resulted in an overflow of vast inter- and intra-individual heterogeneity. This move towards high levels of molecular resolution also means that diagnostics are becoming increasingly refined and precise, with consequences on how cancer is defined, as the categories between cancer as an 'illness' and cancer as a 'disease', seem to dissolve.

Chapter "Crossing the Styx: If Precision Medicine Were to Become Exact Science", by Strand and Chu, also addresses the problems with trying to frame the high complexity of cancer biology in highly sophisticated and exact science. Computational models on which the imaginary of precision oncology relies, promise unachievable levels of numerical precision and conceptual rigour, which would require framing everything from cells to patients as closed and deterministic systems, when they are not. The authors point at the design flaw in precision medicine: it wants to achieve precision and tailoring by relying on exact science, but exact science does not translate into exact technologies that apprehend the complexity of cancer biology. The overflow here, is that the shift to a biology dominated by computational models may reinforce our understanding of life as essentially predictable, understandable and controllable, which in the end supports industrial and economic exploitation. In addition, striving for an unattainable objective will blind us away from what is really at stake here.

Chapters "Assessing the Cost-Effectiveness of Molecular Targeted Therapies and Immune Checkpoint Inhibitors" and "Real-World Data in Health Technology Assessment: Do We Know It Well Enough?", by Cairns and Kang, respectively, direct the analysis towards the details of health technology assessment and specifically, the assessment of the cost-effectiveness of expensive cancer therapies. Their chapters enter into the technical details of such assessments and, by opening these black boxes, show in different ways how what may appear as a "purely" technical frame is, if not necessarily overflowed, co-produced and shaped with social and political concerns. Along such lines, Cairns shows how seemingly parallel innovations, i.e., molecular targeted therapies and immune checkpoint inhibitors, receive subtly but crucially different assessment in terms of the methodologies used. The reader is left with a difficulty to explain these differences except within the more general narratives of the desirability of immune checkpoint inhibitors. Kang offers a similar perspective by discussing how complex and uncertain 'real world data' are incorporated in the relatively rigid frames of health technology assessments, which aim to provide a 'systematic evaluation of short- and long term safety, clinical effects, and cost-effectiveness of health technologies'. The hope with integrating real world data into health technology assessments is more robust clinical and economic decision-making processes. However, managing and making sense of these overwhelming quantities of real world data produced at a very rapid pace is extremely challenging. As a consequence, this will arguably overflow in a much higher degree of complexity when assessing the cost-effectiveness of treatments in health technology assessments.

In chapter "Publication Bias in Precision Oncology and Cancer Biomarker Research; Challenges and Possible Implications", Lotsberg and D'mello Peters explore another overflow that is central to precision oncology: publication bias, i.e., published results are not a representative selection of all results within a study, and not all studies are published, with an imbalance towards reporting 'positive' results. The authors argue that aiming for 'hyper precision' as a general research direction or frame, results in the overflow of publication bias being more present in the fields of precision oncology and biomarker research. Indeed, as the imaginary of precision oncology relies on removing ambiguity, reducing uncertainty and providing molecular certainty, there is naturally less appetite for 'negative' results.

The issues of framing and overflowing are also very stringent in health care priority setting in a context of very expensive drugs. Indeed, it seems like the more efforts are put into establishing a precise cut-off between subgroups of patients for treatment allocation, the more it overflows as heated controversies; with patient subgroups just below the cut-off wanting the unfair situation reframed. In chapter "Reconstruction of Trouble", Dillekås relates the 2012 campaign of the 'Norwegian Breast Cancer Society', who managed to influence policy agendas in order to prioritise immediate breast reconstruction to breast cancer treated patients. This resulted in the dramatic overflow in terms of a resurgence of cleft lip and palate as a public health issue, as plastic surgeons were instructed to prioritise breast reconstructions over this group of patients. This story of frame and overflow is heightened by the fact that Dillekås and her colleagues published a paper pointing at a peak in early relapses in patients who had reconstructed breasts; peak that was not present in patients with similar tumour characteristics that choose not to reconstruct the breast. Their paper is a direct example of 'uncomfortable knowledge', as it points at how the complexities of cancer biology undermine what we think is 'good' prioritisation of health care.

Framing and overflowing also occur in projected priority setting decisions. Fleck analyses in chapter "Just Caring: Precision Health vs. Ethical Ambiguity: Can we Afford the Ethical and Economic Costs?" the argument developed by the oncologist Dr. Raza to abandon paying for targeted therapies for metastatic cancer, in order to rather invest that money for early cancer detection using liquid biopsies. Fleck explains that the apparently simple and 'framed' transaction from handling metastatic cancer to focusing on early detection would result in sacrificing identified lives (those who have metastatic cancer) for the statistical lives of future cancer patients identified through liquid biopsies. This would result in a significant overflow in terms of controversies and heated debates about fairness, compassion, care, and the unjust and unsustainable use of limited health care resources.

Finally, the dynamics of framing and overflowing are found at the level of priority setting institutions themselves, as Tranvåg and Strand explain in chapter

"Rationing of Personalised Cancer Drugs: Rethinking the Co-production of Evidence and Priority Setting Practices". They describe how the priority setting institution in Norway, among other countries, tries to cope with the scientific development of ever finer stratification and smaller patient groups by increasingly refined principles of priority setting (the umbrella values being neutrality, transparency and equal treatment). However, these attempts overflow in the shape of persistent controversies around drug reimbursement decisions as well as novel ways of providing drugs to patients in spite of priority decisions (as by recruitment into trials). The authors argue that the priority-setting frame itself may be due for fundamental reform that also entail a redressing of its umbrella values.

Both relative to the biological complexity and questions of priority setting, we see how all the efforts to frame, control and domesticate biological, ethical and social complexities are extremely resource-intensive, imperfect and often futile, as they result in overflows. It allows us to realise that optimistic discourses around precision medicine have shaky factual foundations.

(iii) The economy of hope and distinction between illness and disease

The resource-intensive efforts put in dismissing or diverting from uncomfortable knowledge, and in attempts to frame biological complexity lead us to the third overarching theme of the anthology. One aspect that contributes to explain why such efforts are developed to shielding at all costs the imaginary of precision medicine from ambivalence and criticisms, is the economic, political, and social interests for sustaining an 'economy of hope' (Rose and Novas 2004). Within the economy of hope, hope is sustained that targeted therapies work, and that every patient will eventually get her or his tailored drug. The limits to achieving these prospects are seen as not being inherent to science: "there are no inherent obstacles or pitfalls of science that could stop the realisation of revolutionary cures" (Brekke and Sirnes 2011, p. 356). Rather, the limits are seen as being exclusively political. It is the politicians who deny suffering cancer patients their life-saving therapies, by not funding them or by not prioritising them. In this economy of hope, patients, or 'somatic individuals' who understand themselves more in biological terms organise themselves into new constellations of 'biocollectives', or alliances with pharmaceutical companies and research groups, in order to influence agendas to promote research on 'their' disease, or enrol themselves actively as research subjects in trials for instance (Brekke and Sirnes 2011).

In chapter "Cancer Currencies: Making and Marketing Resources in a First-in-Human Drug Trial in Denmark", Hillersdal and Svendsen explore the dynamics of the economy of hope by looking at the collaboration between a public hospital in Denmark and a multi-national pharmaceutical company in setting up and running early cancer drug trials for personalised medicine. Notably, they look at how these public-private partnerships stir the direction of research and shape what precision oncology looks like in early clinical trials. They point to the fact that medical advances have become extremely dependent on industrial sponsors and agendas, which has led to less considerations about the real benefit for patients. In addition, they argue that trial qualities such as fast-tracking trial procedures, high-quality data and high compliance of research subjects, were highly demanded by the public-private partnerships, are in fact 'currencies' used in transactions on the global market for drug development. In that way, Hillersdal and Svendsen unravel the ambiguities of the economy of hope, where a demand for a particular targeted drug is grounded in public-private partnerships, further facilitated by the danish welfare state, and finally expected by cancer patients (although the authors argue that participating in an early clinical trial was for some patients a way to give meaning to their disease, by considering that they help research and thus future patients for instance). The tragic irony here, is that these drugs are often too expensive to be prioritised by the same welfare states that contribute to their development.

The economy of hope also runs on fear. The fear of not being able to control one's own last moments of life, the fear of dying 'prematurely' from cancer, the fear of not having the strength or courage to try every extraordinary treatment available, as one has seemingly nothing to lose. In chapter "Filled with Desire, Perceive Molecules", Strand and Engen argue that these fears, and underlying strive, desire and passion to provide immediate help to acute myeloid leukaemia patients leads to losing sight of the important biological questions, such as: 'what is the function of cancer?' Curiosity on the biological, rather than medical questions, would arguably bring important learnings to light. Further, the authors argue that the urgent desire to advance science on AML and help the concerned patients also overshadows the variety of ways to help and accompany patients, in particular by having a better understanding of their illness is for them: is AML an enemy to be defeated, a deficiency to be removed, or an illness to accept? Indeed, the urgency to help may become an obstacle on the road that many of these patients will have to walk, from shock through despair to acceptance of their destiny. In this way the need to act can risk adding to the suffering.

Chapter "Lost in Translation" by Gissum further unravels the distinction between illness and disease that is sustained by the economy of hope. The economy of hope indeed needs to frame cancer as a disease that can be addressed, without ambiguities, by sophisticated technologies and targeted therapies. There is little place for illness in this picture. However, cancer patients 'own' their cancer: it is their illness, and their subjective, personal experience is an important (arguably the most important) consideration to take into account in clinical decision making. However, Gissum point at the mismatches between, on one side, the physician's perception of cancer as a disease that can be measured, and on the basis of which a rational treatment regimen can be established, and on the other side, the patient's experiences of her illness: what it does to her body, her mind, her self-perception, her networks, her activities: in other words, her home-world. The author argues that this mismatch is heightened in a context of precision oncology, where both patients and physicians operate within confusing hopes, realities and temporalities, and where the categories of health, illness and disease are being redefined. Arguably, precision oncology and the strive for hyper-precision and sophistication, both in scientific practices and developed therapies would benefit from being accompanied by a much more prominent place given to cancer as an illness.

This anthology looks at the culture and practice of biomarker research, and how it is powered to a significant extent by the sociotechnical imaginary of precision oncology. The issues at stake and matters of concern are approached with a revealing set of lenses, assembled by a team of authors from fields including fields like oncology, philosophy, STS, anthropology, economics, ethics, and media studies. This anthology is particularly relevant for scholars and practitioners in the many fields that are covered by precision oncology and cancer biomarkers, and for those who want to unpack the timely questions around the feasibility and desirability of precision oncology.

References

Åm, H. 2019. Ethics as ritual: Smoothing over moments of dislocation in biomedicine. *Sociology of Health and Illness* 41 (3): 455–469.

Åm, H., G. Solbu, and K.H. Sørensen. 2020. The imagined scientist of science governance. *Social Studies of Science*. https://doi.org/10.1177/0306312720962573.

Argyris, C., and D.A. Schön. 1974. *Theory in Practice: Increasing Professional Effectiveness*. San Francisco: Jossey-Bass Publishers.

Argyris, C., R. Putnam, and D.M. Smith. 1990. *Action Science*. San Francisco: Jossey-Bass Publishers.

Blanchard, A. 2016. Mapping ethical and social aspects of cancer biomarkers. *New biotechnology* 33 (6): 763–772.

Blanchard, A., and R. Strand. 2017. In *Cancer Biomarkers: Ethics, Economics and Society*, ed. A. Blanchard and R. Strand. Kokstad: Megaloceros Press.

Brekke, O.A., and T. Sirnes. 2011. Biosociality, biocitizenship and the new regime of hope and despair: Interpreting "Portraits of Hope" and the "Mehmet Case". *New Genetics and Society* 30 (4): 347–374.

Callahan, D. 2003. *What Price Better Health? Hazards of the Research Imperative*. Berkeley: University of California Press.

Callon, M. 1998. An essay on framing and overflowing: Economic externalities revisited by sociology. *The Sociological Review* 46 (1): 244–269.

EC. 2015. *European Council Conclusions on Personalised Medicine for Patients*. Luxembourg: Publications Office of the European Union, European Council.

Hesjedal, M.B., H. Åm, K.H. Sørensen, and R. Strand. 2020. Transforming scientists' understanding of science–society relations. Stimulating double-loop learning when teaching RRI. *Science and Engineering Ethics* 26 (3): 1633–1653.

Jasanoff, S. 2015. Future imperfect: Science, technology and the imaginations of modernity. In *Dreamscapes of Modernity: Sociotechnical Imaginaries and the Fabrication of Power*, ed. S. Jasanoff and S.-H. Kim, 1–33. Chicago: Chicago University Press.

Kern, S.E. 2021. Why your new cancer biomarker may never work: Recurrent patterns and remarkable diversity in biomarker failures. *Cancer Research* 72 (23): 6097–6101.

Lakatos, I. 1970. History of science and its rational reconstructions. In *PSA: Proceedings of the biennial meeting of the philosophy of science association*, vol. 1970, 91–136. D. Reidel Publishing.

Loga, J.M. 2004. *Godhetsmakt. Verdikommisjonen – mellom politikk og moral* [The Power of Goodness. The Value Commission – Between Politics and Morality]. PhD thesis for the University of Bergen, Norway.

Rayner, S. 2012. Uncomfortable knowledge: The social construction of ignorance in science and environmental policy discourses. *Economy and Society* 41 (1): 107–125.

Rose, N., and C. Novas. 2004. Biological citizenship. In *Global Assemblages: Technology, Politics, and Ethics as Anthropological Problems*, ed. A. Ong and S.J. Collier, 439–463. New York: Blackwell.
Schön, D. 1983. *The Reflective Practitioner. How Professionals Think in Action*. London: Routledge.
Strand, R. 2000. Naivety in the molecular life sciences. *Futures* 32 (5): 451–470.
Strand, R., and L.A. Akslen. 2017. What is responsible cancer research? *Tidsskr Nor Legeforen* 137 (4): 292–294.
Weinberg, R.A. 2014. Coming full circle – From endless complexity to simplicity and back again. *Cell* 157 (1): 267–271.

Introduction to the Imaginary of Precision Oncology

Caroline Engen

Worldwide, cancer is a major cause of health impairment and premature death.[1] With the rise in life expectancy across the globe, cancer incidence and mortality rates are estimated to increase substantially in the decades to come (Global Burden of Disease Cancer Collaboration et al. 2018; GBD 2017 Causes of Death Collaborators 2018). Efforts aimed at improving clinical management of cancer are extensive (Eckhouse et al. 2008; van de Loo et al. 2012). At the heart of this exertion translational cancer research (Cambrosio et al. 2006) and the imaginaries of precision medicine and precision oncology are taking shape and gaining traction (Hamburg and Collins 2010; National Research Council (US) 2011; Mirnezami et al. 2012; Collins and Varmus 2015; Celis and Heitor 2019).

Precision oncology adheres to the prevailing conceptual understanding of what cancer is: a clonal disease, caused by acquisition and accumulation of genetic alterations in cells, ultimately resulting in disruption of normal cell function (Nowell 1976; Vogelstein and Kinzler 1993; Hanahan and Weinberg 2000; Garraway et al. 2013). On the premise that these molecular events are patient specific and at the core of the causality of cancer precision oncology proposes a change of cancer management along two dimensions: (i) from groups to individuals and (ii) from morphological to molecular classification. Through identification of causally contributing molecular mechanisms the goal is to enable precise disease categorisation, prediction, prevention, early detection and targeted treatment; providing the right

[1] Preliminary versions of parts of this chapter were included in the introduction chapter of the author's PhD dissertation (Engen 2020).

C. Engen (✉)
Centre for Cancer Biomarkers CCBIO, Department of Clinical Medicine,
University of Bergen, Bergen, Norway
e-mail: Caroline.Engen@uib.no

A. Bremer, R. Strand (eds.), *Precision Oncology and Cancer Biomarkers*,
Human Perspectives in Health Sciences and Technology 5,
https://doi.org/10.1007/978-3-030-92612-0_2

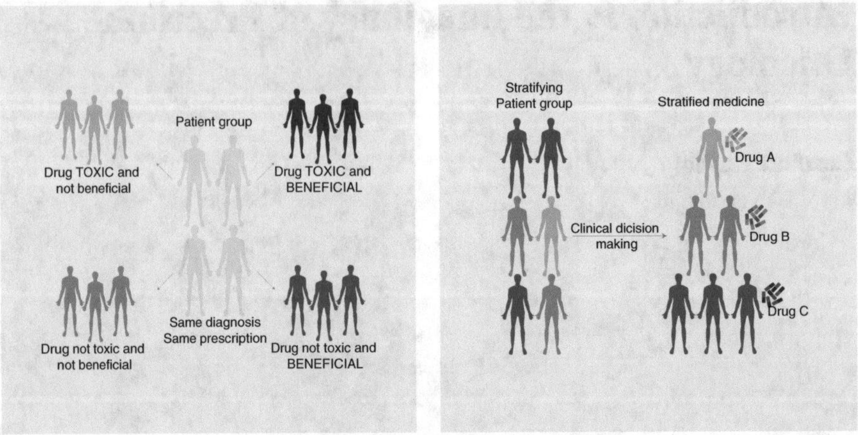

Fig. 1 Precision medicine is a relatively new term overlapping and gradually replacing the preceding term "personalised medicine". According to the Precision Medicine Initiative launched by Barak Obama in 2015 precision medicine is "an emerging approach for disease treatment and prevention that takes into account individual variability in genes, environment, and lifestyle for each person". The approach is founded on the assumption that interindividual heterogeneity result in suboptimal utility of medical interventions. Through disease stratification and tailoring of medical care outcomes can improve. The illustrations are adapted from various online sources based on the google image search "precision medicine"

treatment to the right patient at the right time (Mirnezami et al. 2012; Tsimberidou et al. 2014; Collins and Varmus 2015; Ashley 2016) (Fig. 1).

The integration of precision oncology related approaches in standard patient care is an ongoing process, resulting in shifts in diagnostic thresholds, formation of novel disease subcategories and adaptation of new treatment strategies (Jameson and Longo 2015). Currently, however, only a limited fraction of cancer patients is estimated to benefit from this line of approaches (Marquart et al. 2018). Based on the limited progress so far some investigators and clinicians have even challenged the validity, utility and sustainability of precision oncology all together (Prasad 2016; Prasad et al. 2016; Marquart et al. 2018). The tension between the current status of precision oncology and the optimism related to future benefits of this strategy is an important motivation for this work. In what follows is a brief outline of the emergence of precision medicine and precision oncology.

Precision Medicine – Tradition, Evidence, Reason and Ambition

The advancement towards increased precision in medicine and oncology can be seen as a continuation of the direction modern medicine has had since its conception (Le Fanu 2000). A recent analysis of ancient Hippocratic texts identified that inter-individual heterogeneity was recognised already 2500 years ago. This suggests that individually tailored treatment and medical care always has been a fundamental feature of applied medicine (Konstantinidou et al. 2017). It is, however, only throughout the last two centuries molecular mechanisms underlying this inter-individual heterogeneity have begun to be revealed. Technological progress has allowed a gradual increase in resolution in the exploration of both human physiology as well as pathology. Disease classification systems as well as clinical practices have evolved in close relationship with methodological advancements. This development is characterised by gradual shifts in dimensionality from the clinical and macro-anatomical organisation and understanding of human maladies to tissue centred approaches, followed by increasing attention on cells and subcellular components as the origin of pathology (Keating and Cambrosio 2001).

The concept of "molecular" disease was first put forward in 1949 by Pauling and colleagues in the Science paper "Sickle Cell Anemia, a Molecular Disease". The authors hypothesised the genetic basis for the condition, and experimentally explored the aberrant protein product responsible for erythrocyte "sickling" (Pauling et al. 1949). In the decades that followed and up until the present genotype-phenotype relationships have been confirmed to account for a myriad of human traits and disease phenotypes (Buniello et al. 2019). The idea of precision medicine gradually emerged from this body of knowledge. It was, however, in relation to the planning and execution of the "Human Genome Project", formally commenced in 1990, that the vision of precision medicine was truly articulated (Collins 1999). The Human Genome Project was a milestone in the development of the implicit idea of "precision medicine" into a recognizable sociotechnical imaginary: a shared vision, ambition and commitment, co-created and co- maintained by experts and policy makers (Jasanoff and Kim 2015; Tarkkala et al. 2019). The goal of the "Human Genome Project", providing a complete sequence of the human genome, was ambitious and required considerable financial and intellectual investment. The legitimacy of this publicly funded venture was rationalised through postulations of significant scientific, medical, and societal advancements (National Research Council (US) 1988). Francis Collins[2] put it like this: "Scientists wanted to map the human genetic terrain, knowing it would lead them to previously unimaginable insights, and from there to the common good. That good would include a new understanding of genetic contributions to human disease and the development of rational strategies for minimising or preventing disease phenotypes altogether" (Collins 1999).[3]

Since 1999 the imaginary of precision medicine has matured and expanded beyond its initial scope to propose fundamental changes not only in how diseases are to be managed but also to be categorised and understood. In 2011 the National Research Council (US) released the report "Toward Precision Medicine: Building a

[2] Francis Collins was director of the National Human Genome Research Institute from 1993 to 2008 and was the director of the National Institutes of Health, US, from 2009 until 2021.

[3] Quote from the 1999 Shattuck lecture, titled: "Medical and societal consequences of the human genome project".

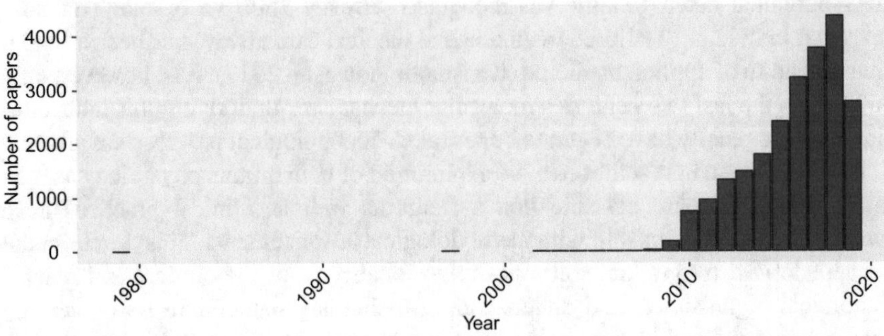

Fig. 2 Histogram of the temporal distribution of papers associated with the term "Precision medicine" in the search-engine Pubmed (10.07.2019)

Knowledge Network for Biomedical Research and a New Taxonomy of Disease". Here the authors commend the development of a new taxonomy of human diseases, predominantly based on intrinsic biology and causal molecular disease mechanisms rather than signs and symptoms (National Research Council (US) 2011). The term "precision medicine" has since then rapidly been integrated in the biomedical and biotechnological scientific literature (Fig. 2) as well as the political, regulatory and public discourse (Blasimme and Vayena 2016). The uptake has been substantially intensified by the launch of the "Precision Medicine Initiative" by Barack Obama in 2015, aimed at accelerating the translation of biomedical science to improved clinical outcomes (Collins and Varmus 2015).

Precision Oncology – Expectations and Realisations

Considered a genetic and molecular disease cancer served as an example of the hypothesised future significance of the Human genome project as well as the transition towards a molecular based disease taxonomy (National Research Council (US) 1988, 2011). Precision medicine in relation to cancer management has been characterised by a strong emphasis on inter-individual variability of genes, and is often referred to as genomics-driven cancer medicine (Garraway et al. 2013). Medical strategies related to precision oncology are profoundly tied to postulations of genetic causality in cancer development. The idea of a monoclonal origin of cancer suggest that the cellular mass of an individual tumour share molecular characteristics involved in pathogenesis. Observations of cellular dependency of mutated or aberrantly expressed gene-products for both initiation and maintenance of malignant phenotypes support this idea and led to the postulation and experimental verification of "oncogene addiction" (Weinstein 2002). This provided a strong rational for the possibility of classifying various cancers with respect to their molecular origin, as well as molecular targeted treatment strategies. The feasibility of this approach was confirmed in the early 2000s based on several unprecedented clinical success stories, including molecular targeted therapy in chronic myeloid leukaemia (Deininger et al. 2005), gastrointestinal stromal tumours (DeMatteo 2002) and a molecular defined sub-group of breast cancer (Slamon et al. 2001).

Identification of shared molecular "drivers" in cancer cells originating from discrete cell-types and diverse tissues led to the hypothesis that this approach may be scalable, perhaps even to all cancers and all cancer patients (Tsimberidou et al. 2014). Instead of managing cancers in accordance with their macro-and microanatomical origin treatment could be guided by genomic profiling (Garraway et al. 2013). Recently, therapeutic compounds based on molecular defined indications, rather than tissue or histology, such as pembrolizumab, were subject to regulatory approval[4] (Lemery et al. 2017; Scott 2019). This development can be seen as a sizeable stride towards making such an approach become standard of care.

While tissue agnostic indications strongly enforce the implementation of molecular profiling of all cancer patients it has been challenging to demonstrate that broad genetic testing followed by rationally selected therapeutic compounds generally lead to superior outcomes compared to current evidence-based practices (Le Tourneau et al. 2015; Stockley et al. 2016; Massard et al. 2017; Rodon et al. 2019; Rothwell et al. 2019). Experience from multiple trials as well as general estimates suggest that currently only a small percentage of cancer patients with advanced stage disease are eligible and will benefit from genome-informed therapy. Furthermore, the magnitude of clinical benefit that can be attributed to biomarker matched interventions is sobering. So far, it is a matter of additional months of life (Marquart et al. 2018; Sicklick et al. 2019), rather than years or decades, as has been achieved in chronic myeloid leukaemia, gastrointestinal stromal tumours and some patients with breast cancer (Slamon et al. 2001; DeMatteo 2002; Deininger et al. 2005).

The limited benefit of precision oncology may in part be accounted for by lack of knowledge as well as restrictions in technology, availability of therapeutic compounds and investigation in suboptimal study populations. Discovery of novel targets, development of better technological solutions, increased availability of therapeutic compounds, improved clinical infrastructure, and therapeutic repositioning to earlier disease stages may all contribute to further progress of this approach. However, more than 20 years have passed since precision oncology related approaches were first projected to result in substantial benefit (Collins 1999). It seems timely to re-explore the theoretical foundations as well as the issues at stakes and matters of concern of precision oncology, as this volume endeavours to do.

---

[4] In 2017 U.S. Food and Drug Administration (FDA) provided approval of a programmed death 1 (PD-1) inhibitor (pembrolizumab) for patients with microsatellite-instability–high or mis-match-repair–deficient solid tumours. This was followed by the authorisation of a tropomyosin kinase receptor inhibitor (larotrectinib) for cancer patients with neurotrophic receptor tyrosine kinase (NTRK) gene fusions regardless of anatomical origin.

Precision Medicine and the Complexity of Biological Systems

The central dogma in molecular biology is the unidirectional flow of information from genes to proteins; the "genotype-phenotype relationship" (Crick 1958). This bottom-up model has dominated the experimental work of biomedicine as well as the interpretation of observational data. While nobody in their right mind might reject the value and validity of the fundamentals of molecular biology, it is becoming clear that a view on cancer as a simple "genotype-phenotype relationship" in linear unidirectional terms is far too simplistic (Bertolaso 2016). I have argued elsewhere that metazoan cell identity, cell state and cell fate are determined by numerous intrinsic *and* extrinsic factors (Engen 2020). For any given cell the selection of potential cell identities and cell states is intrinsically defined by the cell's genetic material, the DNA. Through quantitative or qualitative alterations of complex gene-interactions a somatic mutation can reshape the trajectories of cell fate. Through emergence of new molecular features mutations in the regulatory part of the DNA or mutated gene products can open up unconventional transcriptional states resulting in novel cellular properties. Cell identity and cell state is further strongly influenced by the line of descent of the cell, defining its epigenetic configuration and confining its potential differentiation paths. Fundamentally, metazoan cells are, however, neither self-sufficient nor self-governing. Metazoan cells are collective in nature, and every new cell develops into being profoundly embedded in context. Networks of cells co-produce and co-maintain tissue and organ integrity, and collectively perform plastic transformations in response to perturbations. Through interactions like physical contact, autocrine, paracrine and endocrine signals, the collective of cells co-operate through continuously modifying their individual epigenomes and transcriptomes, in response to their surroundings (Bertolaso and Dieli 2017). Under these premises, cancer, although frequently described as a "genetic" disease, more fundamentally is a manifestation of aberrant cell behaviour. Although mutations can change the boundary conditions for a cell's repertoire of potential phenotypic expression the effect of a mutation on a cell is profoundly relational. As a cell or a line of descending cells phenotypically diversify by expressing non-canonical transcriptional states it is in part the conditions of the environment that defines if the change is beneficial or deleterious. The emergence of novel cellular properties can accordingly never be fully understood or accounted for at the cellular or sub-cellular level. The gene-environment provides a dynamic and relational substrate where the meaning of the gene variant is defined. As neoplastic properties emerge by force of gene-gene-environment alterations the most relevant question may not be how mutations arise and translate to change but how the gene gene-environment relationship restrict the potential translational effect of novel gene-variants. Indeed, the nature of the "phenotype – emergent genotype relation" appears as a highly promising field to explore in cancer. Genotype emergence can under these premises be predictive of disease trajectories through association rather than through causation. The question is what concept of precision medicine one may retain if it is increasingly understood that the disease is a result of stochastic and emergent properties rather than deterministic linear causation.

The Imaginary of Precision Medicine and Unintended Consequences

There is no sharp demarcation between precision and non-precision related medical approaches. Rather, precision oncology is in many regards a continuation of the reductionist biomedical traditions that emerged in the nineteenth century and came to dominate in the twentieth century. As such, precision medicine is not best understood as evolution of practices, but rather as an expansion from medical practice to a techno-scientific imaginary. Although precision medicine has been presented as a societal endeavour, it is not the destination but rather a technical solution to a political oriented objective: namely to improve public health (Fig. 3). Precision oncology is as such a means to this end and not a goal in and of itself. The question is therefore not if precision oncology is feasible, but if it is "feasible enough", as to be both desirable, viable and sustainable within certain frames. This is a combined scientific, medical, political, economic and ethical question. As a medical, political, economic, and societal aim the intended and unintended consequences related to precision medicine far exceed the sum of measurable effects of single medical interventions. The hope, vision and objective of precision medicine shape research objectives, policy agendas (Horgan et al. 2015), legal and regulatory frameworks, health care delivery systems and public expectations across the world (Tarkkala et al. 2019).

A well cited review paper on acute myeloid leukaemia (AML), a severe cancer disease of the blood, summarised the molecular knowledge base with regards to AML pathophysiology and concluded: "Hopefully, this new biological information will contribute to less empirical approaches to treatment" (Ferrara and Schiffer 2013). This statement embodies what seems to be the prevailing mindset of the field. Precision medicine promotes a substantial change in the foundation of clinical decision making, characterised by increased reliance on bio-plausibility and

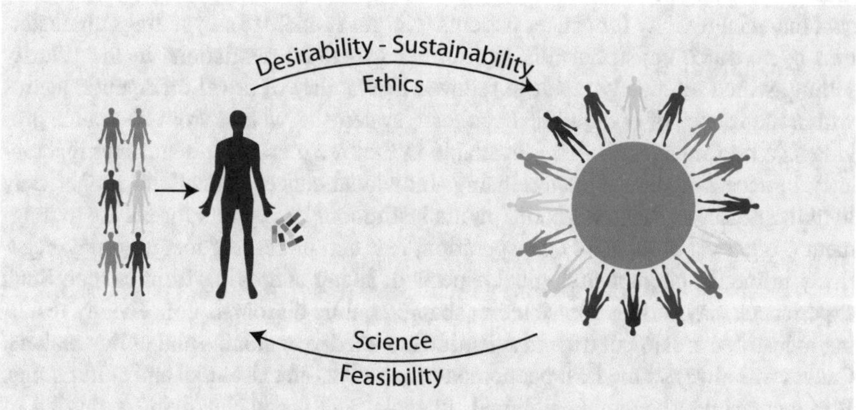

Fig. 3 Precision medicine is a techno-scientific solution to a political problem

de-evaluation of evidence. This is reflected in oncological practice. Stakeholders of precision oncology advocate increased pace in the translation and implementation of novel "promising" agents. This has resulted in deregulation and reduced evidence requirements for marked authorisation of novel drugs, including increased reliance on single arm studies as well as poorly validated surrogate endpoints (Chen et al. 2019; Gyawali et al. 2019; Hilal et al. 2019; Zettler et al. 2019). Based on a systematic evaluation of cancer drugs approved by the European Medicines Agency in the period 2009–2013, Davies et al. demonstrate that the majority of novel oncological agents were approved with no clear evidence of clinical benefit and that evidence of clinically meaningful utility remained unfounded a minimum of 3 years after approval. Quantifiable benefits were marginal, and the estimated median life expansion provided when documented was only 2.7 months (Range:1–5.8 months) (Davis et al. 2017). Early marked-approval is further dis-incentivising for execution of confirmatory well powered randomised studies, resulting in absent quantification of comparative effectiveness. Lack of high-level evidence further affect the possibility and validity of cost-benefit analysis (DeLoughery and Prasad 2018). Low-grade evidence paradoxically increase uncertainty in clinical decision making (Moscow et al. 2018). Despite the failing empirical foundation (Djulbegovic and Ioannidis 2019) uptake of low-grade evidence in clinical care is substantial. There is an increase in use of off-label targeted therapy (Saiyed et al. 2017), despite evidence suggesting inefficiency (Le Tourneau et al. 2015). Some countries and health care delivery systems even have aliquoted funds for such practises, like the National Health Service (NHS, UK) Cancer Drug Funds. A recent analysis of the use of such solutions suggest no meaningful societal or patient value gain (Aggarwal et al. 2017).

These sobering results suggest that there is an increasing distance between the expectations and realisations of precision oncology. Repercussions of this friction are currently materialising across a wide range of medical as well as social domains (Fojo et al. 2014; Bowen and Casadevall 2015; MacLeod et al. 2016). The gradual implementation of low value precision oncology related strategies has contributed to a situation where the total financial burden of cancer treatment and cancer care is rapidly spiralling out of control (Aggarwal et al. 2014). This has resulted in significant financial toxicity for cancer patients (Knight et al. 2018). In settings characterised by resource constraint this has further generated restrictions in the priority setting, which ultimately result in reduced availability of novel therapeutic agents within the frames of both public healthcare systems as well as from insurance providers. With these agents being available in the free market an increasing discrepancy in access to care is materialising. Individual cancer patients no longer only fight their disease, they also battle public institutions or insurers for access to treatment (Aggarwal et al. 2014). Desperation, fear and increasing inequity may negatively influence phenomenological aspects of living and dying from cancer. Such experiences may further contribute in shaping public discourse, conceivably resulting in justified erosion of trust in scientific knowledge, medicine and policy makers. Cancer was always a medical phenomenon as well as an existential and cultural one. It is increasingly becoming political, financial and social, involving a myriad of actors, issues and concerns that literally are matters of life and death.

References

Aggarwal, A., T. Fojo, C. Chamberlain, C. Davis, and R. Sullivan. 2017. Do patient access schemes for high-cost cancer drugs deliver value to society? Lessons from the NHS Cancer Drugs Fund. *Annals of Oncology* 28 (8): 1738–1750.

Aggarwal, A., O. Ginsburg, and T. Fojo. 2014. Cancer economics, policy and politics: What informs the debate? Perspectives from the EU, Canada and US. *Journal of Cancer Policy* 2 (1): 1–11.

Ashley, E.A. 2016. Towards precision medicine. *Nature Reviews. Genetics* 17 (9): 507–522.

Bertolaso, M. 2016. *Philosophy of Cancer: A Dynamic and Relational View*. Dordrecht: Springer.

Bertolaso, M., and A.M. Dieli. 2017. Cancer and intercellular cooperation. *Royal Society Open Science* 4 (10): 170470.

Blasimme, A., and E. Vayena. 2016. "Tailored-to-You": Public engagement and the political legitimation of precision medicine. *Perspectives in Biology and Medicine* 59 (2): 172–188.

Bowen, A., and A. Casadevall. 2015. Increasing disparities between resource inputs and outcomes, as measured by certain health deliverables, in biomedical research. *Proceedings of the National Academy of Sciences of the United States of America* 112 (36): 11335–11340.

Buniello, A., J.A.L. MacArthur, M. Cerezo, L.W. Harris, J. Hayhurst, C. Malangone, A. McMahon, et al. 2019. The NHGRI-EBI GWAS Catalog of published genome-wide association studies, targeted arrays and summary statistics 2019. *Nucleic Acids Research* 47 (D1): D1005–D1012.

Cambrosio, A., P. Keating, S. Mercier, G. Lewison, and A. Mogoutov. 2006. Mapping the emergence and development of translational cancer research. *European Journal of Cancer* 42 (18): 3140–3148.

Celis, J.E., and M. Heitor. 2019. Towards a mission-oriented approach to cancer in Europe: An unmet need in cancer research policy. *Molecular Oncology* 13 (3): 502–510.

Chen, E.Y., V. Raghunathan, and V. Prasad. 2019. An overview of cancer drugs approved by the US Food and Drug Administration based on the surrogate end point of response rate. *JAMA Internal Medicine* 179 (7): 915–921.

Collins, F.S. 1999. Shattuck lecture – Medical and societal consequences of the Human Genome Project. *The New England Journal of Medicine* 341 (1): 28–37.

Collins, F.S., and H. Varmus. 2015. A new initiative on precision medicine. *The New England Journal of Medicine* 372 (9): 793–795.

Crick, F.H. 1958. On protein synthesis. *Symposia of the Society for Experimental Biology* 12: 138–163.

Davis, C., H. Naci, E. Gurpinar, E. Poplavska, A. Pinto, and A. Aggarwal. 2017. Availability of evidence of benefits on overall survival and quality of life of cancer drugs approved by European Medicines Agency: Retrospective cohort study of drug approvals 2009–13. *BMJ* 359: j4530.

Deininger, M., E. Buchdunger, and B.J. Druker. 2005. The development of imatinib as a therapeutic agent for chronic myeloid leukemia. *Blood* 105 (7): 2640–2653.

DeLoughery, E.P., and V. Prasad. 2018. The US Food and Drug Administration's use of regular approval for cancer drugs based on single-arm studies: Implications for subsequent evidence generation. *Annals of Oncology* 29 (3): 527–529.

DeMatteo, R.P. 2002. The GIST of targeted cancer therapy: A tumor (gastrointestinal stromal tumor), a mutated gene (c-kit), and a molecular inhibitor (STI571). *Annals of Surgical Oncology* 9 (9): 831–839.

Djulbegovic, B., and J.P.A. Ioannidis. 2019. Precision medicine for individual patients should use population group averages and larger, not smaller, groups. *European Journal of Clinical Investigation* 49 (1): e13031.

Eckhouse, S., G. Lewison, and R. Sullivan. 2008. Trends in the global funding and activity of cancer research. *Molecular Oncology* 2 (1): 20–32.

Engen, C.B. 2020. *Exploring the Boundaries of Precision Haemato-Oncology – The Case of FLT3 Length Mutated Acute Myeloid Leukaemia*. PhD dissertation. University of Bergen.

Ferrara, F., and C.A. Schiffer. 2013. Acute myeloid leukaemia in adults. *Lancet* 381 (9865): 484–495.

Fojo, T., S. Mailankody, and A. Lo. 2014. Unintended consequences of expensive cancer therapeutics-the pursuit of marginal indications and a me-too mentality that stifles innovation and creativity: The John Conley Lecture. *JAMA Otolaryngology. Head & Neck Surgery* 140 (12): 1225–1236.

Garraway, L.A., J. Verweij, and K.V. Ballman. 2013. Precision oncology: An overview. *Journal of Clinical Oncology* 31 (15): 1803–1805.

Global Burden of Disease Cancer Collaboration, C. Fitzmaurice, T.F. Akinyemiju, F.H. Al Lami, T. Alam, R. Alizadeh-Navaei, C. Allen, et al. 2018. Global, regional, and national cancer incidence, mortality, years of life lost, years lived with disability, and disability-adjusted life-years for 29 cancer groups, 1990 to 2016: A systematic analysis for the global burden of disease study. *JAMA Oncology* 4 (11): 1553–1568.

GBD 2017 Causes of Death Collaborators. 2018. Global, regional, and national age-sex-specific mortality for 282 causes of death in 195 countries and territories, 1980–2017: A systematic analysis for the Global Burden of Disease Study 2017. *Lancet* 392 (10159): 1736–1788.

Gyawali, B., S.P. Hey, and A.S. Kesselheim. 2019. Assessment of the clinical benefit of cancer drugs receiving accelerated approval. *JAMA Internal Medicine* 179 (7): 906–913.

Hamburg, M.A., and F.S. Collins. 2010. The path to personalized medicine. *The New England Journal of Medicine* 363 (4): 301–304.

Hanahan, D., and R.A. Weinberg. 2000. The hallmarks of cancer. *Cell* 100 (1): 57–70.

Hilal, T., M.B. Sonbol, and V. Prasad. 2019. Analysis of control arm quality in randomized clinical trials leading to anticancer drug approval by the US Food and Drug Administration. *JAMA Oncology* 5 (6): 887–892.

Horgan, D., M. Lawler, and A. Brand. 2015. Getting personal: Accelerating personalised and precision medicine integration into clinical cancer research and care in clinical trials. *Public Health Genomics* 18 (6): 325–328.

Jameson, J.L., and D.L. Longo. 2015. Precision medicine – Personalized, problematic, and promising. *The New England Journal of Medicine* 372 (23): 2229–2234.

Jasanoff, S., and S.-H. Kim. 2015. *Dreamscapes of Modernity: Sociotechnical Imaginaries and the Fabrication of Power*. Chicago/London: The University of Chicago Press.

Keating, P., and A. Cambrosio. 2001. The new genetics and cancer: The contributions of clinical medicine in the era of biomedicine. *Journal of the History of Medicine and Allied Sciences* 56 (4): 321–352.

Knight, T.G., A.M. Deal, S.B. Dusetzina, H.B. Muss, S.K. Choi, J.T. Bensen, and G.R. Williams. 2018. Financial toxicity in adults with cancer: Adverse outcomes and noncompliance. *Journal of Oncology Practice* 14 (11): e665–e673. https://doi.org/10.1200/JOP.18.00120.

Konstantinidou, M.K., M. Karaglani, M. Panagopoulou, A. Fiska, and E. Chatzaki. 2017. Are the origins of precision medicine found in the corpus hippocraticum? *Molecular Diagnosis & Therapy* 21 (6): 601–606.

Le Fanu, J. 2000. *The Rise and Fall of Modern Medicine*. New York: Carroll & Graf Publishers.

Le Tourneau, C., J.P. Delord, A. Goncalves, C. Gavoille, C. Dubot, N. Isambert, M. Campone, et al. 2015. Molecularly targeted therapy based on tumour molecular profiling versus conventional therapy for advanced cancer (SHIVA): A multicentre, open-label, proof-of-concept, randomised, controlled phase 2 trial. *The Lancet Oncology* 16 (13): 1324–1334.

Lemery, S., P. Keegan, and R. Pazdur. 2017. First FDA approval agnostic of cancer site – When a biomarker defines the indication. *The New England Journal of Medicine* 377 (15): 1409–1412.

MacLeod, T.E., A.H. Harris, and A. Mahal. 2016. Stated and revealed preferences for funding new high-cost cancer drugs: A critical review of the evidence from patients, the public and payers. *Patient* 9 (3): 201–222.

Marquart, J., E.Y. Chen, and V. Prasad. 2018. Estimation of the percentage of US patients with cancer who benefit from genome-driven oncology. *JAMA Oncology* 4 (8): 1093–1098.

Massard, C., S. Michiels, C. Ferte, M.C. Le Deley, L. Lacroix, A. Hollebecque, L. Verlingue, et al. 2017. High-throughput genomics and clinical outcome in hard-to-treat advanced cancers: Results of the MOSCATO 01 trial. *Cancer Discovery* 7 (6): 586–595.
Mirnezami, R., J. Nicholson, and A. Darzi. 2012. Preparing for precision medicine. *The New England Journal of Medicine* 366 (6): 489–491.
Moscow, J.A., T. Fojo, and R.L. Schilsky. 2018. The evidence framework for precision cancer medicine. *Nature Reviews. Clinical Oncology* 15 (3): 183–192.
National Research Council (US). 1988. *Mapping and Sequencing the Human Genome*. Washington, DC.
National Research Council (US), Committee on A Framework for Developing a New Taxonomy of Disease. 2011. *Toward Precision Medicine: Building a Knowledge Network for Biomedical Research and a New Taxonomy of Disease*. Washington, DC: National Academies Press.
Nowell, P.C. 1976. The clonal evolution of tumor cell populations. *Science* 194 (4260): 23–28.
Pauling, L., H.A. Itano, S.J. Singer, and Ibert C. Wells. 1949. Sickle cell anemia, a molecular disease. *Science* 110 (2865): 443–448.
Prasad, V. 2016. Perspective: The precision-oncology illusion. *Nature* 537 (7619): S63.
Prasad, V., T. Fojo, and M. Brada. 2016. Precision oncology: Origins, optimism, and potential. *The Lancet Oncology* 17 (2): e81–e86.
Rodon, J., J.C. Soria, R. Berger, W.H. Miller, E. Rubin, A. Kugel, A. Tsimberidou, et al. 2019. Genomic and transcriptomic profiling expands precision cancer medicine: The WINTHER trial. *Nature Medicine* 25 (5): 751–758.
Rothwell, D.G., M. Ayub, N. Cook, F. Thistlethwaite, L. Carter, E. Dean, N. Smith, et al. 2019. Utility of ctDNA to support patient selection for early phase clinical trials: The TARGET study. *Nature Medicine* 25 (5): 738–743.
Saiyed, M.M., P.S. Ong, and L. Chew. 2017. Off-label drug use in oncology: A systematic review of literature. *Journal of Clinical Pharmacy and Therapeutics* 42 (3): 251–258.
Scott, L.J. 2019. Larotrectinib: First global approval. *Drugs* 79 (2): 201–206.
Sicklick, J.K., S. Kato, R. Okamura, M. Schwaederle, M.E. Hahn, C.B. Williams, P. De, et al. 2019. Molecular profiling of cancer patients enables personalized combination therapy: The I-PREDICT study. *Nature Medicine* 25 (5): 744–750.
Slamon, D.J., B. Leyland-Jones, S. Shak, H. Fuchs, V. Paton, A. Bajamonde, T. Fleming, et al. 2001. Use of chemotherapy plus a monoclonal antibody against HER2 for metastatic breast cancer that overexpresses HER2. *The New England Journal of Medicine* 344 (11): 783–792.
Stockley, T.L., A.M. Oza, H.K. Berman, N.B. Leighl, J.J. Knox, F.A. Shepherd, E.X. Chen, et al. 2016. Molecular profiling of advanced solid tumors and patient outcomes with genotype-matched clinical trials: The Princess Margaret IMPACT/COMPACT trial. *Genome Medicine* 8 (1): 109.
Tarkkala, H., I. Helén, and K. Snell. 2019. From health to wealth: The future of personalized medicine in the making. *Futures* 109: 142–152.
Tsimberidou, A.M., A.M. Eggermont, and R.L. Schilsky. 2014. Precision cancer medicine: The future is now, only better. *American Society of Clinical Oncology Educational Book* 34: 61–69.
Van de Loo, J.W., D. Trzaska, K. Berkouk, M. Vidal, and R. Draghia-Akli. 2012. Emphasising the European Union's Commitment to Cancer Research: A helicopter view of the Seventh Framework Programme for Research and Technological Development. *The Oncologist* 17 (10): e26–e32.
Vogelstein, B., and K.W. Kinzler. 1993. The multistep nature of cancer. *Trends in Genetics* 9 (4): 138–141.
Weinstein, I.B. 2002. Cancer. Addiction to oncogenes – The Achilles heal of cancer. *Science* 297 (5578): 63–64.
Zettler, M., E. Basch, and C. Nabhan. 2019. Surrogate end points and patient-reported outcomes for novel oncology drugs approved between 2011 and 2017. *JAMA Oncology*. Published ahead of print, July 3, 2019.

Precision Oncology in the News

Mille Sofie Stenmarck and Irmelin W. Nilsen

Introduction

Opportunities and costs of cancer treatment are on the increase (WHO Technical report 2018). Opportunities are often framed as medical and costs as economic. This creates an imbalance between medical opportunities, economic realities and public expectations. Globally, national health expenditures have risen considerably in recent decades, with the US seeing an increase in its GDP healthcare spending from 5.2% in 1960 to 17.9% in 2017 – and is projected to reach 20% by the end of 2020 (CMS 2018). Concurrently, annual global oncology drug costs have exceeded 100 billion US dollars, and were projected to exceed 150 billion US dollars in 2020 (IMS Institute 2016). As prices on new cancer drugs continue to rise, they claim an increasingly large proportion of health budgets – and are likely to do so at an increasing rate, as we face ageing populations more prone to cancer disease.

Following this development, the issue of cancer and cancer drugs holds a considerable presence in the news media. The public interest on the issue has been especially persistent in recent years with the rise and alleged promise of precision medicine. National and global costs of cancer medicines affect entire populations of patients, and the issue of priority-setting in relation to expensive cancer drugs has thus become a highly relevant and much-debated issue. As a result, in several countries there are ongoing debates in the public on medical priority-setting related to the introduction of new and expensive cancer drugs and treatments.

In both the public, political and medical discourse, a dominant frame of the issue of expensive cancer drugs is that of tragic choices, where suffering and death by cancer is considered an intolerable evil for the patients, while drug prices are

M. S. Stenmarck (✉) · I. W. Nilsen
Centre for Cancer Biomarkers CCBIO, Centre for the Study of the Sciences and the Humanities, University of Bergen, Bergen, Norway
e-mail: mille.sofie.stenmarck@ahus.no

A. Bremer, R. Strand (eds.), *Precision Oncology and Cancer Biomarkers*, Human Perspectives in Health Sciences and Technology 5,
https://doi.org/10.1007/978-3-030-92612-0_3

considered intolerable to society as a whole (Fleck 2013). In the most polarised expressions of these controversies, as presented in hundreds of Norwegian newspaper articles, cancer patients and their representatives experience the situation as one of the government threatening their lives by denying them access to the newest and most costly drugs, over concerns for the public health budget. These polarised expressions, along with other expressions of similarly extreme framings, have become so typical of the debate that it seems to have stagnated, with a continuous reproduction of virtually the same positions. This suggested to us that there was a need for research on public media framings on the issue of priority-setting in relation to expensive cancer drugs.

Further, the news media devotes considerable attention to cancer research and what the future of cancer treatment will hold. Precision medicine has been particularly prevalent in the news, and it has been associated with great opportunities and high expectations. The portrayal of precision medicine often includes strong normative visions regarding future treatments. Adding to this, precision medicine is linked to strong optimism and is depicted in the news as the ultimate breakthrough in cancer research, leading the public to believe the cure of cancer is nearing – 40 years after President Nixon declared war on the disease (National Cancer Institute, National Cancer Act of 1971). When institutions and actors in society make important decisions regarding the direction of science and future research, it should be based on an informed and critical basis concerning the issue at hand. One of the media's central roles is to provide a platform where such debates can take place and to separate information and facts from opinions and beliefs. This, therefore, raises questions concerning the possible implications such an optimistic and determined depiction of precision medicine might have for society, and calls for an exploration of how cancer research is framed in the public news media.

In this chapter, we will briefly present two empirical studies of the Norwegian media debate. The first is a study by Mille Sofie Stenmarck, a medical doctor and PhD candidate, on the media debate and public discourse concerning expensive cancer drugs (Stenmarck et al. 2021). The second is by Irmelin Nilsen, who has a background in media studies, focusing broadly on the role of the media in today's society. The studies were undertaken from 2017–2019 and they addressed news material on cancer spanning from 2013 to 2017. We focus on media as the central actor in presenting and providing information on precision oncology, thus holding a media centred perspective. These two studies, although performed separately, are well suited for a joint presentation as they highlight two different but important aspects of the field of cancer research. The first study explores the public discourse on the current issues surrounding modern cancer treatment, whilst the second study examines the scientific community's conception of the future of cancer treatment – both in light of the framing these issues are presented with in the media.

We will display various frames but above all highlight how there is a striking lack of polyphony in the media debate, leaving out important views and stories from the public discourse. We will also present and discuss four unquestioned assumptions that we identified in the material, which we find highly questionable from the perspective of medical ethics, as well as from a social and political critique. We will subject these to critical analysis and discuss how the framing of cancer and cancer research in the news affects public understanding of those issues. Through this discussion, we aim to gain a better understanding of the structure of the debate on cancer and cancer drugs, with the hope of moving thinking beyond that solely of tragic choices.

Framing Within the News Media

Before presenting our studies, we will briefly introduce some basic understandings of the operations of the news media and their use of frames, an understanding which underpins the following studies. The news media serves the citizens with a certain worldview, and much of our understanding of certain topics is shaped by how the news media chooses to present this. The central role of the media is to provide unbiased information, ensure that those in power do not abuse their power and ensure diversity, in the context that different views of society and political directions are visible in the public debate.[1] Adding to this, the press is institutionalized – meaning journalists and editors follow a specific set of established ideals, practices, and routines in their work (Eide 2011). For instance, the definition of what is regarded as a novelty, is constituted by a set of news criteria, or news values. More precisely, if an incident is close in time, space, and culture, and if the incident or issue was unexpected, unusual, and meaningful, these are considered characteristics that may qualify as a novelty, and noteworthy enough to report in the news media (Galtung and Ruge 1965).

However, news value can be added by the journalist, and does not necessarily need to characterize the incident or issue in itself (Eide 2011). By doing this, journalists can add a specific *framing*. Framing is often understood as selecting one aspect of the issue and amplifying this to create a frame that the incident or issue is presented within (Entman 1993). In turn, this can affect how an issue is understood by the news consumer (Schudson 2003). The sources can also play a significant role in deciding how the issue is framed. If the sources have a clear agenda which they wish to present, they can present it in a way that fits the established news criteria. Gamson and Modigliani (1989) address that sources can use sensational words and phrases to lead the journalist to choose a specific framing of the issue. At the same time, journalists can use certain sources to amplify their own framing of the issue. Nevertheless, previous media studies have addressed the fact that the so-called political and societal elites and their views seem to dominate the news, and journalists often choose framings that support these views, rather than challenge them (Bennett et al. 2006).

[1] This is part of the ethical guidelines of the Norwegian news press named *Vær varsom-plakaten* (Pressens faglige utvalg 2021).

Identifying Frames Within the Discourse on Priority-Setting and Cancer

As exploratory research suggested a clear lack of nuance in the Norwegian public discourse on priority-setting in relation to cancer and expensive cancer drugs, we saw the need for an in-depth analysis of this debate and how it is framed within the public discourse. We wished to explore both how the issue itself is framed, and if there are central stories that remain untold. At a superficial glance, the discourse appeared un-nuanced and overly simplified, with an abundance of stories highlighting – rather coarsely – the tragedy that cancer entails, but with a simplistic presentation of both the realities of living with a cancer diagnosis, and of the realities, possibilities and constraints of cancer research. The issue of priority-setting and cancer appeared to be continually portrayed as one of tragic choices with the framings enhancing – or perhaps even creating – a dichotomy between winners and losers in the battle over budget allocation and scarce health resources. We therefore wished to gain an overview of the public discourse and the stories that are told within it – and to explore if there were missing aspects to this discourse, which, if brought to light, could perhaps challenge the persisting view that priority-setting in relation to cancer is always and indubitably an issue of tragic choices.

In order to study the framings within this debate, we applied framing theory. We used a methodological approach based on the work of Erving Goffman, who argued that frame analysis is a study of the cognitive organisation of experiences (Goffman 1974). He argued that we cannot understand the issues in front of us without consciously or subconsciously framing them in our minds, and that each and every issue we consider is consciously or subconsciously placed within a framework that makes it understandable to us. Through framing theory, one can identify the lenses through which we view society, and also identify the frames that are chosen for us in story-telling, and in the media. Framing theory thus allows us to consider not only *what* is said in the news media, but also *how* it is said. Through applying this theory, we can thus also – importantly – begin to understand *why* it is said.

In order to identify frames in our material on the discourse on priority-setting and cancer, each article was structurally examined along the following four considerations: (1) what is the problem, (2) who are the actors, (3) where is the allocation of power, and (4) what sources have been used. In order to collect relevant data for our studies we used the Norwegian newspaper database *Atekst*. We included all articles published on the issue of priority-setting and cancer from the eight largest Norwegian national newspapers, spanning from January 1st 2013 to December 31st 2016. This provided us with a total of 439 newspaper articles. We chose the above-mentioned time frame due to the developments in the public debate on cancer and priority-setting, an issue that gained an increasingly large presence in the news media around this time. This was largely due to a controversy surrounding the melanoma drug Ipilimumab, which took place in 2013. Ipilimumab is a particularly costly cancer drug and at the time, the organ which considers the use of new drugs in the public healthcare system, the Decision-Making Forum (Beslutningsforum), recommended that the drug should not be made available in Norwegian public hospitals, due to an unfavourable cost-benefit analysis. This gave the issue considerable media attention, with a large number of newspaper stories featuring individual patients who would consequently be denied the new drug, and highlighting the tragedy and alleged inherent unfairness of this decision. Following considerable news coverage and a loud public outcry, the health minister in Norway at the time, Jonas Gahr Støre, chose to overrule Beslutningsforum's recommendation and allow for the use of this drug in public hospitals. This kind of political intervention in the process of decision-making and distribution of health resources was unprecedented at the time. One could argue that this incident gave birth to a new wave of public engagement surrounding priority-setting in relation to cancer treatment in Norway, as it demonstrated that public and political uproar had the power to alter recommendations and decisions made by the existing, and well-established, framework for priority-setting. As the study was led in 2017, the data collection cut-off was the end of 2016.

The pervasive theme in the identification of the framings of these articles was a striking lack of nuance. Although we identified nine separate frames as the most common ones, in almost every frame the problem was the same – namely that of the suffering patient denied an expensive cancer drugs, and the apparent injustice of these decisions. The constellation of actors was similarly unvaried, with the patient/doctor/patient organisation being pitched against the priority-setting system/health authorities/politicians. The allocation of power was determinedly placed with the system/authorities, whilst patients/doctors were presented as virtually powerless. And finally, the sources appeared to be surprisingly unvaried, mainly consisting of statements from patients, doctors or the Norwegian Cancer Society. Following a closer study of the articles, we gradually identified not only the lack of nuance of this discourse, but also what appeared to be four major assumptions underlying every newspaper article. These assumptions were so prevailing in the material that they arguably form the basis of the discourse itself. They read as follows:

1. Cancer drugs are de facto expensive, and one does not and should not question why.
2. These cancer drugs work, and there is no need to question the validity of this claim.
3. Prolongment of life for a cancer patient is an absolute and unproblematic good, and any gained time – whatever shape or form that time has – is a blessing.
4. Patients, and their doctors, own the truth about cancer and cancer drugs, and "outsider" views on these are irrelevant, or unwelcome.

Once we had identified these four assumptions, we were surprised by how very prevalent they were in the material. The adherence to these assumptions thus suggests that they are indeed four central *premises* that underlie the public discourse on cancer and cancer drugs. Furthermore, our study of these articles covered a timespan from 2013 to 2016, yet there was very little evidence to suggest that the discourse went through any meaningful development during that time period. There did not seem to be any progression in either its structure or in its content, nor any real evolution to the arguments within it. It seems to us that there existed, purposefully or not, an apparent unwillingness to challenge these premises that underlie the discourse, and this suggested to us that the continued adherence to these premises is indeed a root cause to the stagnation of the discourse itself.

Precision Medicine as a Medical Revolution

While the first study mainly considered the dominant framings of the priority-setting debate around cancer and expensive cancer drugs, the second study considered how the future of cancer research is imagined by cancer researchers and other prominent actors in the field of cancer. In addition, we examined how research and scientific perspectives on cancer treatment is framed in the editorial news. Based upon a qualitative exploration, we examined the dominant framings of cancer research coverage in three national Norwegian newspapers: Aftenposten, Dagbladet and Verdens Gang. The reason for choosing these three papers is that they are three nationwide newspapers, thus covering and representing stories from the whole country, as well as being the three most read papers in the years of the study, namely 2016–2017 (Medienorge 2019). We wanted to focus on the papers that target the general public, hence we chose to not include news sources that were specifically targeted to people associated with the health field.

As the study commenced in 2018, we wanted to use the most recent material available. Adding to this, we also wanted to cover a fairly broad period. The material was thus gathered from *Atekst* in the period from January 1st 2016 to December 31st 2017, and the search resulted in 464 texts. This number was reduced to 53, to ensure that the material included only the texts that mentioned cancer research and cancer drugs and excluding those that did not.

We applied a qualitative content analysis and studied expressions, opinions and argumentations that allowed us to identify different future imaginaries within the news content. The theoretical framework applied to the study was based on the analytical term *sociotechnical imaginaries*. This theoretical tradition argues that there exist strong collective visions of a desired societal, scientific and technological future (Jasanoff and Kim 2009, 2015). They are considered particular visions for a future that tends to be presented as optimistic and positive. These imaginaries can influence the choices made in research, and accordingly the future that is created based on these visions. In the field of cancer research, previous international literature has established that there exists a future vision on precision oncology in the Western society, and that this consists of strong normative aspects (Blasimme 2017; Blanchard and Strand 2017).

We wished to examine whether there existed strong future visions surrounding cancer research and treatments in the Norwegian media discourse – and if so, *how* they were articulated and argued for. In the material, we found that the vision of precision oncology is strongly present in the Norwegian media discourse. By exploring the material, which included both editorial and non-editorial news content (e.g. debates, chronicles, reader letters), we identified three variants of the sociotechnical imaginary concerning precision medicine. The three main visions and imaginaries which proved prominent, all related to precision medicine and the future of cancer treatments, were as follows:

1. Precision medicine, particularly immunotherapy, will revolutionise cancer treatment.
2. Artificial intelligence will make diagnosis and treatment more effective.
3. The cancer-industry will become Norway's next billion-dollar industry.

These three future visions, all tied to precision medicine, thus comprised (1) a scientific aspect, (2) a technological aspect, and (3) a socioeconomic aspect. In the first imaginary, immunotherapy was imagined as revolutionising cancer treatment by securing a targeted treatment for each and every patient. The second imaginary praised the use of emerging technologies, and how artificial intelligence and machine learning in particular were going to be used as a central part of future diagnostics and prognostics, and would make treatments more effective. The third and final imaginary revolved around harvesting the socioeconomical benefits of building a Norwegian cancer industry, and the idea that Norway will lead the way in the development of precision medicine. In addition, these visions were articulated as the *only* natural direction that the future of cancer research and treatment should move towards. These imaginaries were stated by oncologists, researchers and leading figures in the cancer industry, mainly through opinion pieces in non-editorial news content.[2]

In this study, we also considered how the topic of cancer research was handled by journalists, and if the sociotechnical imaginaries identified were ever questioned or challenged. Through examining the editorial news content, we found that the news articles were characterised by the same future imaginaries as the actors in the field of cancer had stated in their own pieces. The news articles further tended to use words such as "revolution", "miracle" and "breakthrough" when presenting new cancer research, particularly immunotherapy.

The prevalent frame that journalists used to cover cancer research was the *sensation/information* frame, identified through the journalists' use of sensationalistic headlines and lead paragraphs, as well as the neutral and informational body text. Here, they usually cited a cancer researcher or an oncologist, and used almost exclusively only a handful of prominent actors in the field of cancer as their sources. By letting the same sources comment repeatedly, they both limit the diversity that the media should strive for, and also contribute to amplifying their opinions and views at the expense of others. Thus, few voices are shaping the framework and understanding of an entire field. The future vision on precision medicine, as outlined above, was prevalent throughout the timespan of the material and was not met with

[2] Some examples of headlines from the opinion pieces: "Future cancer treatment – bring your own data" Aftenposten, 21.12.17", "Open the Nordic boarders for research and treatment" Aftenposten, 18.05.17", "Machines must learn to detect how dangerous the cancer is" Aftenposten, 17.08.17" [our translations].

a single critical question by the journalists, and had no counter-visions to challenge it.

Further, our study shows that the editorial frames on cancer research is characterised by optimism concerning precision medicine. The analysis demonstrated that the news articles tend to favour research related to precision medicine, by referring to this research as continually producing highly promising breakthroughs in cancer research. We found exceptionally few texts that challenged this vision, and even fewer that presented an alternative future vision of cancer treatments and cancer drugs. Thus, we found that journalists tended to frame cancer research in the same way as the actors in the field of cancer – arguably overly optimistic, and undeniably lacking in nuance. This demonstrates the lack of variety in the coverage of the cancer research in the news, both in terms of what research is referred to in the news media, but also in *how* the research is presented – with an (overly) optimistic and deterministic framing. It also demonstrates that cancer researchers' visions seem to be reproduced and reinforced by journalists and editors in the Norwegian news media – lending these visions a validity they cannot always rightfully claim.

Discussion

The common and general findings from our studies are (1) the lack of nuance in the discourse on cancer and cancer drugs, and (2) the lack of variation in terms of how the issue of cancer and cancer research is covered in the news media. This lack of nuance has in turn arguably led to a lack of progression in the discourse and public debate. We would argue this stagnation continues to hinder a true understanding of the issues of cancer disease, cancer treatment and cancer research amongst the general public, with an overly simplistic and un-nuanced presentation of what are arguably complex and ambiguous issues. Our most noteworthy findings are therefore the aspects that remain *absent* from the discourse.

In the following section, we wish to highlight and address some of these absent aspects, and further consider how and why the discourse has become what it is today. We thus hope to further challenge the adherence to the premises that underlie the discourse, and in doing so, to confront the notion that the issue of cancer and cancer drugs must solely be an issue of tragic choices.

Why Have the Premises Remained Unchallenged?

Cancer treatment is one of the largest research fields in medicine, and cancer illness is increasingly common and thus also an increasingly relevant aspect of public health. Yet, as we have demonstrated, the discourse on these issues seems to have stagnated, remaining virtually unchanged both in form and substance throughout the years we have studied it. Thus, despite the size and alleged promise of the field of cancer research, the issue of cancer treatment remains framed as one of tragic choices. This paradox is arguably explained in part by a stubborn adherence to the four abovementioned premises – an adherence that monopolises the discourse on priority-setting and expensive cancer drugs to the extent that it inhibits any other understandings of the issue beyond that of tragic choices. After identifying these premises, apparently so deeply ingrained in the discourse, the question therefore soon arose: why have they not been challenged?

One explanation for this unwillingness to challenge the four premises, and to us the most salient, stems from the standpoint of what is, or appears to be, morally acceptable. It seems, from our study of the material, that these premises claim a moral and ethical superiority. They speak to the rights of the suffering cancer patient to be both heard and saved, as well as to the insistent hope that the field of cancer research will indeed provide that salvation, through curative drugs. Because the premises thus appear to stand on moral high ground, the prospect of challenging them in contrast becomes unethical and cold-hearted. The sociological perspective offered by Norwegian sociologist Jill Loga helps us to understand the intricacies of such a mechanism, through her analysis of the interface between morality and politics in the Norwegian public sphere (Loga 2004). Loga proposes the idea of *godhetsmakt*, which we have translated into the term "power of goodness". She considers how professing goodness affects discourse, and argues that by openly claiming to represent goodness, an argument gains a superior stance that makes it difficult for an alternate stance to claim legitimacy. She states that:

> [...] open goodness has the ability to define its opponent. This creates a great discursive power. It becomes impossible to oppose against the open goodness, because one would appear as evil, cynical or egotistical. If one can administer and preach from the position of goodness, one becomes unimpeachable and immunized against criticism. Goodness needs only be expressed in order to become a conclusion (Loga 2004, p. 323).

This notion is well suited when considering the underlying mechanisms of the discourse on priority-setting and cancer drugs, where each of the four premises arguably claim a power of goodness. We will demonstrate this with a brief consideration of each premise, understood through the power of goodness:

1. Cancer drugs are expensive, but that is irrelevant because their administration saves lives.
2. Cancer drugs work, and questioning this is indefensible because that amounts to questioning if cancer patients' lives can be saved at all – which is painful to accept and cruel to consider.
3. Any prolongment of life for a cancer patient is an absolute good, because it is evidence that we are trying to save them – and failing to try is wrong, no matter the pain or suffering our efforts might cause them.
4. Patients, and their doctors, own the truth about cancer and cancer drugs – and questioning their ownership of this is disloyal to them, and both undermines the academic proficiency of medical professionals and enhances the suffering cancer patients are already subjected to.

These premises, by supposedly supporting the cause of cancer patients, claim moral superiority, rendering all other arguments invalid as they in contrast represent an immoral stance. Indeed, by opposing the goodness of the four premises, counterarguments become unethical. Thus, these premises gain discursive control, and through their incontestability maintain a monopoly of the discourse. Loga further suggests that through lack of opposition, the goodness discourse becomes self-reinforcing as 'discursive power forces one down a goodness spiral' (Loga 2004, p. 324). The discourse thus loses its nuance and stifles the many and varied opinions one intuitively would have expected these issues to give rise to. The four premises become four truths, as the public for lack of opposing arguments accepts them as foregone conclusions. This demonstrates, as Loga argues, 'the power of definition in the incontestable' (Loga 2004, p. 323).

The notion of the "power of goodness" becomes prominent also in the seemingly insufficient ability of the news media to challenge the current future imaginaries of cancer treatment – exemplified by the inherent optimism surrounding immunotherapy. The actors in the field of cancer, and their normative visions, are rarely (if ever) challenged by journalists, and one could argue that this reflects the idea of one party claiming the position of goodness, and the other thus rendered unethical. By challenging experts who are trying to do something indisputably good – treating, or at times curing, cancer – one is portrayed as the party opposing the inherent goodness of helping these patients. Journalists seemingly accept the optimistic narrative surrounding precision medicine without hesitation, and indeed present it as the solution we have all been waiting for. The discourse thus becomes oversimplified and unilateral – an injustice to a field as complex as that of cancer research.

Cancer and Cancer Research: An Issue of More Than Tragic Choices?

It remains surprising to us that the issue of priority-setting and cancer drugs is so stifled in the public discourse, particularly because our experience from both private conversations and public debates suggests that it is an issue that ignites impassioned opinions – arguably for several reasons. First, the story of a suffering cancer patient is relatable; as populations age and incidence rates of cancers increase, most of us have either had cancer or know someone who has. The issue is relatable because it is personal. This also means that the issue is important to us because we know that the decisions made top-level today have a direct impact on our own future health and treatment options, or on that of our loved ones. Thus, we are deeply invested in the decisions made by authorities, and the public debate surrounding these decisions – although it is not our health that is threatened today, or our story that is portrayed in the newspaper right now, we know it could be us tomorrow.

Secondly, this issue arguably stirs us deeply precisely because of its tragic nature. Our study of the material suggests that the general public – unsurprisingly – holds the view that there is something inherently wrong and unjust in denying a suffering patient the right to treatment, if we believe that treatment might help them. It seems to go against our nature to accept death – particularly deaths we believe might have been prevented, or at least delayed. However, we would argue that these views are partly based on understandings of the issue that are misinformed, largely due to the discourse's adherence to the abovementioned four underlying principles. In order to achieve a meaningful debate on the issue of cancer and cancer drugs, it is essential that arguments within this debate are based on accurate understandings of the issues it concerns itself with. This seems particularly important in relation to the second and third premise. Without a genuine and critical consideration of the actual efficacy of these drugs, it seems impossible to have any meaningful debate on the decisions made in relation to these drugs' implementation – and only serves to further polarize the debate. Further, without a genuine understanding of what the life and time gained through treatment with these expensive drugs indeed looks like, we are doing cancer patients a disservice. If we ignore the possible negative outcomes of failed treatment and harmful side-effects, we are uncritically arguing for costly treatments that potentially rob patients of whatever quality of life they might have had in their final months or years. Herein lies real injustice.

Because adherence to these premises undermines the discourse, as well as any meaningful progression in the debate, the issue of cancer and cancer treatment remains one of tragic choices. Further, even in the world's richest countries, as highlighted by the covid-19 pandemic and the very real and tough choices that have had to be made in emergency rooms and intensive care units all over the world, our health care expenditure has a limit, and healthcare resources are exhaustible. One could argue that the persistent tragic choices framing of the issue of cancer and cancer drugs further enhances the imbalance between financial constraints and public expectations, and blocks our path not only to a more constructive public discourse, but also to attaining an overall better outcome and improved care for cancer patients.

Alternative Imaginaries

Having stressed the importance of challenging the current discourse on cancer and cancer research in Norway, a natural next step is to consider what the discourse could look like when the premises that underlie it are questioned – and if this might alter the current framing of the issue as one solely of tragic choices. We have done this by considering alternative future imaginaries. Whilst frames say something about the past and the present and how we understand reality today, imaginaries are a particular way of "futuring", or imagining what could be. Interestingly, one could argue that at least one of the premises that underlie the discourse, and thus one of the framings that are prominent in it, is in fact more resemblant of an imaginary than of a framing. This pertains to premise number one, namely that new and expensive cancer drugs work and that their efficacy need not be questioned. In studying the discourse more closely, it becomes apparent that this framing often points not to the current status of cancer treatment and cancer research, but more towards the hope of what they might become. There are countless articles pointing to the idea that the drugs currently being developed *will* work, that they *will* have less side effects, and that they *will* be cheaper – and that this justifies both their use today and the investments we place in pharmaceuticals in order to allow for their enhancement. This is a sidenote to our following exploration of imaginaries, but nonetheless worth noting.

The imaginaries we now present are intended to complete each other in terms of providing an interplay between various political and philosophical levels of the issue at hand. Some are concerned with institutional change and political reform. Others suggest the need for a cultural shift in perceptions of how we as a society view illness and health. We have also considered philosophical aspects of society's views on life and death, and how future imaginaries on this level could challenge current perceptions. Though we will not provide comprehensive analyses of these imaginaries, we propose them in order to challenge the current framings of the issues relating to cancer and cancer research, and the premises that underlie the discourse on these issues (Stenmarck et al. 2021).

A cornerstone to the issue of priority-setting and cancer care is the immense costs of new cancer drugs, and the price tags they are often associated with. Within a market economy and with fiscal policies that necessarily put constraints on health care budgets, this makes healthcare a commodity that can be bought – at a price. One alternative imaginary is institutional change, whereby one, through major changes to social policies and financial models, changes the system through which we provide healthcare. This would entail a comprehensive analysis and a considerable review of the current healthcare model, which allows Big Pharma to set the prices they want rather than the payment they need (WHO Technical report 2018).

A second imaginary relates to current public perceptions of illness and health. It would seem that current cultural perceptions of health deem it the opposite of any discomfort – indeed, the WHO defines health as 'complete physical, mental and social well-being' (WHO FAQ 2020). This suggests that any deviation from a perfect state of health equals disease, and arguably enhances the perception we hold that any deviation from this state must therefore be treated. This coincides poorly with healthcare budget limitations, as well as with ageing populations, and we therefore wish to propose an alternate imaginary where perceptions of disease and health are challenged: where we accept health as a continuum, and where absence of disease is not the only element that defines it. Doing so would also challenge the premises that underlie the current discourse, particularly the premise that any prolongment of life for a cancer patient is an unequivocal good. If that prolonged period involves a level of suffering, we are arguably not providing the patient with an overall improved health outcome.

A third imaginary which we wish to briefly explore is perhaps controversial, but nonetheless interesting as it highlights how pervasive the notion that disease *must* be treated has become. In this imaginary we consider a scenario where society as a whole comes to the conclusion that certain cancers have become so prevalent, their treatments so costly, and the treatment results often so marginal that we decide to abandon the notion that these cancers can and should be *cured*, and rather take the approach that the healthcare system aims to provide *care*. That we choose to no longer provide treatment as such, but rather to help the affected individual by alleviating their pain, and to cope with the suffering associated with a terminal illness. In doing so, we accept disease as a necessary evil of life – rather than denying death as a basic condition of it. This imaginary may seem both radical and crude – which arguably highlights how deeply entrenched healthism has become in modern society (Stenmarck et al. 2021).

A fourth and final imaginary is related to the power given to single future visions by the media. The prominent vision of precision oncology is a vision that extends beyond just the medical, and should, in our opinion, be considered as such. As this vision is founded on normative concepts and has the potential to form and alter fundamental societal structures, it should be critically examined in line with other political issues. Thus, understanding that cancer research and the alleged promise of precision medicine is more complex than what is being portrayed in and communicated through the media, is an understanding that should be taken seriously. To achieve a responsible media discourse on cancer and cancer research requires both that journalists and editors are more critical of established truths in the field of the research they cover, and that researchers themselves dare to communicate potential challenges and weaknesses in their own visions of the future.

Lack of Nuance in the Field of Cancer and Cancer Research: Why?

Thus far, we have attempted to illustrate the stagnation of the discourses on cancer drugs and cancer research. We have explored why the four identified premises of the discourse remain unchallenged, particularly in light of the notion of the power of goodness. We have questioned if perhaps the issues of priority-setting and cancer could be framed through other lenses than solely that of tragic choices, and attempted to highlight the disservice one does both to cancer patients and to the discourse itself by employing this lone view. Finally, we suggested some future imaginaries, as a way of exploring other potential futures of cancer treatment and cancer research. Throughout, we have attempted to highlight the obvious and common finding of our research: *the complete lack of nuance* on the discourses of both priority-setting and cancer, and cancer as a research field. As this chapter draws to a close, we wish to pose the obvious question: why? Why is a field such as cancer and cancer research, so obviously complex, reduced to a presentation in the news media that so poorly represents the realities of its complexities?

In the material we have studied, all of which was sourced from the news media, we made the interesting observation that while studying the over 500 articles we had collected, it was surprisingly difficult to distinguish between them in terms of style of writing, type of article and newspaper outlet. Not only was there a very stereotypical presentation of cancer as a scientific field, with a clear focus on the black and white "facts" of its scientific findings and a lack of appreciation of the ambiguities inherent to it. But more surprisingly, whether the article was a news report, an opinion piece, an editorial, a column or a feature article, the style was markedly similar and the angling – here too – un-nuanced. In other words, we could not from the style of writing or layout clearly distinguish which pieces were written by professional journalists, and which were commentaries or opinion pieces from the general public. We will not delve deeply into what the causes of this may be, but simply question if this is suggestive of failings in the state of journalism generally. It seems one might deduce that the news media holds itself to sub-par standards in its presentation and framing of cancer and cancer research – which is startling, considering the vast interest it has shown on these issues.

That being said, it would be an over-simplification to point only to faults of the news media as an explanation for the lack of nuance in the public discourse on cancer drugs and cancer research. We would argue that the lack of appreciation for these nuances is evident not only in the media, but in broader understandings and a wider context pertaining to the field of cancer more generally. Considering the scope both of the disease that is cancer, and of the research field that seeks to obliterate it, it is indeed peculiar that there is not a greater acknowledgment or appreciation for the complexities within it. A central purpose of this book is to attempt to reduce some of the ambiguity in the field of cancer research – a first step must therefore be to acknowledge that there is, for whatever reason, considerable resistance to accepting this ambiguity. Perhaps this is in part due to a faulty public discourse, oversimplified and unappreciative of the complexities of the issues it concerns itself with. Perhaps it is caused in part by an unwillingness to accept these complexities, because it suggests that cancer researchers cannot easily eradicate this illness, and some of the deaths caused by it are thereby unavoidable. Perhaps it has to do with the imaginaries we have started to accept as realities – that the future of cancer treatment, precision oncology, will indeed provide us with a molecular precision that obliterates ambiguity. Perhaps researchers are doing too poor a job of communicating the enigmatic character of their research object, fuelling hopes that cancer is a fully understood, soon-to-be always-curable disease. Perhaps there are other forces, political or financial, that have something to gain from oversimplifying our understanding of the field of cancer research, feeding our hopes that it can cure the diseases it researches, and thus also fuelling our willingness to pay for it. These are questions we will not attempt to answer – but our analyses and findings suggest they are worth contemplating.

Conclusion

The issues of cancer, cancer treatment and cancer research are complex, and will unfortunately continue to involve a certain degree of tragedy: they concern the loss of health, or ultimately life, for millions of people. Whatever progress precision medicine and cancer care makes in the coming years, there is no cure to all cancer disease – and we remain mortally vulnerable to its advance. We face a reality where patients will continue to be denied access to drugs they want, whether for lack of any expected positive outcome or due to financial restraints.

Having performed studies that considered the public discourse on these issues from 2013 to 2017, it therefore alarms us to see that the discourse fails to accurately depict the realities of priority-setting and cancer, as well as the promise of cancer research. Further, it is regrettable that the discourse continues to so heavily rely on premises that lack vigilant questioning and consideration. We believe the adherence to these premises enhances the perception that cancer and cancer treatment is an issue *solely* of tragic choices – and thus inhibits alternative perspectives that would better do justice to the complexity of these issues. We argue that a wholesome and sober approach, through a more responsible media discourse, might reveal that these issues need not *solely* be framed as issues of tragic choices. Cancer disease, cancer care and cancer research are more than that, and undermining their complexity does neither patients, doctors, researchers or society as a whole any favours.

References

Bennett, W.L., R.G. Lawrence, and S. Livingston. 2006. None dare call it torture: Indexing and the limits of press independence in the Abu Ghraib Scandal. *Journal of Communication* 56: 467–485.

Blanchard, A., and R. Strand. 2017. *Cancer Biomarkers: Ethics, Economics and Society*. Kokstad: Megaloceros Press.

Blasimme, A. 2017. Health research meets big data: The science and politics of precision medicine. In *Cancer Biomarkers: Ethics, Economics and Society*, ed. A. Blanchard and R. Strand, 95–110. Kokstad: Megaloceros Press.

Centers for Medicare and Medicaid Services, CMS. 2018. *National Health Expenditure Data*. https://www.cms.gov/research-statistics-data-and-systems/statistics-trends-and-reports/nationalhealthexpenddata/nationalhealthaccountshistorical.html. Accessed 20 May 2019.

Eide, M. 2011. *Hva er journalistikk*. Oslo: Universitetsforlaget.

Entman, R. 1993. Framing: Toward clarification of a fractured paradigm. *Journal of Communication* 43 (4): 51–58.

Fleck, L.M. 2013. "Just Caring": Can we afford the ethical and economic costs of circumventing cancer drug resistance. *Journal of Personalized Medicine* 3: 124–143.

Galtung, J., and M.H. Ruge. 1965. *The Structure of Foreign News. The Presentation of the Congo, Cuba and Cyprus Crises in Four Norwegian Newspapers*. Oslo: Peace Research Institute.

Gamson, W.A., and A. Modigliani. 1989. Media discourse and public opinion on nuclear power: A constructionist approach. *American Journal of Sociology* 95: 1–37.

Goffman, E. 1974. *Frame Analysis: An Essay on the Organisation of Experience*. Massachusetts: Harvard University Press.

Jasanoff, S., and S. Kim. 2009. Containing the atom: Sociotechnical imaginaries and nuclear power in the United States and South Korea. *Minerva* 47: 119–146.

———. 2015. *Dreamscapes of Modernity. Sociotechnical Imaginaries and the Fabrication of Power*. Chicago/London: University of Chicago Press.

IMS Institute for Health Informatics. 2016. *Global oncology trend report*. https://morningconsult.com/wp-content/uploads/2016/06/IMS-Institute-Global-Oncology-Report-05.31.16.pdf. Accessed 2 July 2020.

Loga, J.M. 2004. *Godhetsmakt. Verdikommisjonen – mellom politikk og moral* [The power of goodness. The value commission – Between politics and morality]. PhD thesis for the University of Bergen, Norway.

Medienorge. 2019. *Lesertall for norske papiraviser – resultat*. http://medienorge.uib.no/statistikk/medium/avis/273. Accessed 1 Feb 2021.

National Cancer Institute. National Cancer Act of 1971. https://dtp.cancer.gov/timeline/flash/milestones/M4_Nixon.htm. Accessed 22 Mar 2021.

Schudson, M. 2003. *The Sociology of News*. New York/London: W.W. Norton & Company.

Stenmarck, M.S., C. Engen, and R. Strand. 2021. Reframing cancer: challenging the discourse on cancer and cancer drugs—a Norwegian perspective. *BMC Med Ethics* 22, 126 (2021). https://doi.org/10.1186/s12910-021-00693-5

Pressens faglige utvalg. 2021. *Vær Varsom-plakaten*. https://presse.no/pfu/etiske-regler/vaer-varsom-plakaten/. Accessed 25 Feb 2021.

WHO. 2018. *Technical report: Pricing of cancer medicines and its impact*. https://apps.who.int/iris/bitstream/handle/10665/277190/9789241515115-eng.pdf?ua=1. Accessed 1 Sept 2020.

———. 2020. *Frequently asked questions*. https://www.who.int/about/who-we-are/frequently-asked-questions. Accessed 26 Sept 2020.

Cancer Currencies: Making and Marketing Resources in a First-in-Human Drug Trial in Denmark

Line Hillersdal and Mette N. Svendsen

Vignette: Trial Qualities on Display

It is early morning. Only the day before, Line received her final clearance to participate in a 'site initiation visit' between a cancer clinic in a Danish public hospital and a sponsor from a multi-national pharmaceutical company. The meeting's purpose is to settle and formalise agreements that will enable and kick off a new cancer drug trial. The trial to be started is a 'first-in-human trial,' which means that the drug has so far only been tested on animals. The new drug is hypothesized to affect a specific genetic mutation in colon cancer tissue. Still, at this early drug development stage, it is only the toxicity of the drug that will be assessed. If the sponsor finds the hospital unit to be a suitable place to locate the study, the drug will be included in the unit's targeted therapies offered to incurably ill patients who have exhausted all standard treatment options.

Line arrives at the office building just across the hospital, and two of the trial nurses from the unit, Lena and Sarah, greet her and invite her to take a seat at the table in the small room. They will be responsible for coordinating the trial's clinical procedures, supervising the nurses in the clinic and schedule treatment with the enrolled patients if the sponsor decides to locate the study at the unit. Present at the meeting are also a sharply-dressed medical doctor flown in from the US-based pharmaceutical company sponsoring the drug, the overseas lead, a research project nurse from their Danish subsidiary company, and 'the monitor' – an external consultant – whose job is to control and assess the quality of data collected during the trial. Lena,

L. Hillersdal (✉)
Department of Anthropology, University of Copenhagen, Copenhagen, Denmark

Centre for Medical Science and Technology Studies, Department of Public Health, University of Copenhagen, Copenhagen, Denmark
e-mail: line.hillersdal@anthro.ku.dk

M. N. Svendsen
Centre for Medical Science and Technology Studies, Department of Public Health, University of Copenhagen, Copenhagen, Denmark
e-mail: mesv@sund.ku.dk

A. Bremer, R. Strand (eds.), *Precision Oncology and Cancer Biomarkers*, Human Perspectives in Health Sciences and Technology 5,
https://doi.org/10.1007/978-3-030-92612-0_4

the trial nurse, shows Line the day's program, which lists time slots for the next 5 h with new people dropping in and out of the meeting every half hour. Lena explains that the purpose of the meeting is to go over all procedures to reassure the sponsor: that all relevant staff have been trained according to the regulatory standards that the hospital can recruit the right patients for the trial, and all clinical procedures will follow protocol standards.

During the first 30 min, the unit's principal investigator, Doctor Mathew, joins the meeting to go through the book-thick trial protocol, the details of the quality assessments, and the examinations to be performed in the trial. At one point, the industry collaborator voices his concern about recruitment. He doubts that the unit will be able to enrol enough patients within the timeframe of the trial. Doctor Mathew promptly responds, *"We already have relevant patients waiting, we can start tomorrow!"* During the next hours, staff from across the hospital arrive at their allocated time slots: 'pharmacy', 'lab', 'economic officer' – all to ensure the sponsor that their part of running the trial will be smooth and efficient as logistics of recruitment or data handling are brought forward again and again. Showing their support, the nurses noddingly confirm the staff's short presentations and assure everyone in the room by saying: *"you couldn't get a more qualified person for this task*".

The above vignette describes an event from Line Hillersdal's ethnographic fieldwork exploring the negotiations and practical work involved in setting up new cancer drug trials for personalised medicine in a public hospital in Denmark. As described, the meeting unfolded as a neatly orchestrated demonstration of the unit's professionalism, reliability and procedural excellence. What struck us in the situation was how it unfolded as a business meeting in which the hospital staff presented themselves as a unit worth investing in, an ideal research site containing the state of the art infrastructure of patients, clinical staff and equipment needed to run a high quality trial. The way the unit 'sold' itself and its capacity to recruit relevant patients and to execute clinical trials according to the highest standards made us curious. What, in effect, were they selling, and how do collaborations with pharmaceutical companies intersect with the research and care practices at the unit?

Introduction

The Danish healthcare system builds on universalism and tax-financed health care principles and provides free and equal access to healthcare for all citizens. At the same time, recent developments within welfare state service delivery have shown an increasing reliance on public-private collaborations and a push towards turning core welfare provisions into a profitable business working on an international scale (Larsen and Stone 2015). Particularly within cancer treatment development, the cost of medical research is increasing, and many policymakers see partnerships between private and public partners as mandatory to sustain public welfare services (Ministry of Finance 2015). In Denmark, personalized medicine has become an important focus. The Danish Regions governing Danish healthcare and the Danish government announced a national strategy on personalized medicine emphasizing both economic gain and an improved treatment based on genome sequencing and tumour profiling in oncology (Ministry of Health 2016; Danish Regions 2015). The claim seems to be to sell welfare in order to save welfare. However, research shows that financial interest may also involve competing interests between public and private actors when publicly funded research receives fewer funds and innovation in treatment increasingly comes to rely on industry support (Healy 2004; Sismondo 2018). Globally, pharmaceutical companies have taken the front seat to develop new targeted therapies within basic cancer research. Consequently, industrial concerns about, for instance, marketability come to shape how research is conducted and what kinds of questions are explored and which kinds of hypotheses are tested (Fisher 2009; Rajan 2017).

In this chapter, we follow this development on the ground by investigating the daily work of setting up and running so-called early cancer drug trials in collaboration with pharmaceutical companies in a public hospital in Denmark. We draw on fieldwork conducted, from November 2019 through March 2020. During fieldwork, Line followed the oncologists' daily activities, observing patient consultations, and enrolment meetings in which patient allocation in trials was discussed. She also followed the nurses as they administered the treatment, filled out trial protocols, and reported data to industry partners. In addition, she interviewed clinical staff, research nurses, data consultants, industry partners and participating patients. Based on this ethnographic research, we analyse how practices of competition, investment and exchange shape how welfare resources for personalised medicine are defined, produced and offered. We argue that qualities facilitated by the welfare state – i.e. fast-tracking trial procedures, high-quality data and high compliance of research subjects – become *currencies* transactable on the global market for drug development.

By exploring the day-to-day collaboration between public institutions and private industry actors invested in developing personalised medicine for treating cancer, the article contributes to the field of political economies of health care markets. Our case from a Nordic welfare state in this regard offers a unique perspective. As cancer research becomes increasingly entangled with big pharma interests, it becomes crucial to understand how welfare state practices and values intersect with commercial interests. We show that this development exposes the inherently contested character of the current welfare state, aiming to secure the health of citizens and the wealth of the state.

Cancer Currencies and Trial Qualities

When Tarkalla and colleagues argue that the promise of personalised medicine is determined "more in terms of wealth than health" (Tarkalla et al. 2019: 149), they point to the presence of a discourse of investment and potential economic gain going hand in hand with the promise of new discoveries in cancer treatment. Analysing the Finnish national policies and strategies, they show how justifications related to science and health care are being eclipsed by economic and business rationales. Similarly, anthropological work on the global medical industry has unravelled the role of pharmaceutical companies in defining clinical research, determining available treatment, and setting drug prices (Sismondo 2007; Petryna et al. 2006; Dumit 2012). In his work on drug trials in India, Sunder Rajan provides critical analytical attention to the confrontation between global economic structures and local democratic institutions, in particular when he unfolds how national policy changes have been implemented to facilitate corporate interests with major consequences for public health in India (Rajan 2017). In general, research on bio-economies, including the economy of trialling, has exposed the exploitation of vulnerable populations (Petryna 2009) when tissue and biological samples become commodified in transnational relations of exchange (Cooper 2011; Mitchell and Waldby 2010; Rose and Novas 2005; Rajan 2006).

In contrast to these studies of transnational bio-economies, our case is situated in the Danish health care system, where care and treatment are delivered within the frame of a social contract between state and citizen. In this context, entering a clinical trial and becoming a research subject is inseparable from relationships of state-citizen reciprocity and questions of social belonging. In stepping into a Danish hospital unit, we analyse the practical and collaborative work that goes into running the trials between the clinic and the pharmaceutical company in a welfare state that values egalitarianism and comprehensive universal welfare services.

By analysing how the entanglement and collaboration between the pharmaceutical industry and the clinic shape and stir research and care practices, we draw out specific 'trial qualities,' such as a strong bureaucratic infrastructure, well-described and compliant patients, and accessible public biobanks. Notably, we highlight how the 'gold-standardness' or procedural excellence of the trial (Timmermans and Berg 2003) is facilitated by these local infrastructures, patients, and biobanks yet becomes recognized as a universal value that can be bought and sold on a global market. With the notion of 'cancer currencies', we aim to capture how 'trial qualities' get an economic life of their own when transacted beyond the local clinic and enter a global market. We argue that the competencies and resources already made available by a strong welfare state condition what can be 'made in Denmark' and sold to other countries to make money on a global market.

Unit One[1]: A Research Business at the Heart of a Public Hospital

Unit One at the national hospital in Denmark specialises in early phase one drug trials of new personalised or targeted therapy, which targets specific genetic mutations in cancer tumours. The unit offers treatment to patients with advanced cancer

[1] The names of the hospital unit, staff and patients are all pseudonyms.

disease who have exhausted all standard treatment. The unit receives approximately 500 patients each year, with around 100 being included in trials. Upon entering the unit, all patients will be offered a whole-genome sequencing of cancer tissue to identify whether they have a specific genetic mutation that a trial drug can target. If their cancer tissue expresses the genetic mutation, they are allowed to enter the trial. Only one in five will have mutations leading to enrolment in such a trial. Most other patients will be redirected to trials testing other new drugs though not selected based on genetic targets. All trials at the unit are in the earliest drug development phase (phase 1) testing the dosage's safety and toxicity level (Smith and O'Donnell 2006). The aim of the early trials is to identify at what dosage the drug is likely to have an effect without causing severe side effects. At this stage, it is not expected that the drug will have an effect on the individual person's cancer disease.

Unit One has a unique history. From the start, it was designed to be a research business at the forefront of current efforts to develop and deliver personalised cancer medicine to a global market. It was established in 2001, and contrary to other public phase 1 units run as an integrated part of the hospital's clinical work. Four doctors and ten nurses are responsible for coordinating and running the trials, along with recruiting patients. As the head of the unit recounts in an interview held at his office:

> *We said from the start that we want to take on those tasks, but primarily we want to make a professional unit for phase 1 trials, that is similar to what I had seen in Europe. We closed a bed section and said: here we make a phase 1 unit. Here's how we could attract foreign trials from pharmaceutical companies....to begin with it was quite impossible to get any of the companies to say, well, maybe we should also try trials in Copenhagen. It was hard, so it required a lot of work. Both for me, but also for the Danish subsidiaries for all companies, it became a bit of a success and a prestigious project for the hospital. It was an investment that returned. So, it was not something that cost money; on the contrary, it was a good business, plus it was research merit, and the companies invested a lot, the offices grew, and ... So there was growth in many areas.* (Medical lead of the unit)

Elaborating further, he stressed how the timing was right as the unit was established around the last financial crisis. The large pharmaceutical companies were shutting down their regional offices, and many people were losing their jobs. He saw a strong incentive for everyone, publicly or privately employed, to work hard to attract more trials to keep their local business running. Moreover, there was a strong political agenda pushing Danish initiatives on personalised medicine onto a global market. Together, this momentum secured the unit's rare popularity among global pharmaceutical companies that saw multiple benefits of setting up trials in Denmark despite the number of patients being lower and cost per patient higher than larger countries.

Today, the unit comprises a small clinical ward with six beds located next to the standard oncology treatment wards. The unit's business of running experimental drug tests for the pharmaceutical industry is hardly visible to visitors. Only the small whiteboard hanging in the staff office lists all the current company 'sponsors' and protocols. In addition, consent forms and information material mention the companies when introducing the potential trial to the patients at their first consultation. During the day, patients enrolled in trials come in and receive their treatment. Their health status is meticulously followed by the project nurses doing measurements of, e.g. ECG, blood samples or assessing symptoms of their cancer disease or documenting if side effects are experienced from the new drug. Most patients have a personal oncologist whom they will see weekly to get a status on their wellbeing or the results from their latest scan, being performed monthly.

Located near Unit One is the Clinical Research Unit (CRU), taking care of the administrative tasks required to conduct the drug trials. On average, the unit has about 150 open protocols receiving or treating patients with new medicine. CRU has approximately thirty employees (project nurses, research assistants, lab technicians, IT staff, secretaries, service staff and doctors), all working on clinical trials of new and known drugs for adult patients with cancer. Their core expertise is to collect and document data and coordinate the administrative and clinical tasks required in conducting drug trials. In this space, the concurrent aims of securing better treatments for patients, doing good research, and sustaining local budgets by earning money on drug development intersect to produce personalized medicine. In the following section, we unpack the daily trial practices in the clinic in which the doctors have to balance a variety of considerations simultaneously.

Fast-Tracking Trialling at the Unit

> *You have to be able to act quite quickly to get these phase 1 studies. It's not like big phase 2 and 3 studies where it is possible to say that you start [recruitment] two months later than those in Poland and those in Germany. Here, time is such a crucial factor ... We have no advantages in Denmark other than time.* (Head of the oncology unit)

Doctors in Unit One have to balance their concern for the patient, the business, and the studies' scientific quality. In this context, speed is a key and defining trial quality. Getting patients ready for enrolment to deliver high-quality data fast is central to the daily work in the unit. To this end, many of the unit's operational procedures have been through a process of optimization. Such forms of fast-tracking are of direct economic value since a delay in the production and marketing of what might turn out to be a blockbuster drug may very well cost a drug company millions of dollars per day (cf. Rajan 2003). However, the ideal of speediness potentially conflicts with other temporalities, such as the time needed for the clinic to sort out new symptoms reported by the patients, which unsolved will not allow patients to stay in the trials, thus confronting the doctors at the unit with the question of how to produce excellent and ethical science *fast*.

Competitive Enrolment

Most discussions in the unit revolved around recruitment. The intense international competition among the trial centres to quickly find the right patients and deliver high-quality data was a constant and daily issue to handle. On the wall in the doctors' shared office hangs a small notice board with the unit's recruitment statistics depicted in bar charts. Next to the chart, someone has written: *"recruitment is everything!"* Every week the unit doctors and a project research nurse responsible for the practical enrolment of patients meet here to do a status on the allocation of patients to the many trials running in the unit. On spreadsheets lying on the table are listed all the trial protocols currently running at the unit, the number of open or upcoming slots to be filled, and the names of patients ready to be screened or enrolled. Some of the protocols recruit only patients with specific genetic mutations, and it can be hard to find patients matching, other protocols include more broad diagnoses of cancers ("all-comers"), and they fill up more easily.

Running the recruitment by keeping up the right flow of patients is crucial to the unit's work. Indeed, the pharmaceutical company starting up a phase 1 trial would often do so in several countries simultaneously, but only a few patients were included in the first cycles. The competition to get a 'slot' (a place in the trial) for a patient means that the unit's staff will line up patients even before the trial opens, to be able to report immediately that they have a patient ready. This procedure means that the unit doctors screen the list of patients waiting. They look for patients with only little signs of disease, as the ideal research subject is what they term "healthy sick" (cf. Bogicevic and Svendsen 2021). This person is well enough to stay in the trial for at least a month and able to report on side effects. Moreover, the doctors select patients with different diagnoses of cancer to present a broad 'catalogue of patients waiting', making them able to respond to new studies opening up.

Even if the doctors optimize their procedures and try to secure a diverse group of patients waiting to be enrolled, the daily work is still unpredictable. Patients and procedures are continually changing, and the doctors need to adjust accordingly to make ends meet. A patient's condition may suddenly and rapidly deteriorate. Because the unit runs early drug tests, the protocols receive many amendments, which mean more waiting time and postponement of the enrolment. Sometimes, trials are even put to a stoppage. At one status meeting, the unit doctor Sarah mentions that a patient she had hoped to include has developed some strange tics, hypertonia: *"He has got an appointment with the neurologist. If it is some strange side effect, he will not be ready to enrol."* Similarly, another study to which they have patients waiting was suddenly closed due to: *"acute adrenal insufficiency in the French arm"*. The discovery of possible adverse side effects in one arm (site) of the study shuts down all sites across countries at once while amendments to the protocol are written and approved. Esben, another of the unit doctors, compares that situation to Formula One racing. When an engine is down with oil leaking on the track and all cars are pulled in to wait until a new go-ahead is given. The contingencies of open and closed slots are worrying, and at the status meeting, the unit's lead physician Doctor Matthew concludes that they need to be ahead of things: *"We have to be on the ball ourselves. We cannot wait for them to call us. We also say there is an open slot; we have to register when we have some. And also ask around, at unit A [the regular oncological unit] if there are patients in late line [the last chemotherapy cycles], who are likely to fit."*

Producing Good Care and Good Science Fast

The unit's chief medical lead had urged the company to get rid of competitive enrolment. It occasionally led to "screen failure" instances where patients were lined up for a trial to secure an open slot but without all the necessary assessments finished. The time allocated to collect all the data needed to enrol a patient was often insufficient. Talking to one of the nurses collecting the data for the "eligibility package" – a standard package containing the selected patient's data – she told us that many of the blood samples could not be more than 72 h old when the patient entered the trial. Similar measures, i.e. EGC, had to be taken immediately before the enrolment. If the patient experienced new symptoms in the week leading up to the trial, there was simply not enough time to treat and keep them under control while preparing and securing the final regulatory green light.

At the site initiation visit described at the beginning of the chapter, the company representative had mentioned "fairness" as the principle guiding the allocation of slots. She had proposed a waiting list to give the different study sites in Denmark, France and Britain an equal opportunity. In discussing enrolment, the unit's medical lead, Mathew, underlined how the patients' needs should be guiding the allocation of slots, rather than the waiting list. This would guarantee a high-quality selection, and enrolment of the right patients, rather than patients who are first on the waiting list.

As Doctor Mathew pointed to, on the one hand, the competition to get the open slots does not always benefit the patient as the time to deliver the correct data might lead to the exclusion of the patient. On the other hand, enrolling patients and providing data fast is valuable to their collaborators and benefits the business. As the trial proceeds, commercial pressures come into play and reveal how the marketing of new drugs and innovative research at public hospitals go hand in hand (cf. Lakoff 2007). Doing things fast remains operative for the commercial life of the product. Many clinical trials are conducted not merely to assess efficacy and safety, but to secure regulatory approval at the least possible risk, and to bolster marketability. We see in the Danish clinic that the pace and temporality of enrolment become a transactable trial quality paving the way for patients into transnational trials while funding drugs and staff locally.

The Genomic Project: An Investment in Potentiality

Initiatives aimed at developing personalised medicine rely on data collection and data pooling (Prainsack 2015). Concurrently, storing valuable patient data in a publicly owned biobank to develop and deliver treatment for future patients (Hoeyer 2019) has been central to the strategies supporting public funding. At the unit, the project, 'Copenhagen Prospective Personalized Oncology' – in daily conversation referred to as 'the genomic project' – is a particular research protocol collecting tumour biopsies from patients and mapping the cancer genetics of their tumour (see Tuxen et al. 2018). The hospital funds the biopsies, data collection and storage. The unit considers this heavy public investment in biomarker-driven cancer research as crucial to their business. As explained by the head of the oncology unit:

> *The genomic project is a critical condition for the growth we have seen. At a very early stage, we could go out and say, well, we know everything about our patients; we can do a full gene sequencing on our patients. We were among the first in Europe to do it on such a large scale. After all, it costs money, it's expensive, and I was lucky to get some public funding to do it. And the other big sites in Europe, they said 'how can you do that? How can you afford it?' Because they did not have the opportunity. Even the places that were much bigger than us couldn't do the things we could. And that meant that we were often preferred over other sites, too.* (Head of the oncology unit).

At the site initiation visit we described at the beginning of this chapter, Line noticed how discussions centred on the role of mandatory biopsies upon entering a trial at the unit (see also Peppercorn et al. 2010). On discussing the first cycles of the trial and the first patients to be enrolled in the trial, the industry partner voiced the possibility of not taking a research biopsy, to which the medical lead immediately refused by saying: "We follow the protocol". This tension points to the doctor's focus on securing the resources for doing research, which will ensure the unit a competitive market position. In contrast, the company does not need the biopsy data at this early stage of drug development. For them, securing data through the fast and successful inclusion of a patient is more important. Getting data as fast as possible is the company's main objective, allowing them to move on to the subsequent development cycles.

In the weeks following the site initiation visit, Line observed the consultations with the first patient to be enrolled in that particular first-in-human trial as described in the following field note:

> Doctor Sarah makes herself ready to meet Bryan for his baseline examination, the final assessment before being enrolled in the trial. She gathers the papers she has just finished looking though, squeezes a booklet under her arm and heads for the patients' waiting area. On the way, she mentions that Bryan will be the first in the world to receive this treatment. *"He is really tough. But I am not sure if he is aware of how sick he is"* – she pauses with a sad look, *"I just saw his scan, and he has many metastases in his liver"*. Bryan greets us smiling, throwing a heavy rucksack on his back. He is in his late 50s and has never been ill before he was diagnosed with cancer, he says. Sitting down in the consultation room, Doctor Sarah and him discuss the recent biopsy procedure last week. He had seven biopsies taken from tumours in his liver, which caused him to throw up and then faint. Doctor Sarah wants to know whether he is still in pain: "*do you take Panodil [painkillers]?*" *"Nah, one*

or two", he answers. *"You are allowed to take more, you know"*. She asks him if there is anything new since they spoke together last week. Bryan says, no. She gives him a status: *"It is mostly your blood count I was worried about. It was a bit low, but it is up again, I can see*". Bryan confirms, *"Yes, this is also what I sense when I'm out for a run."* Sarah continues: *"Your blood count is related to the number of days since your last Chemo, so that it will come up again. I cannot see anything that would stop you from entering the trial. Then I just need to have a look at you*", she points to the couch in the room, and puts on her stethoscope, saying: *"What do the kids say?"* Bryan: *"they just say go for it!"* Doctor Sarah goes through the last test result needed to be included in the 'eligibility packet', a data package to be sent off to the company. Looking up at Bryan, she clarifies: *"the company needs to look at all these numbers and approve. As soon as we have their approval, we can get started*". Bryan nods, and she rounds off the consultation: *I will call you on Friday to see if you are still in pain. Then I will tell you if we can start on Monday*.

Shortly after the biopsy procedure and Bryan's consultation with Doctor Sarah, Bryan developed a fever lasting for more than a week, which was difficult to get under control and ultimately led to Bryan's exit from the trial. The inclusion of mandatory tumour biopsies as a condition for enrolment in early phase studies of little direct benefit for the patient is critiqued for not letting the choice of biopsy be up to the patients. Despite the low prevalence of complications, most patients would say no to biopsies if they had the choice (El-Osta et al. 2011). More generally, Barbara Prainsack has pointed to how personalised medicine intensifies data collection and pooling to deliver the promise of personalised medicine, a data practice that puts additional responsibility on the patients for contributing personal data to shared resources (Prainsack 2015). In Bryans's case, the unit doctors would argue that the biopsy was essential to assess whether the patient's cancer tumour had the genetic mutations matching the experimental drug. To them, careful selection of patients based on genetic profiling showed promise of better clinical results even in early phase trials (Tuxen 2019). Furthermore, the research biopsy was also important in getting more knowledge on tumour specific variation for the biobank targeted future drug testing (Green et al. 2021). In contrast, the company representative suggested the possibility of getting this knowledge at a later stage because the company's first priority was to get data on safety and toxicity. Had the clinic not taken the biopsies, Bryan might have stayed in the trial – yet he would not have contributed to the unit's research.

The unit is interested in pursuing research that they hope eventually will lead to better treatment for future patients. This entailed both profiling the patients to match treatments as a way to attract more industry targeted trials and an ambition to build national resources for developing better treatment in the future. Because the unit entered the market early and was able to deliver genomic profiling, it was able to attract many of the new targeted trials. Therefore, by investing in the precise selection of patients based on publicly funded genetic profiling, the unit aimed to deliver yet another valuable trial quality to the industry while ensuring that patients entering trials had a chance of benefiting from trials drugs. In this way, the genomic project underlines the role of public ownership of data in legitimizing investments in biobanking (cf. Salter and Salter 2017) and shaping economic exchange trajectories or specific partnerships to sustain research and development.

Research Bodies: High Patient Compliance

A national selling point in attracting international research investments and research to Denmark is reliable research subjects. The Danish civil registration system allocates a personal identification number to all citizens making potential trial participants easy to track over time and across data sources, as exemplified in the following quote from the branding material sent out from the Ministry of Foreign Affairs and the Regions in Denmark:

> *With its unique social security number system, longstanding tradition for patient and population registration and access to comprehensive biobanks, Denmark is an ideal location for medical and clinical research. Furthermore, Denmark has a homogeneous and compliant population, and individual patients are easily traced, which makes for a low lost-to-follow-up rate. In addition, Danes are very open to participating in clinical trials.* (Start with Denmark 2016)

Here the Danish population is promoted as a particularly resourceful population with high compliance and willingness to participate in trials. Furthermore, the welfare state's lifelong registration and tracking of its citizens has proven an excellent trial quality. In contrast, the companies running trials in other countries experience many "lost-to-follow-up," i.e. cases where patients are unreachable, and data cannot be obtained due to lack of a central registration (Dettori 2011).

At Unit One, the patients volunteering for trials have been referred by their general practitioner or local oncologist. The patients are invited to an introductory visitation to assess their possibilities of participating in a trial. The ideal patient should have a solid tumour from which a biopsy can be drawn and tested. Moreover, the potential patient needs to be well enough to endure the many weekly trips to the clinic, and has to be able to report back on side effects of the drug and not mixing up side effects with the progressed disease (cf. Tuxen 2019).

Anne, a former schoolteacher and patient we interviewed, had participated in several trials at the unit. She had been ill with bowel cancer for 9 years. She described the importance of being in a trial and the effort she put into reporting back and keeping herself as healthy as possible:

> *I was flattered when they told me that I had a good performance status. Because then you might be able to take part in an experiment. Because you cannot do that when you are poorly. They [doctors] can see if you are feeling well or badly. Of course, you can overplay or underplay how much headache you have or how much of the one and the other. But I do not. I am honest about that. But the thought is there. If you get too poorly, you are out. And that's why I'm exercising all the time, and that's why I eat healthily and live a sensible life because I know well that if I suddenly lie down and I am not in shape... It's for my own good. So I am top motivated for it. I have a little notebook where I write what medicine I take. I also write if I get any side effects because I have to answer whether there is a connection [to symptoms] or not. If I say I have had a headache, they will ask what day I had a headache. So I have trained myself to be a professional patient. There was not much choice for me in saying yes to a trial. For me, it was hope. Yes, there are some tough things you read in the papers you need to read and sign. So, I am well aware of the likelihood of side effects and I have tried many side effects. But I must try it. I am not ready to leave.* (Anne, 67 years old)

Like Anne, the patients enrolled at the unit were primarily well-educated and with will and resources put into being a research subject. They were all diagnosed with incurable cancer but in good health, showing no sign of disease. Because they had incurable cancer but were not yet terminal, they did not receive any treatment and had been discharged from their former oncology units. They all had a deep wish to receive treatment, even if it meant treatment with unknown effects. As one patient said: *"if you are a part of the unit, then you do not feel you have been abandoned."* Another patient voiced the despair she felt of sitting at home waiting as: *"there can't be nothing,"* with which she expressed her need for receiving treatment but also an expectation to be treated, pulled back into to being seen by the health care system, and thereby being asserted as deserving of care and resources (see also Dam et al. 2022). The fact that patients in the unit experience trial participation as a form of care is very clear and has also been shown to be the case for patients elsewhere (cf. Will and Moreira 2016; Keating and Cambrosio 2012; Kaufman 2015).

One of the unit doctors explained the unit as a specific space of care where data collection created more time with the patients than in standard oncology treatment:

> *It takes a huge amount of data collection to do these phase 1 studies, so when we see the patients, there is time set aside, both to talk to them of course, where we have to ask a little more about all sorts of things and examine them, more than we might otherwise want, and so also to report all the data through the various systems and such. So more time has been set aside. And the time set aside is, in principle, something that the pharmaceutical companies help pay for. If we did not take a high price from the companies to do these studies, both to start the studies and to include and treat each patient, then it would not be possible at all.* (Unit doctor)

The patients value the attention and being in a trial connects them to the social contract and exchange with the state. The patients are a valuable resource but not in the narrow sense of delivering bio-resources. Rather, their wish to be in an exchange with a health care system and be part of a national collective secures high compliance and thus the qualities of trialling that the unit is capable of providing. The social contract with the welfare state is meaningful to patients, and they are willing to give back by participating in trials and doing so with high compliance and trust but also with the expectation of being seen as deserving of care by the state. This commitment to the collective is also what becomes fragile when the exchange is not reciprocated.

During autumn of 2019, news spread at the Unit One that a drug being trialled to treat ovarian cancer showed very promising results could now be offered in treatment. Many patients were hoping to start on the new drug.

Due to the rise in drug prices, a national council, the Danish Medicine Council, was founded in 2017 to provide guidance about new medicines for use in the Danish hospital sector. In this case, the medicine council decided that only women with *BRCA* mutations, representing just 20% of patients with advanced ovarian cancer, were eligible to be treated with the drug, even though patients without the specific mutation had also shown to have some effect.

The Danish Medicines Council's reason for recommending the new drug only to patients with a BRCA-mutation, associated with hereditary breast and ovarian cancer, was that *"the council found that there is a reasonable relationship between the drug's clinical added value and the cost of treatment with Zejula [new drug] compared to Lynparza, which is Danish standard treatment. The reason for not recommending Zejula to patients without a BRCA mutation, on the other hand, is that the council considers that there is no reasonable relationship between the clinical added value of the drug and the cost of treatment with Zejula compared to placebo*" (Danish Medicines Council 2019, translation by the authors). The argument of not granting all patients access built on a comparison of the price of the placebo pills compared to the price of the new drug, which made the relatively little benefit of the drug look very costly compared to cheap calcium pills. The Zejula case illustrates the economy of prioritisation as a consequence of limited resources in public healthcare. In the end, the collaboration between pharma and public clinics may develop new medicine too expensive for the public health care budgets to pay for. This tension challenges the promise of personalised medicine, which is sought to deliver more precise medicine "*to the benefit of patients*," as the first Danish national strategy of precision medicine is named (Ministy of Health 2016). In the case of Zejula, only a very small subpopulation came to benefit from the national precision medicine investment (see also Day et al. 2017; Tannock and Hickman 2016; Marquart et al. 2018). Moreover, considerable doubts have been raised about whether targeted treatments would be affordable in practice, and the fairness of resource allocation compared with other priorities, such as cancer prevention (see Vineis and Wild 2014, a.o.). There is a lot of pressure from patients to get new drugs on the market. If the development of the drugs – conditioned on the collective resources delivered by citizens – does not result in citizens' access to the new medicine, public trust may be challenged (cf. Petryna 2011).

Discussion

The chapter has aimed to analyse how public-private entanglements shape and stir the research and care practices at a public hospital in Denmark. We have unpacked the concrete ways the exchange and interdependence between the public hospital unit and the pharmaceutical industry shape what personalized medicine becomes in clinical practice. Where social science studies of transnational bio-economies have focused on how clinical trials exploit disadvantaged populations, our study shows how clinical trials operate in wealthy nations already providing comprehensive and equal access to health care. Here, the contract between citizens and the state becomes an important context for understanding how trial qualities gain traction as currencies ready to be transacted on the global market for drug development.

Internationally, concern has been raised about whether medical innovation has become too dependent upon industrial sponsors and whether medical innovation creates sufficient benefit for the patients. Most of the new cancer medicines only promise a small 'survival increase', which at best means slowing the growth of cancer for a few months. As these drugs are expensive, the pharmaceutical industries receive a considerable profit in delivering a small survival increase. At present, the success of personalised medicine is debated, as the endpoint measured in trials is often not 'overall survival', but typically 'progression-free survival'. Furthermore, the current focus on precision responses to cancer has led scientific and policy communities to eschew primary prevention measures that might ultimately prove a more effective and cost-efficient way to fight cancer since a third to a half of all cancers are deemed preventable based on present knowledge (Plutynski 2020). The patients we met represent a highly selected group of citizens who want treatment and will accept a drug that might only give them a few months more to live in. However, letting hugely expensive drugs with only very little promise in terms of overall survival hit the market testifies to the role of the market in defining available care. Patients demand these products, yet what about the tax-financed health care services, which are to pay for them? Paradoxically, maybe, the Danish welfare state comes to both profit from and help produce a demand for drugs, which it is not willing to pay for.

References

Bogicevic, I., and Svendsen, M. N. 2021. Taming time: Configuring cancer patients as research subjects. *Medical Anthropology Quarterly*, 35 (3), 386–401.

Cooper, M. 2011. TRIAL BY ACCIDENT: Tort Law, Industrial Risks and the History of Medical Experiment. *Journal of Cultural Economy* 4 (1): 81–96. https://doi.org/10.1080/17530350.2011.535374.

Dam, M.S., S. Green, I. Borgicevic, L. Hillersdal, I. Spanggaard, K.S. Rohrberg, and M.N. Svendsen. 2022. *Precision Patients: Selection Practices and Moral Pathfinding in Experimental Oncology*. Sociology of Health and illness: Epub ahead of press.

Danish Medicines Council. 2019. *Niraparib (Zejula) – Kræft i æggestokkene*. https://medicinraadet.dk/igangvaerende-vurderinger/laegemidler-og-indikationsudvidelser/niraparib-zejula-kraeft-i-aeggestokkene#phase_7746. Accessed 22 Mar 2020.

Danish Regions. 2015. Handlingsplan for Personlig Medicin: PIXI-udgave [Plan of Action for Personlised Medicine: PIXI-version]. https://www.regioner.dk/media/1280/handlingsplan-forpersonlig-medicin.pdf.

Day, S., R.C. Coombes, L. McGrath-Lone, C. Schoenborn, and H. Ward. 2017. Stratified, precision or personalised medicine? Cancer services in the "real world" of a London Hospital. *Sociology of Health & Illness* 39 (1): 143–158.

Dettori, J.R. 2011. Loss to follow-up. *Evidence-Based Spine-Care Journal* 2 (1): 7–10.

Dumit, J. 2012. *Drugs for Life: How Pharmaceutical Companies Define Our Health*. Durham: Duke University Press.

El-Osta, H., D. Hong, J. Wheler, S. Fu, A. Naing, G. Falchook, M. Hicks, S. Wen, A.M. Tsimberidou, and R. Kurzrock. 2011. Outcomes of research biopsies in phase 1 clinical trials: The MD Anderson Cancer Center Experience. *Oncologist* 16 (9): 1292–1298.

Fisher, J.A. 2009. *Medical Research for Hire: The Political Economy of Pharmaceutical Clinical Trials*. New Brunswick/London: Rutgers University Press.

Green, S., M.S. Dam, and M.N. Svendsen. 2021. Patient-derived organoids in precision oncology – Towards a science of and for the individual? In *Personalized Medicine in the Making. Philosophical Perspectives from Biology to Healthcare*, ed. C. Beneduce, and M. Bertolaso. Dordrecht: Springer.

Healy, D. 2004. Shaping the intimate: Influences on the experience of everyday nerves. *Social Studies of Science* 34: 219–245.

Hoeyer, K. 2019. Data as promise: Reconfiguring Danish public health through personalized medicine. *Social Studies of Science* 49 (4): 531–555.

Kaufman, S.R. 2015. *Ordinary Medicine: Extraordinary Treatments, Longer Lives, and Where to Draw the Line*. North Carolina: Duke University Press.

Keating, P., and A. Cambrosio. 2012. *Cancer on Trial: Oncology as a New Style of Practice*. Reprint edition. Chicago: University of Chicago Press.

Lakoff, A. 2007. The Right Patients for the Drug: Managing the Placebo Effect in Antidepressant Trials. *BioSocieties* 2 (1): 57–71. https://doi.org/10.1017/S1745855207005054.

Larsen, L.T., and D. Stone. 2015. Governing health care through free choice: Neoliberal reforms in Denmark and the United States. *Journal of Health Politics, Policy & Law* 40 (5): 941–970.

Marquart, J., E.Y. Chen, and V. Prasad. 2018. Estimation of the percentage of US patients with cancer who benefit from genome-driven oncology. *JAMA Oncology* 4 (8): 1093–1098.

Ministry of Finance. 2015. *Økonomisk fordelagtighed ved offentlige-private partnerskaber*. [Economic Advantages In Public-Private Partnerships]. Copenhagen: Ministry of Finance

Ministy of Health. 2016. *Til gavn for patienterne*. Danish National Strategy of Precision Medicine 2016–2020. https://www.regioner.dk/media/4352/national-strategi-for-personlig-medicin.pdf. Accessed 29 Mar 2021.

Mitchell, R., and C. Waldby. 2010. National biobanks: Clinical labor, risk production, and the creation of biovalue. *Science, Technology, & Human Values* 35 (3): 330–355.

Peppercorn, J., I. Shapira, D. Collyar, T. Deshields, N. Lin, I. Krop, H. Grunwald, et al. 2010. Ethics of mandatory research biopsy for correlative end points within clinical trials in oncology. *Journal of Clinical Oncology* 28 (15): 2635–2640.

Petryna, A. 2009. *When Experiments Travel: Clinical Trials and the Global Search for Human Subjects*. Princeton and Oxford: Princeton University Press.

———. 2011. Pharmaceuticals and the right to health: Reclaiming patients and the evidence base of new drugs. *Anthropological Quarterly* 84 (2): 305–329.

Petryna, A., A. Kleinman, and A. Lakoff. 2006. *Global Pharmaceuticals: Ethics, Markets, Practices*. North Carolina: Duke University Press.

Plutynski, A. 2020. Why Precision Oncology is Not Very Precise (and Why This Should Not Surprise Us). *Philosophical Issues in Precision Medicine*.

Prainsack, B. 2015. Is personalized medicine different? (reinscription: The sequel) A response to Troy Duster. *The British Journal of Sociology* 66 (1): 28–35.

Rajan, K.S. 2003. Genomic Capital: Public Cultures and Market Logics of Corporate Biotechnology. *Science as Culture* 12 (1): 87–121. https://doi.org/10.1080/0950543032000062272.

———. 2006. *Biocapital: The Constitution of Postgenomic Life*. Durham: Duke University Press.

———. 2017. *Pharmocracy: Value, Politics and Knowledge in Global Biomedicine*. Durham: Duke University Press.

Rose, N., and C. Novas. 2005. Biological citizenship. In *Global Assemblages: Technology, Politics, and Ethics as Anthropological Problems*, ed. Aihwa Ong and Stephen J. Collier, 439–463. Malden: Blackwell.

Salter, B., and C. Salter. 2017. Controlling new knowledge: Genomic science, governance and the politics of bioinformatics. *Social Studies of Science* 47 (2): 263–287.

Sismondo, S. 2007. Ghost Management: How Much of the Medical Literature Is Shaped Behind the Scenes by the Pharmaceutical Industry? *PLOS Medicine* 4 (9): e286. https://doi.org/10.1371/journal.pmed.0040286.

———. 2018. *Ghost-Managed Medicine: Big Pharma's Invisible Hands*. Manchester: Mattering Press.

Smith, C.G., and J.T. O'Donnel. 2006. *The Process of New Drug Discovery and Development*. 2nd ed. New York: Informa Healthcare.

Start with Denmark. 2016 report published by the Danish Health Regions in collaboration with the Ministry of Foreign affairs. https://www.regioner.dk/media/3759/270916-startwithdenmark2016-fullreport.pdf Retrieved 23 November 2020. Accessed 29 Mar 2021.

Tannock, I.F., and J.A. Hickman. 2016. Limits to personalized cancer medicine. *New England Journal of Medicine* 375 (13): 1289–1294.

Tarkkala, H.A.U., I.A. Helen, and K. Snell. 2019. From health to wealth: The future of personalized medicine in the making. *Futures* 109: 142–152.

Timmermans, S., and M. Berg. 2003. *The Gold Standard: The Challenge of Evidence-Based Medicine and Standardization in Health Care*. Philadelphia: Temple University Press.

Tuxen, I.E.V. 2019. *Precision Oncology – The Clinical Utility and Potentials of Molecular Profiling in the Early Trial Setting*. PhD-thesis, University of Copenhagen.

Tuxen, I.E.V., K.S. Rohrberg, O. Oestrup, L.B. Ahlborn, A.Y. Schmidt, I. Spanggaard, J.P. Hasselby, et al. 2018. Copenhagen Prospective Personalized Oncology (CoPPO) – Clinical utility of using molecular profiling to select patients to phase I trials. *Clinical Cancer Research.* 25 (4): 1239–1247.

Vineis, P., and C.P. Wild. 2014. Global cancer patterns: Causes and prevention. *Lancet.* 383 (9916): 549–557.

Waldby, C., and R. Mitchell. 2006. *Tissue Economies: Blood, Organs, and Cell lines in late Capitalism*. Durham, NC: Duke University Press.

Will, C., and T. Moreira. 2016. *Medical Proofs, Social Experiments: Clinical Trials in Shifting Contexts*. Abingdon: Routledge.

"Reconstruction of Trouble"

Hanna Elisabet Dillekås

Introduction: The Special Status of Breast Cancer in Society

The imaginary of precision oncology could be said to be a cancer treatment so highly efficient that it completely eradicates all traces of the disease without any damage to healthy tissues or processes, and thus causing no side effects. Surgical excision of a tumour, although increasingly precise and limited in extent over the past decades, is nowhere near this. In fact, for most cancer surgery, it is a goal in itself to remove some healthy tissue surrounding the cancer to secure tumour-free margins, to ensure that all cancer cells have been removed from the organ. When this tumour and tissue removal is applied to the female breast, a body part of immense focus in society regarding shape and size, important to sexuality, femininity and body image, the desire to reconstruct the breast is evident. In addition, for some patients, the lacking breast serves as a daily reminder of the cancer and a source of fear of a recurrence. This fear is actually quite rational, breast cancer is infamous for its propensity of late relapses, even decades after apparently successful treatment of the primary tumour. These late relapses are assumedly caused by early dissemination of cancer cells that subsequently enter a state of dormancy at a distant site, and later escape dormancy to produce overt metastatic disease (Phan and Croucher 2020). As of yet, we are not able to reliably detect these dormant cancer cells and have no prognostic biomarkers of neither dormancy nor escape from dormancy (Yadav et al. 2018).

Surgery, maybe apart from the most minimal procedures, has systemic effects in the body. The tissue trauma initiates a cascade of reactions ranging from blood clotting through inflammation, wound healing and tissue remodelling (Shaw and Martin 2016). These systemic biologic effects aim to restore tissue integrity and

H. E. Dillekås (✉)
Department of Oncology, Haukeland University Hospital, Bergen, Norway
e-mail: Hanna.elisabet.dillekas@helse-bergen.no

A. Bremer, R. Strand (eds.), *Precision Oncology and Cancer Biomarkers*, Human Perspectives in Health Sciences and Technology 5,
https://doi.org/10.1007/978-3-030-92612-0_5

homeostasis and to protect the body from intruding microorganisms. These are tightly regulated, precise reactions, as can be deducted from the major health problems caused when they are disturbed, such as wound healing deficiencies in diabetes and cardiovascular disease, or over active wound healing resulting in keloid scars (Martin 1997). Cancer has an interesting dual identity: it is a part of our body, many tumour lumps are mainly comprised of normal cells, connective tissue and blood vessels, and even the cancer cells themselves originate from normal cells with the same DNA although with some mutations. At the same time the cancer is an intruder, not adhering to any of the rules regulating cell growth and movement. Even though it has for long periods in history been viewed as such, in our efforts to understand and overcome this disease, we must remember that cancer is not an island. As a part of the organism, the cancer cells can benefit from systemic signalling in the body, for instance the stimulation to grow and move that is necessary to heal a wound (Dillekås and Straume 2019; Antonio et al. 2015; Martins-Green et al. 1994).

In his chapter, I will look at the different roles of the various actors involved, in the context of a priority debate around breast reconstruction after surgery. In order to do so, I will first, in section "A case study of how breast reconstruction was prioritised, and the unintended consequences this had for breast cancer patients and other patient groups", describe a highly mediatised campaign for increased and faster availability of breast reconstruction for breast cancer patients, and the political decision-making and scientific discussions that followed from that. In section "The actors and their roles in the prioritisation game", I will outline the roles of the various actors involved, and in section "Debate, decision-making and their consequences", I will discuss the desirable and non-desirable consequences of the decisions being made, some of which were not foreseen.

A Case Study of How Breast Reconstruction Was Prioritised, and the Unintended Consequences This Had for Breast Cancer Patients and Other Patient Groups

In 2012, the patient association 'Norwegian Breast Cancer Society' campaigned for improved access to breast reconstruction. At this time, breast reconstruction after mastectomy for breast cancer was government funded and performed in public hospitals, but waiting times were often several years. The campaign was well orchestrated by the Norwegian Breast Cancer Society, and quite dramatic with women standing outside parliament showing their mastectomy scars. Naturally, media caught on to this and it was on both national television and newspapers with headlines such as "Cancer treatment is not completed until we get our bodies back" (Tv2, 31.05.12). It did not take too long before the minister of health declared an extra grant of 150 million NOK earmarked to breast reconstruction and a directive to plastic surgery departments across the country to prioritise these procedures (Bakke 2012). As numbers of plastic surgeons cannot be increased overnight, some unintended effects were later discovered. Children born with cleft lip and palate are another patient group treated by plastic surgeons, and while the most pressing procedures on infants not able to breast feed were still prioritised, the later corrections, usually performed on teenagers to achieve better facial cosmetic result were not, and waiting times increased (Bordvik 2014). This patient group have an interest organisation that is more focused on advising and supporting patients and their parents than influencing politicians. It is also considerably smaller than the breast cancer society and lacks a professional communication machinery. Into this discussion came a paper published as a part of my thesis, evaluating relapse dynamics after breast reconstruction. Relapse rates after breast reconstruction have been studied previously, with contradictory results. Some have demonstrated increase relapse rates in reconstructed patients and other reduced (Geers et al. 2018; Isern et al. 2011; Svee et al. 2018). These studies, however, suffered from different underlying risk factors in the groups compared and did not explore the time dynamics of relapses. In our study, we discovered a peak in early relapses in the reconstructed patients, the first two years after reconstruction, not present in controls with similar tumour characteristics that choose not to reconstruct the breast (Dillekås et al. 2016). We also found what looked like a dose-response relationship: the more extensive reconstructive procedures resulted in a higher peak of relapses compared to simple implant surgery. Consequently, the question was asked whether the campaign to increase the capacity for reconstructive surgery could have caused an increase in early metastatic relapses. This could, as we explained in the paper, only indirectly be suggested. Still, our results were strongly supported by some and equally refuted by others, as will be discussed further in the discussion section. The Norwegian Breast Cancer Society, when asked the same question, rightly explained that this was not a known hypothesis at the time of the campaign, and immediately demanded another 150 million NOK for research into this possible connection (Bordvik 2016). Decision-makers were, however, not as ready to leap into action this time. Had we or others had a pre-planned study protocol to explore this, ready at that time point, and a media strategy to push political decisions, or even teamed up with the breast cancer society, perhaps we could have utilised the momentum and received similar funding. Current waiting times for reconstruction after breast cancer in Norway span from 12 weeks to 3 years, depending on hospital and type of reconstruction, with longer waiting lists for autologous flap procedures and shorter for simple implant surgery.

The Actors and Their Roles in the Prioritisation Game

The roles of the different actors in this drama are shifting as the situation evolves. The breast cancer patients and their advocacy group play dual roles of both victims suffering from the missing breast that politicians are denying them by not prioritising and funding these procedures sufficiently. In addition, they take on a role as warriors and heroes, battling to change this. The politicians on their side are first labelled the villains as described above. As this is a very unsatisfactory role, they eventually buy themselves the role of the hero, for 150 million NOK, giving the patients what they want and getting praise in the media in return. A couple of years later comes the backlash, and they are again the villains as they are accused of both having caused suffering to young people with birth defects and of having accelerated breast cancer relapses. Briefly, a small attempt may be said to have been made to label the breast cancer patients the villain, by confronting them with the possibility that their campaign was what spurred the rise in reconstructions, and thus indirectly may have caused earlier relapses. This was rapidly refuted by the simple statement that this potential link was neither known nor suspected at the time.

The journalists sometimes claim that they have a neutral position, merely conveying the message of others, but I must argue that they also take on the role of the hero. They are definitely trying to influence politicians, giving attention to causes they deem good and worthy, like breast reconstruction to cancer treated patients in this case. If we try to imagine a scenario where women were campaigning for aesthetic plastic surgery to be government funded and performed in public hospitals, I cannot imagine the media reporting this in a similar way. With the velocity of the media landscape, issues are presented without understanding of the underlying depth and complexity. As opposed to a true hero, however, the media refuses to take responsibility for the unforeseen consequences of their actions.

In general, plastic breast surgery is an elective form of surgery, in its nature cosmetic, and thus not considered a priority for public funding. Some have even proposed that cosmetic plastic surgery is in opposition to the Hippocratic oath, unethical and should not be performed by any physician (Vogt and Pahle 2018). Naturally, it has been in the interest of patients and surgeons to frame reconstruction after mastectomy for breast cancer as something completely different, an integrated part of the breast cancer treatment rather than an aesthetic procedure. I believe most people would agree that it is indeed something quite different to reconstruct a breast after mastectomy for breast cancer compared to aesthetic breast surgery.

At first, the plastic surgeons had a neutral position, when instructed from the highest political level to prioritise reconstructions, they did. When the problematic side of this was presented, both regarding cleft lip and palate surgery and breast cancer recurrences, the field was divided. Some were genuinely concerned that they had been too eager to perform reconstructions, and, although prematurely as the study was quite small, began changing their clinical practice, advising a larger proportion of women against reconstruction, or at least to take the possibility of a stimulating effect on dormant cancer into account. Others were, perhaps as a reaction to protect their own field of work, deeply sceptical of our findings. Pharmaceutical companies did not get involved in the discussion, neither did manufacturers of breast implants or private plastic surgical clinics. Thus, financial gain did not seem like a driving force in this debate. Although our results were quite convincing, due to the fact that this was a retrospective study no causal relation can be established, and thus everyone can push their own agenda. The complexity of tumour biology as well as systemic responses to tissue trauma and wound healing, not to mention when these are combined, makes the effect even more difficult to determine.

Debates on cancer care almost inevitably become emotional at some time point. The debate was indeed mainly evolving around the emotional and psychological effects on women treated for breast cancer of living with only one breast. The feminist argument of diseases affecting women historically being less prioritised was also used. There is nothing glamourous about cancer but if one cancer form could be said to have some glamourous status, it would probably be breast cancer. The pink ribbon campaign, celebrity survivors and the fact that the disease is not strongly associated with life-style risk factor such as smoking may contribute to make breast cancer considered a "worthy cause" for politicians and philanthropists. As opposed to lung cancer and head and neck cancers where it is perhaps easier to put some blame of the disease on patients' lifestyle choices.

Norway has a long tradition of systematic work regarding priority settings in the health care sector. The government document "Open and fair - priority setting in the health service" was unanimously approved by parliament in 2014 (Norheim et al. 2014). The underlying principles of good priority setting are stated to be a fair distribution of as many years of good life quality as possible. Distinct criteria for priority setting are applied when making these decisions: health gained from an intervention, health lost without it and resources demanded. The health care priority settings are founded on systematic work with openness and user involvement and comprehensive implementation by efficient means. Even if the 2012 "right to reconstruction"-campaign preceded this document, its precursor was founded on the same principles. For sure, in none of these priority setting documents is it stated that the ones screaming the loudest and with the largest media attention should be prioritised. Ignorance of these forces increases the risk of unknowingly being influenced by them and thus of allocating resources to the most visible instead of where they could give the greatest benefit. The case described here is definitely not the first time these noble principles are forgotten or ignored. Just recently, the Norwegian parliament decided that all women should get free access to early prenatal diagnostic procedures, without any evaluation of the cost or what amount of health could be gained. Further, it seems like the criteria are more readily applied to medical treatment and less when implementing surgical techniques or prioritising between surgical interventions. Perhaps this is a reflection of the difficulty in obtaining the highest levels of scientific evidence through randomized clinical trials for surgical interventions compared to medical substances. Surgical procedures are more difficult to standardize, both due to patients' factors like anatomical variability and surgeon factors like experience and technical skills.

Debate, Decision-Making and Their Consequences

In this chapter, I have described a real-life case from Norway of how patients' advocacy groups and media dictated medical prioritisations, and the consequences this had. Even though it is not surprising that politicians yielded to the massive and highly emotional pressure, my argument is that they must explore the consequences of such decisions before making them. This should be done from more angles than the will of patients directly involved, the scientific angle of biological effects and resource reallocation from other patients' groups would in my opinion be a minimum requirement. To describe a case evolving around surgery, a non-precise form of treatment, and breast reconstruction, an at best only indirect part of oncology, in an anthology on precision oncology may be considered unorthodox. Still, my claim is that is has its place in this book. The precision in our approach to cancer must, in addition to the molecular level traditionally associated with the concept, also be applied to the societal level when we decide what treatment should be available for whom. If we are to be able to make good decisions, we must make the highest effort to explore the broader consequences our decisions may have, for the patient group immediately affected, as well as for those indirectly involved. Where to stop this exploration of consequences quickly becomes a dilemma, should we weigh in the patient's relatives? Other parts of the health sector, or even other sectors in society? While it undoubtedly would be naïve to presume that we would ever be able to predict all consequences of a political decision, on the healthcare system and society at large, adhering to established criteria for priority settings would seem to be our safest option. There may always be dissatisfaction and an experience of unfairness in groups not getting what they want and feel entitled to, but as long as resources are limited there will always be a need to prioritise. The priority setting criteria in Norway are implemented through democratic processes and should not be ignored. However, as these criteria do not address in detail every situation where prioritisation choices must be made, different agendas and values of what is (most) important will inevitably create debate. As we strive for precision in oncologic treatment, with the noble intention to effectively treat the cancer without negative side effects, we must also keep in mind the inherent complex nature of this disease. The heterogeneity of cancer, with different clones in the same patient at the same time and the capability to change and adapt through a high mutational rate may escape narrowly targeted therapeutic approaches. Indeed, such an imprecise treatment form as surgical excision of the tumour still remains a cornerstone in cancer treatment, with demonstrated benefit on both relapse free and overall survival (Fisher 1985). As a clinician, treating breast cancer patients, I was of course not unaware of the strong emotions surrounding breast reconstructions when I started on the work presented in the paper on recurrence dynamics. Still, my main focus was on the enigma of tumour dormancy and late recurrences, and what could provoke escape from dormancy, with a hope to contribute to preventing metastatic relapse, rather than aiming at influencing health setting priorities. When we first were to present our results, we realized that the finding of a possible connection between surgical procedures of breast reconstruction and accelerated relapses could be both provocative and scary. Certainly, I soon became an actor in in the debate myself, with interviews in national media as well as invitations to present and discus our findings both in the annual meeting of Norwegian Surgical Society and that of the Norwegian Oncology Society. In these fora, questions and feedback were in general coloured by concern for the patients. The uncertainties around the results were acknowledged, and utilised by some to support their own view, but mainly spurred curiosity and discussion regarding the way onward to a potential clinical implementation. A deeper understanding of the systemic biological consequences of this treatment form, and all of the associated physiological reactions, for the biological organism in general and the cancer especially, may direct us to different interventions in the perioperative time window with a greater oncologic impact. However, we have seen in this chapter that the discussions around what to prioritise and when, are much politicized, and values are strongly at play here. Therefore, we should also be careful about thinking that more knowledge will always lead to more straightforward medical and political decision-making processes.

References

Antonio, N., M.L. Bønelykke-Behrndtz, L. Ward, J. Collin, I.J. Christensen, T. Steiniche, H. Schmidt, Y. Feng, and P. Martin. 2015. The wound inflammatory response exacerbates growth of pre-neoplastic cells and progression to cancer. *The EMBO Journal* 34: 2219–2236.

Bakke, K.A. 2012. 100 Millioner Til Brystrekonstruksjon. *Dagens Medisin.*

Bordvik, M. 2014. Ungdom Må Vente I 4 År På Operasjon. *Dagens Medisin* (2014-03-19 2014).

Bordvik, Målfrid. 2016. Brystkreftforeningen Krever Nye 150 Millioner. *Dagens Medisin* 26 (08): 2016.

Dillekås, H., and O. Straume. 2019. The link between wound healing and escape from tumor dormancy. *Surgical Oncology* 28: 50–56.

Dillekås, H., R. Demicheli, I. Ardoino, S.A. Jensen, E. Biganzoli, and O. Straume. 2016. The recurrence pattern following delayed breast reconstruction after mastectomy for breast cancer suggests a systemic effect of surgery on occult dormant micrometastases. *Breast Cancer Research and Treatment* 158 (1): 169–178.

Fisher, B. 1985. The revolution in breast cancer surgery: Science or anecdotalism? [In eng]. *World Journal of Surgery* 9 (5): 655–666.

Geers, J., H. Wildiers, K. Van Calster, A. Laenen, G. Floris, M. Vandevoort, G. Fabre, I. Nevelsteen, and A. Smeets. 2018. Oncological safety of autologous breast reconstruction after mastectomy for invasive breast cancer. [In eng]. *BMC Cancer* 18 (1): 994.

Isern, A.E., J. Manjer, J. Malina, N. Loman, T. Mårtensson, A. Bofin, A.I. Hagen, I. Tengrup, G. Landberg, and A. Ringberg. 2011. Risk of recurrence following delayed large flap reconstruction after mastectomy for breast cancer. *The British Journal of Surgery* 98 (5): 659–666.

Martin, P. 1997. Wound healing – Aiming for perfect skin regeneration. *Science* 276 (5309): 75–81.

Martins-Green, M., N. Boudreau, and M.J. Bissell. 1994. Inflammation is responsible for the development of Wound-Induced Tumors in Chickens Infected with Rous Sarcoma Virus. [In eng]. *Cancer Research* 54 (16): 4334–4341.

Norheim, O.F., et al. 2014. *Åpent Og Rettferdig – Prioriteringer I Helsetjenesten (Open and Fair – Priority Setting in the Health Service).* Official Norwegian Reports 2014:12 Oslo: Departementenes sikkerhets og serviceorganisasjon; 2014.

Phan, T.G., and P.I. Croucher. 2020. The dormant cancer cell life cycle. [In eng]. *Nature Reviews. Cancer* 20: 398–411.

Shaw, T.J., and P. Martin. 2016. Wound repair: A showcase for cell plasticity and migration. *Current Opinion in Cell Biology* 42: 29–37.

Svee, A., M. Mani, K. Sandquist, T. Audolfsson, Y. Folkvaljon, A.E. Isern, A. Ringberg, et al. 2018. Survival and risk of breast cancer recurrence after breast reconstruction with deep inferior epigastric perforator flap. [In eng]. *The British Journal of Surgery* 105: 1446–1453.

Vogt, H., and A. Pahle. *2018*. Legeprofesjonens (Kosm)Etiske problem. *Tidsskr Nor Legeforen* 2018: 13.

Yadav, A.S., P.R. Pandey, R. Butti, N.N.V. Radharani, S. Roy, S.R. Bhalara, M. Gorain, G.C. Kundu, and D. Kumar. 2018. The biology and therapeutic implications of tumor dormancy and reactivation. [In eng]. *Frontiers in Oncology* 8: 72.

Lost in Translation

Karen Rosnes Gissum

A Short Introduction to Health, Illness and Disease in the Context of Ovarian Cancer

What is health? And why is this concept important to define? Throughout history several perceptions and concepts of health have been designed and discussed. With the progression in medicine, there is an urgent need to take into consideration what health is today, how we talk about health, who is declaring a status of 'good health', and the limitations of modern medicine in being able to declare 'good health'.

> Health is the harmonious functioning of the organs (Pindar).

Health was first understood as a divine responsibility held by the one who created Man; The Eternal One Himself. In the ancient Greece, the demand of reality being explained through natural causes arose. Dualism: the separation of mind and body, the connection between a person and the environment and the nature of disease, are perspectives brought to life by poets and philosophers like Pindar, Plato, Aristoteles and Hippocrates.

> Health is a state of being in complete harmony with the 'universe', a universe never affected by old age and disease due to the harmonious synthesis of the four fundamental elements (fire, earth, water and air) providing its sub-stance (Plato).

In the modern world, the perspectives from the ancient Greece are still valid, however new perspectives and concepts of health have emerged. The World Health Organization (WHO) defines health as "A state of complete physical, mental and social well-being and not merely the absence of disease or infirmity" (World Health

K. R. Gissum (✉)
Centre for Cancer Biomarkers CCBIO, Department of Clinical Science, University of Bergen, Bergen, Norway
e-mail: Karen.gissum@uib.no

A. Bremer, R. Strand (eds.), *Precision Oncology and Cancer Biomarkers*, Human Perspectives in Health Sciences and Technology 5,
https://doi.org/10.1007/978-3-030-92612-0_6

Organization 1946). The definition has not been amended since it was entered into force in 1948. However, the advances in medicine and science have led us into a new era; *the era of precision and personalised medicine*, challenging the WHO definition and former philosophical theories of health with new understandings of disease, study of diseases (pathology) and medical phenomena (The Lancet editorial 2009). Other attempts to define health introduce several other elementary ideas, all considered as "paradigm objects of medical concern": (1) Value; (2) Treatment by physicians; (3) Statistical normality; (4) Pain, suffering and discomfort; (5) Disability; (6) Adaption and (7) Homeostasis (Boorse 1977). All these ideas are part of different conceptual frameworks, or models, of thinking about health, disease and illness. Among these models, we have the biomedical model; the biopsychosocial model; the holistic (religious) model; the ideal model; the humanistic (holistic) model; the psychosomatic model and the existential model (Tamm 1993; Hofmann 2005). All these models present health from different perspectives, both individual and societal.

Perspectives and Concepts of Disease and Illness in Health

Disease is not the opposite of health, however, the understanding of disease and how to define disease is as tricky as defining health. There are several different perspectives and concepts of what disease is, even among health professionals as well as what good health is not, one being disease understood as a generalized phenomenon (the psychological concept) and another one is disease understood as entities (the ontological concept) (Cassell 1991, 4). The medical model provides criteria for something to be a disease, and these criteria are stricter than "just" being un-healthy (Hofmann 2005).

> Disease, then, is something an organ has; illness is something a man has (Cassell 1976, 27).

The easy way would be to define disease as a deviation from what is normal, in traditional medicine also called pathophysiology, diagnosed and treated by doctors (Pool and Geissler 2005). However, there are subscales of disease, subscales that have certain normative features reflected in the institutions of medical practice – illness (Boorse 1975). Patricia Benner (1989, 303) once defined illness as "the human experience of dysfunction". What is important to highlight in that definition is the 'human experience'. Illness is something that is subjectively experienced by an individual. Let's look further at the difference between disease and illness.

Health professionals involved in patient treatment relate to concepts, models and perspectives of health, disease and illness. The clinical gaze, the perspective of seeing the patient through a biomedical lens where biology causes disease, was introduced by Michel Foucault in *The birth of the clinic* in 1963 (Foucault 2003). In the perspective of Hippocrates, the father of medicine, disease was caused by biological factors. He conveyed the dualistic view of mind and body making way for the biomedical model: viewing different aspects of the human body rather than man as a totality – a reductionistic view (Tamm 1993). The biomedical model has made enormous improvements in medical care focusing on physical health, genetic vulnerabilities and drug effects (Tamm 1993), however this model has been criticised for the exclusion of social, psychological and behavioural dimensions of illness introducing the biopsychosocial model (Engel 1977).The biomedical model explain health as defined in the biophysiological world, while the holistic model represents the world of phenomena, promoting the lived experience of health, and dealing with illness and disease (Benner 1985). The holistic model, the expression of wholeness and the care for the whole person; the spiritual, physical, mental and social needs of a person are all related to human good. The criteria of health being the absence of disease is set aside, introducing criteria for health in wellbeing, happiness, human flourishing, ability to realise goals and to promote human functioning as a whole (Hofmann 2005). Compared to the biomedical model, the clinical gaze, focusing on disease, the holistic model gives an important place to illness. As a comparison, in the ideal model of health disease is seen as the absence of health and health as the absence of disease.

The different models of health, disease and illness show that there are different framings of these concepts, and different priorities given to a holistic perspective or a more reductionist one. This starts to outline the challenges of communicating the understanding of health, illness and disease between the patients and the clinicians. Arguably, this issue is even more stringent for cancer, as it is an extremely complex disease that evolves over time, and where the patient's 'home-world' is fundamentally changed. Cancer patients live in both the biophysiological world and the phenomena world with their lived experience of health, dealing with illness and disease; two levels of discourse, calling for different explanations (Benner 1985).

How do physicians and patients communicate and understand health, the disease of cancer, the illness caused by cancer and the reality of cancer treatment? In Susan Sontag's book *Illness as Metaphor and AIDS and Its Metaphores* (2005), she writes about metaphors of cancer, especially metaphors regarding cancer and illness. She urges us to take illness literally, it is not a metaphor although it is often treated as one. Illness is a difficult word to understand. At first it sounds negative much due to personal interpretations and misunderstanding of the word, but it can also be healthy and normal when used in its right shape. You may experience illness, but you will return to wellness.

In a context of high biological complexity and uncertainty, where medical decision-making is not guided by clear, full-proof facts, it is easy to be misled by the assumption that the concept of "good health" is not a fixed entity.

In his book *Nature of Suffering and the Goals of Medicine* (1991) Eric J. Cassell claims the hallmark of modern medicine to be its dependence on science and technology and the conflict these two creates when dealing with suffering will be a predicament for medicine. The science of modern medicine, in its new meaning, is the science of normal and abnormal biology. Why is the disease the centre of focus in modern medical science instead of the sick patient, what happened?

> By defining disease or disability in terms of genetic loci, the relationship to experience is made a step more distant: removed not just from the lived experience of the phenotype, but from the development of the phenotype itself (Scully 2004, 653).

The rapid advances in medicine and the era of precision medicine urge us to provide the ultimate explanation of disease, often to the expense of the more superfluously trapping of "human culture and language" (Benner 1985). In the perfect world of personalised medicine, physicians would help patients adapt to their unique prevailing conditions (The Lancet editorial 2009). But are doctors and patients living in the same world, communicating and understanding by the same language of medicine and health, or are they all lost in translation?

The aim of this chapter is to provide an understanding of how the biophysiological, and phenomena worlds interact with each other in the decision-making process and management of ovarian cancer. This chapter discusses changes in health, disease and illness over time, the trajectory, and the discrepancies in knowledge translation, understanding and eventually decision-making in ovarian cancer from a patient perspective and from a doctor's perspective. In order to explore these questions and concepts, the author will guide you through two fictional journeys, based on literature, focus-group interviews with patients, physicians and the author's own experience. But first and foremost, there is a need for a short understanding to what ovarian cancer is.

A Brief Introduction to Ovarian Cancer – "The Silent Killer"

In the western world, epithelial ovarian cancer (EOC) is the leading cause of death from gynaecological malignancies (Allemani et al. 2015; Mor and Alvero 2013). EOC is the 8th most common neoplasm with 295,414 new cases in 2018 worldwide, and the 8th leading cause of cancer deaths in women (Bray et al. 2018). The 5-year survival rates range from 29% to 89%, depending on histological group and advancement of the disease (Timmermans et al. 2018; De Angelis et al. 2014). The cellular origin and pathogenesis of EOC is not well known or understood (Desai et al. 2014; Kurman 2014). EOC can be subdivided into at least five different histological subtypes, each with different aetiologies and genetic, phenotypic, and clinical features (Kurman 2014; Desai and Soon-Shiong 2014).

Symptoms of EOC are non-specific. Women may experience gastrointestinal symptoms like nausea, loss of appetite, early satiety, abdominal distension, bloating, pain, tenesmus and constipation (Goff et al. 2004; Fitch et al. 2002). Others report of symptoms of unusual fatigue, weight loss, urinary symptoms and gynaecological symptoms (Bankhead et al. 2005). All of these symptoms could be related to several other diagnosis, leading to the inevitable result: the majority of women are diagnosed at late and advanced stages with metastases disseminated throughout the peritoneal cavity (Berek et al. 2018a; Kurman 2014). Advanced disease at time of diagnosis is the main reason why the overall survival rate still is <45% (Morgan 2011), giving EOC the reputation of being a "silent killer".

Staging is the description of the size of the cancer and where it is located. EOC is staged according to the International Federation of Gynaecology and Obstetrics (FIGO) staging system from 2014 (updated in 2018) (Berek et al. 2018b). The stage is set based on perioperative judgements in combination with histopathological evaluation. There are a variety of different EOC subtypes, the high-grade serous ovarian cancer (HGSOC) is the most lethal one of them all as it often presents itself in a high stage (Vang et al. 2009; Lisio et al. 2019). The main treatment of HGSOC, both early and late stages, has been the same for decades; advanced surgery followed by platinum- and taxane-based chemotherapy. Macroscopic visible residual disease has been shown to be the single most important, independent, negative and prognostic factor for survival. At recurrence, which is common, patients still receive multiple therapeutics (primarily chemotherapy and/or targeted drugs), but the attention will change from cure to palliation.

Biomarkers in Epithelial Ovarian Cancer

Cellular biology and cancer biomarkers are the unique molecular signature of your cells or a protein, like a fingerprint. This is a simple explanation of a research area, that has been unveiled since President Obama announced the research initiative towards a new era of precision medicine in 2015 (Collins and Varmus 2015). Biomarkers are still a young area of research and the era of precision medicine in the landscape of ovarian cancer biomarkers is yet to come (Ueland 2017).

For many years one believed the origin of the HGSOC to be from the surface epithelium or epithelial inclusions in the ovary, but recent studies indicate its origin to be the fallopian tube (Labidi-Galy et al. 2017; Berek et al. 2018a). The pathogenesis of HGSOC differs from the low-grade serous carcinomas with a high level of chromosomal instability, mutated TP53 and a rapid tempo of tumour development (Vang et al. 2009). This pathogenesis causes diagnostic challenges in HGSOC, making this histological subtype responsible for 70-80% of all deaths in ovarian cancer (Lisio et al. 2019). Identifying biomarkers that can identify the disease at an early stage is of the outmost importance to reduce deaths in HGSOC. HGSOC, biomarkers that are identified shows to have an impact of diagnostic, predictive and prognostic clinical value as in: symptomatic disease/asymptomatic disease; tumour invasion; ascites; imaging; general health status; carbohydrate antigen (CA125); decrease in CA125; platinum response; BRCA status single molecules versus signatures molecules. As one can see biomarkers have different functions: *diagnostic*, *predictive*, *prognostic* or *therapeutic* (Hennessy et al. 2008). Diagnostic biomarkers are used in the determination of whether a patient has HGSOC, indicating the treatment the patient should be offered (FDA-NIH Biomarker Working Group 2016). While predictive biomarkers can predict if the patient will respond to the treatment, prognostic biomarkers will give information of the outcome of the disease (Italiano 2011). Main types of predictive and prognostic biomarkers are *clinical markers*, *histopathologic markers*, *molecular markers* and *imaging markers*. However, predictive and prognostic biomarkers can be exchanged causing confusion mix-up (Oldenhuis et al. 2008). Therapeutic biomarkers in ovarian cancer can be proteins as target for therapy. Few, or none, are related to early diagnosis and/or patient-centred treatment. Up until recently, ovarian cancer biomarkers have consisted of single – biomarkers like the CA125, reliable but not very sensitive (Dochez et al. 2019). In the present, biomarkers in ovarian cancer are going towards algorithms consisting of several single – biomarkers like the Multivariate Index Assay, Risk of Malignancy Index (RMI) and the Risk of Ovarian Malignancy Algorithm (ROMA) in order to improve the characteristics of single-biomarkers (Ueland 2017; Dochez et al. 2019).

The Journey and Trajectory of Ovarian Cancer

The word journey is often used as a metaphor by cancer patients when describing their illness experience (Semino et al. 2017) while trajectory is used when describing the course of the disease. You are now to be presented to Anne her journey, her journey of illness: through the curative and the palliative journey of ovarian cancer. Starting with who Anne is then leading to the psychological and cognitive suspicion of something being medically wrong, not knowing what it is or where it will lead her, involving existential and spiritual experiences.

> Cancer is a journey, but you walk the road alone. There are many places to stop along the way and get nourishment – you just have to be willing to take it (Emily Hollenberg, cancer survivor, 2020).

Anne – The Suspicion of Something Being Wrong

Anne is living in the twenty-first century, the year is 2020. She was born in 1964, the same year as Martin Luther King Jr. received the Nobel Prize and seven years before President Richard Nixon signed the National Cancer Act later known as his War on Cancer. She spent her childhood and teens in a small industrial community in the western part of Norway, being somewhat happy and fearless, unaware of what the future will bring. Then her mother dies of cancer when Anne is 17 years old, an adolescent now feeling both angry and sad by the loss of her mother, everything she had taken for granted was now thrown into the air. They said the cancer was in her "tummy", nobody asked for more information and Anne does not want to upset her father by asking. He, her father, is a man of few words. Every day she sees the grief in his face, the same face Anne looks at in the mirror, but without her mother her father becomes the cornerstone in her life, no words are needed.

Years passes, and Anne meets John and becomes pregnant, a boy. She becomes a teacher but studying and caring for her family comes with a cost, Anne and John drifts apart however still connected with a deep desire to keep the family together in marriage. Anne finds comfort in her job, she loves teaching small children, but lately she has been feeling tired and for the first time she is feeling low on energy. She has just turned 57 years old. She can't remember when the changes started, they probably came gradually. Anne relates it to lack of exercise, she hasn't found the time or effort to exercising and she is getting older and sensing more stress to most things in life. Her stomach feels "filled up", her appetite is poor at the same time she struggles with constipation, and she has to admit that her pants are getting too tight. Anne doesn't like complaining, she thinks the changes in her body are probably normal for her age, her friends are experiencing some of the same changes, however her husband advises her to see her general practitioner (GP), but the GP doesn't find anything abnormal when examining her, but he wants her to take some blood samples "just to be safe". Being reassured by her GP, Anne continues her everyday living, accepting that nothing was found but she doesn't understand why her physical state is not improving, it should improve if nothing is wrong shouldn't it? Should she return to her GP? Besides, she remembers being to the gynaecologist taking the ultrasound and other tests just six months ago, everything was normal then, things can't change so fast, or can they?

Now, Anne has been feeling ill for a while, without her GP detecting a disease. She is feeling more tired and is experiencing more undesirable bodily complaints such as weight loss, nausea and shortened of breath. She is however still undermining her symptoms, relating them to bodily changes, maybe more a hope of normality rather than logic normality.

Intermission – Moving from Illness to Disease: Classifying Ovarian Cancer

A survey performed by the World Ovarian Cancer Coalition (World Ovarian Cancer Coalition 2018) found that more than 90% of women diagnosed with ovarian cancer experienced multiple symptoms prior to diagnosis, but even so, almost half of the women waited three to six months before going to the doctor. Increased abdominal size is the most commonly reported symptom in recognising ovarian cancer, but women in general have not heard of ovarian cancer when experiencing the symptom (World Ovarian Cancer Coalition 2018), the women don't see what's coming so they don't react. In Anne's case, the GP did not find any sufficient physical cause when she consulted him the first time. However, one of the blood samples from Anne's GP's appointment turns out to be elevated to 325 kU/L (normal <35 kU/L) – the CA125. This elevation could be caused by several non-malignant processes in her body like endometriosis, infections or heart failure, but Anne doesn't know that CA125 is very often is increased in patients with EOC and the blood test is used as part of diagnostics for tumour in the abdomen (Rustin et al. 1996; Buamah 2000; Bast et al. 1983). The GP is aware of the uncertainty of the CA125, and doesn't want to give Anne a preliminary cancer diagnosis, causing her tremendous agony if not true, but how will he be able to inform Anne of the test result without introducing the word of cancer? Anne will come to trust this indicator, this biomarker, as a cue in which direction her diagnosis is taking, unaware of the knowledge that the CA125 is still considered to be neither a sensitive nor a specific biomarker (Ueland 2017; Moss et al. 2005; Coticchia et al. 2008).

Anne – The Preliminary Diagnosis

The elevated CA125 was not sufficient for a cancer diagnosis, but in combination with US findings and the fact that Anne is menopausal, all single-biomarkers involved in the RMI algorithm, results in a strong suspicion of ovarian cancer (Javdekar and Maitra 2015). Anne is referred back to her gynaecologist. She dreads it, it's so intimate being examined in the most private part of her body, but she knows she has to go through with it – there is no other option. New ultrasounds (US) both transvaginal and abdominal are performed by the gynaecologist within two weeks after being to the general practitioner, giving her new unpleasant information: the suspicion of ovarian cancer has been reinforced by the detection of ascites in her abdomen and by abnormal findings on both her ovaries.

Now Anne is scared. She understands the seriousness of the findings, but she doesn't understand all the words or the meaning of them, they are all representing something new, something frightful. Her assumptions that these changes in her body were normal has been proven wrong, the experience of illness was not her mind playing with her they were all vague symptoms of a malignant disease she had never heard of up until now, now they're suddenly equal to death. The guilt of not being familiar with her own body hits her like a stone, is she herself to blame for not taking the bodily symptoms for something being wrong? will she get through this? How will she cope? Even though more people survive cancer, more people die of cancer than ever before. How can she tell her family and friends that she has cancer? Her husband, whom is more like a friend than a husband, and her son… Even though he is an adult he is still her boy and she feel close to him. Her father who has experienced the evil of cancer before, how and who will take care of him? She has had the role of being the caretaker for all of the men in her life, now all of a sudden, she feels the need to be taken care of be somebody else. Now, the burden of her own thoughts seems unbearable, but putting them on her husband, son or father seems even more awful. So, she keeps her thoughts and feelings to herself, remaining a wife, mother and daughter as nothing has happened, the roles are maintained although her experience of illness grows. Her home-world as she knows it will forever be changed.

Anne's gynaecologist sends a referral for a computed tomography (CT) and to the Gynaecologic Oncology unit at the nearest University Hospital. But in spite of the information from her gynaecologist Anne doesn't understand the seriousness of her disease, not even her gynaecologist understands. None of the examination's performed, the tests, the biomarkers can predict the stage, prognosis or outcome of her disease – ovarian cancer.

Anne – Her Experienced Symptoms Confirmed into a Diagnosis

Anne receives an appointment for a preoperative computed tomography (CT), a clinically relevant and imaging biomarker for staging and distinguishing between primary cytoreductive surgery and neo-adjuvant chemotherapy in ovarian cancer (Chang et al. 2015; Suidan et al. 2014). She has a friend who has taking an MRI, telling her of the claustrophobic experience of lying inside the machine, not able to move. This is all new to Anne, lying still in an enclosed space with fear of claustrophobia, feeling the contrast fluid moves its way through her veins giving her a sense of flushing in her pelvic. Above all is the fear she is feeling- the fear of the result of the CT scan, to her it's just another test, an imaging test, and she is not able to differ between the different modalities. She senses something being wrong, it's almost like she can feel something growing inside of her, and she is afraid. She remembers her mother, dead at the age of 48 from cancer in her tummy. At this time on her journey Jenny, a gynaecological oncologist at the University Hospital, receives the results of Anne's CT scan, showing a macroscopic peritoneal metastasis beyond the pelvis with metastasis to the retroperitoneal lymph nodes.

Jenny – The Gynaecological Oncologist Meeting Her Patient Anne

Jenny meets Anne at the hospital. She has 30 min to her next appointment. Thirty minutes to read Anne's history, perform a clinical examination, answer questions from Anne and to establish a physician-patient relationship. By the first glimpse of Anne, Jenny notices the big tummy, the shortened of breath from going from the elevator to the examining room. At first, they talk. Jenny knows the questions to ask, to ask open-ended questions rather than the closed questions, but she also knows open-ended questions takes time and for setting the diagnose she needs also to ask closed questions, questions with answers that may lead to a diagnose when interpreted and translated into the biomedical world. She senses the tension in Anne, her fear. Jenny has to perform a gynaecological examination and palpate Anne's abdomen. It takes time, she has to do the US and manual examination, translating her understanding of what her eyes can see, and her hands can feel into pathology. She has already seen the CT scan and the CA125 result.

Even though Jenny and her colleagues are able to set Anne's diagnosis at an early point of the disease due to diagnostic biomarkers as CA125 and CT, the outcome of the disease has not improved the last decades (Chandra et al. 2019). Every single test and finding affect what treatment Anne will get and decide her chances for the future, but even so, Jenny knows that according to statistics Anne has a 20% chance of 5 years life expectancy, should she tell her, or not? What will Anne's reaction be? Jenny and Anne have just met and the power-balance between them are not equal: Jenny has the power in her knowledge, training and experience, Anne is depending on Jenny, literarily laying her life in her hands. The relationship is unbalanced, the doctor-patient relationship is dependent on trust, from both sides. Anne will come to cherish the affective quality of the consultations, both now and in the consultations to come. Jenny seeing Anne as the person she is, not just the disease resulting in a higher quality of life and satisfaction for Anne (Ong et al. 2000).

Anne – Accepting the Decision of Surgery

The positive side of the CT scan is that Anne is eligible for surgery. Jenny informs Anne of the side effects of surgery, bodily changes and potential complications of surgery. In Anne's mind surgery is the only treatment that can cure her, removing the "disease" leaving her without cancer. The anaesthesiologist examines her and informs her about the anaesthesia while the physiotherapist gives her instructions of how-to breath while protecting her midline incision. The nurse gives her information of everything that will happen before and after surgery while the laboratory technician is taking blood samples from the vein in her arm. This is all too much. Anne's head is full, she is tired, she is afraid thinking this must be serious. All these people, informing her about everything that might go wrong. Anne is feeling stressed. She finds it hard to cope with her new situation it's unknown to her, she hovers not knowing where she will land navigating the uncertainty. She cannot control this, or anything, or so it seems. What could be the meaning of this, did she deserve it? Was it something she did or didn't do? Is this Gods intention? These questions are tearing her apart – WHY HER? Was it not enough that her mother was taken from them at an early age? Time is running fast, too fast. She struggles to look ahead, fearing the future, fearing what the surgeons will find, fearing never to wake up. Not being able to think she goes into surgery three weeks after her appointment with her gynaecologist, four weeks after being to the general practitioner, and approximately five months after sensing the first symptom.

Jenny – Posing a Final Diagnosis

Jenny knows by her training, research and experience that cytoreductive surgery is the most important cornerstone in the treatment of ovarian cancer (Chang and Bristow 2012; Chang et al. 2015) and that going through with the surgery is the best medical advice she can give to Anne. She will perform the surgery herself, being a skilled gynaecologic oncologist at a university hospital, and this will increase Anne's survival rate (Paulsen et al. 2006). The CT scan has provided Jenny information of what to expect when they "open" Anne, but the complete answer will not be revealed before Anne is lying on the operation table. Jenny was able to give Anne a complete cytoreduction, a major impact on survival along with the chemotherapy that is to come. The final diagnosis, histology, staging and prognosis of ovarian cancer depends on operative findings (Berek et al. 2018a). Jenny receives the histological report from Anne's surgery showing a high-grade serous ovarian carcinoma and the presence of mutated *TP53* strengthen the diagnosis. The germline test for mutations in the BRCA1 and BRCA2 genes were negative, however the pathologist found a somatic BRCA1 mutations when her primary tumour was examined. This finding makes her eligible for Olapraib, approved for this population by The National System for Managed Introduction of New Health Technologies. The findings of somatic mutations in BRCA1/2 predicts she may benefit from PARP inhibitors targeting DNA repair, combined with chemotherapy prolonging her progression-free survival (Ledermann 2016; Moore et al. 2018).

Jenny concludes that Anne has an overall poor prognosis. Findings from the CT scan and surgery indicates that Anne's disease is at FIGO stage IIIC: Macroscopic peritoneal metastasis beyond the pelvis more than 2 cm in greatest dimension, with or without metastasis to the retroperitoneal lymph nodes (includes extension of tumour to capsule of liver and spleen without parenchymal involvement of either organ) (Berek et al. 2018b). Relapse maybe unavoidable, it will come, but the timing of it is essential for further treatment and prognosis. How will Jenny inform Anne of these devastating findings?

Anne – The Start of Chemotherapy

Anne starts chemotherapy four weeks after her surgery. She has recovered from surgery and her performance status is 0 being fully active and able to carry on all pre-disease performance without restriction (Oken et al. 1982). She will receive six courses of paclitaxel 175 mg/m^2 body surface and carboplatin AUC = 5, a platinum-based chemotherapy (6 treatments cycles, 3 weeks interval), followed by olaparib, a PARP inhibitor for 2 years. The chemotherapy is being administered at an oncological day unit. Women bearing ovarian cancer and other gynaecological diseases are sitting next to each other, waiting for their doctor appointment, talking to each other comparing treatment end experiences. Anne starts talking to Linda. Linda has ovarian cancer, apparently the same as Anne, but Linda is receiving bevacizumab and not olaparib. Why so? They have the same disease, should they not receive the same treatment? The differences make they them both stress – Anne can't help but thinking "I wonder if I am receiving the best treatment for my disease?" Anne is reassured by her doctor's choice of treatment. During her surgery complete cytoreduction was archived. This means that her risk for recurrence is reduced. In Norway bevacizumab consolidation therapy is only approved for patients with high-risk for recurrence. She read about bevacizumab in the newspapers, but not olaparib.

CA125 dropped to 125 kU/L after the surgery, and it dropped further already after one course of chemotherapy indicating a therapeutically effect and her hope for survival amplifies. She is experiencing some side-effects of the treatment, but they are minor details in her goal of conquering the disease. Her sufferings fell meaningful. When ending her sixth course of chemotherapy Anne continues olaparib maintenance monotherapy hoping for longer overall survival (Ledermann et al. 2016). Anne is not naïve, she knows the relapse probably will come, of course it will, but she doesn't know when especially since the survival data for use of PARP inhibitors are not yet known. Twenty-seven months after she ended her first line of chemotherapy Anne starts to feel some discomfort, it feels like a fibroid or something similar in her back. She relates it to previous discomfort in her back, not to her cancer disease, and with no immediate plans to contact her doctor, she not out of control. Instead, she gets an appointment at her physiotherapist, but the discomfort stays and the inevitable occurs – Anne relapses.

Intermission – The Dichotomy Between Illness and Disease at the Diagnosis Stage

Anne feels unhealthy with a reduced capacity; she is experiencing illness (Niebrój 2006). She is diseased with ovarian cancer and Jenny is the physician who professes to be able to heal her disease. Anne's feelings of health and illness are a result of her lived experiences; her perceptions, her beliefs, her skills, her practices and her expectations (Benner 1985). The changes in her body, the vague bodily symptoms are often not recognised by health professionals, and women like Anne, may confuse them with normal bodily changes (Fitch et al. 2002). These changes are not detected by the GP or the laboratory, but are all part for Anne's illness experience, before the diagnose is suspected. There is no connection between the cellular level and her experience of illness, but even so, the experience is real to Anne although no one else can proof them. Should she be angry that her illness experience was not enough, those changes were not seen as alterations in a disease process; they were proof of a disease to her, but not for the biomedical model of disease. The illness Anne is experiencing becomes a disease when "the underlying biological abnormalities that cause the symptoms and signs of the illness are clarified" (Komaroff 2019). The diagnose frees Anne from prejudice, she has a disease not "only" an illness. Her assumptions are no longer signs of something vague, it has left her with an explanation. Although receiving a diagnose rearranges reality to Anne, it also reduces her to a diagnose and not the woman, mother, wife, daughter and friend she is. Being diagnosed with gynaecological cancer may be stigmatizing. Ovarian cancer is a sex- specific disease – only women can have ovarian cancer. The sex-linked factor (the biological differences) like the ovaries and the fallopian tubes, are decisive in ovarian cancer. Is the inevitable status of being a woman itself one of the reasons to why Anne's disease was diagnosed at a late stage? Do women like Anne have less trust in their own body or a lack of knowledge of how the female body functions? Or is it the intimacy of the disease? It almost seems like the knowledge of anatomy in ones one body is too vague, leading to the inevitable fact: women diagnosed with ovarian cancer have little understanding of what ovarian cancer is or what it will mean to them (Simacek et al. 2017).

Anne's horizon of her illness meets Jenny's horizon of disease in the consultation at the hospital, a meeting characterised by misunderstanding, Anne's understanding of illness and Jenny's understanding of disease. Their relationship is based on trust; however, their communication of understanding is not straightforward. Together, they are to share and discuss on the basis of the best available evidence, to provide Anne the support she needs to consider her options (Elwyn et al. 2010). But are there really any options? Accept treatment or face death.

Being a physician, Jenny is a natural scientist, trained according to the biomedical model of health and disease, but she is also a mother, a wife, a daughter and friend. Jenny has obtained the knowledge of human organisms in health and disease to cure diseases, diseases as classified by the WHO 11th International Statistical Classification of Diseases and Related Health Problems (ICD-11). Jenny regards Anne both as an object and human, a human organism by which she has the knowledge of. She makes a deduction based on her theoretical knowledge and the object she is seeing in front of her. The decision of which treatment Jenny is to offer Anne is based on several different reasonings, both objective and subjective, reasoning on theoretical knowledge and past experience, but also reasoning by understanding Anne as well as ethical considerations (Wulff 1999). Jenny's understanding of ovarian cancer is that it has a biological reason, her assignment is to focus on the treatment of the disease rather than a treatment for Anne as a sick person affected by disease (Svenaeus 2005, 97).

Anne – The Meaning of Relapse

Many EOC women will experience relapse of their ovarian cancer, it's almost inevitable, and it will happen more than once. The fear of recurrence has been lurking in Anne's thoughts ever since she received the diagnosis. She has found ways of coping with her fear, but now all of a sudden, her fear has become real. When Anne first got the diagnosis and formed an understanding of the diagnosis treatment, she counted the days and weeks till it would be over, it was her way of coping until she would be back to normal, at least her normal. Anne has been tackling cancer as a to-do list, ticking of items and moving on to the next one. During the chemotherapy she developed a new identity of stoic optimistic tenacity, she adjusted her lifestyle making chemotherapy bearable even when she lost her hair, vomited and felt the muscle weakness and the numbness in her feet. Some of them, especially the abnormal heartbeat she fails to tell Jenny, she is afraid she will have a reduced dose of chemotherapy making her odds of cure reduced. What now? Is this it, the point of no return to normal or is this her new normal? Accepting the unacceptable means accepting the unknown, this disease will be the end of her. Living with cancer is her involuntarily new normal, there is no need for holding her breath waiting for cancer to be over, she will not conquer it. Cancer and cancer treatment will be her life. How will she adapt? How will she cope with this? How *can* she cope with this? She is fighting an internal fight; regard the symptoms hoping them to be normal or facing her fears and contacting Jenny. Contacting Jenny wins, and she is scheduled with an appointment with Jenny after taking a new CT scan.

Jenny – Telling the Truth or Giving False Hope?

CT scan leaves Anne with no hope of the relapse being misinterpreted. Jenny has been anticipating this to happen, although nothing would give her more joy than if Anne would have been one of the few ones, not experiencing a relapse of this ugly disease. This is it. Jenny knows it by her training and by her experience, it is only a matter of time. Anne will never be cured from her ovarian cancer. She wonders if Anne understands the full meaning of the relapse. She has come to know Anne better from the chemotherapy consultations. She thinks of Anne as a woman capable of receiving the truth of her diagnosis and prognosis, although the relapse itself will cause her agony, leaving Anne with no hope. How can Anne find the strength to another round of treatment if there is no hope? Jenny knows that many women experiencing relapse of ovarian cancer have unrealistic expectations of the treatment they are offered, despite the symptom burden, clinging to hope even when reality leaves them with no hope of cure or a long life. Should Jenny be direct informing Anne of her prognosis now or should the reality of her prognosis remain unspoken and rather fight death to the bitter end. In the end there is death, sooner or later, but for Anne it will be sooner rather than later. Postponing death is the new goal, not cure. Jenny advises Anne to have a secondary cytoreduction based on the absence of ascites and her performance status being 1, she has some restrictions in physically strenuous activity, but still, cytoreduction is an opportunity. But does Anne want a new cytoreduction?

Anne – Towards the End of Her Journey

Her husband and son advise Anne to act on what Jenny advises her to, this is her only hope for recovering from this disease. Anne doesn't want to let them down. If she doesn't go through with it, what then? Is it a real choice? Choose surgery or choose death, now. Anne is aware of the risk of surgery, she knows it will be tougher this time and that this operation will not cure her, it is just another item to tick off on her cancer journey. But somewhere in her subconscious there is a glimpse of hope that maybe, maybe there will be a cure, a revolutionary new treatment for her so that she will have the opportunity to see her grandchildren grow up and to enjoy the life she had planned for. Now, she is standing at the train station, watching her life passing without knowing if she should jump on.

Once again, Anne has a complete cytoreduction leaving her tumour free although Jenny recommends more chemotherapy, a platinum containing regimen. Will this journey ever end, all this time at the clinic? Her life is being lived at the hospital, being dependent on physicians like Jenny. Coming to the clinic feels like a doubled-edged sword; sometimes it feels like going on a vacation leaving the house, meeting other women undergoing the same treatment as herself, sharing some kind of a bond, then of course it's the smell of hospital that sometimes makes her stomach twist, the fear of blood, values not being compatible with treatment and last but not the least; Jenny's veridiction of treatment. Sometimes the women share their experiences of their cancer journey, but more the side-effects than the thoughts. There is a sense of optimism at the clinic, the nurses have the time to talk to them, however the sensitive topics, the thoughts closest to her heart remain inside her. There is never the time nor place to share these thoughts. Sometimes she feels ashamed for not feeling the optimism, but not all days are the same. Is there the slightest amount of hope for her, for her outliving her destiny? Perhaps a clinical trial she can participate in, not with the intention to be cured, but to be given the opportunity of receiving a few more months, maybe then another medicine has been approved and she may gain time postponing the inevitable.

Intermission – The Dichotomy Between Illness and Disease at the Stage of Relapse

What is the dichotomy between illness and disease at the stage of relapse? Are EOC women aware the differences between the experience of illness and the symptoms of disease indicating relapse, what to look for? Are the women fully aware of their diagnosis, even after relapse? The answer is not unanimous, there is not a yes or no answer to it. Some women will say they are not fully aware of which symptoms to look for, their disease, nor the treatment options, but some women would say yes, however, within the yes lies death as imminent despite incurability (Finlayson 2017).

Anne has been experiencing illness for a very long time, long before the disease was a fact. The disease could come back any place in her body, causing her vague symptoms and/or problems of pain, bloating, nausea or constipation. Anne has been through two major surgeries in her abdomen, and she had twelve courses of chemotherapy. These symptoms could be just a result of her previous cancer treatment, not necessarily the symptoms of relapse. How can she be sure whether the illness she is experiencing is part of the disease or as a result of cancer treatment, not just something she is sensing or is normal? In the physician-patient communication there is a risk of talking past each other's purposes, talking but not reaching the other, not as a result of bad practice from the physician, but rather different realities and relations to illness (Toombs 1987). Again, their horizons are not the same, they are not living in the same world.

Anne is in a vulnerable situation expecting alleviation and the best possible service from Jenny. Jenny professes her willingness to help and heal Anne, by doing so she promises to act to benefit Anne rather than harming her; to do good. However, for Jenny doing 'good' is not necessarily the same as doing biomedical good when considering Anne's values and her experience of illness. Jenny's actions of medicine should lead to a 'correct healing decision'. As a result of the advances in modern medicine as well as in medical research, and the consequences of these advances, Jenny is not in a position of control to always act in Anne's interest, to protect her 'good life' and facilitate for her to make value choices in decision-making (Pellegrino 1981). Jenny is controlled by the administrative burden placed upon her, legal restraints, industrial seduction and implicit rationing making her unable to act in the best interest of Anne (Bircher 2005). Jenny considers herself being a realist not bound by theorists or philosophers, believing in what works in clinical practice rather than theories of clinical practice leading to unhappy results, removing the *why* and potentially causing discomfort for the patient (Cassell 1991).

When suffering from a life-threatening illness the timeframe, needs and interactions with health-services become important, even more and yet unspoken when "how long have I got" hides the "what will happen" for both patients and carers (Murray et al. 2005). Physicians, like Jenny, may be reluctant in speaking of the prognosis of a disease even though this unspeaking of prognosis may leave to patients and relatives with hope and a drive to fight death with palliative treatment that are unlikely to benefit them (Murray et al. 2005).

Anne – Accepting the Inevitable

In the depth of her mind, Anne knows this disease will be the end of her. The surgery, the chemotherapy, another surgery and another chemotherapy, they were all part of a hope, a hope for cure. She raises the question of how long time she has left, but the unspoken question was: what will happen to me? Anne doesn't fear death, she is more afraid of not being able to breathe because her lungs are filled up with fluids, or her stomach being large and hard also making it hard to breath. Will this happen to her? Her legs so filled with water making the skin feel like it should blast? Anne knows it's an ugly death dying of ovarian cancer and it takes time, but when death comes, she will be free of her sufferings.

As time has gone by, the acceptance of this outcome as become clearer to her. She keeps these thoughts inside, not sharing them with her husband, her son or father, she doesn't want to cause them more pain and misery. She is alone. She wonders if her mother felt the same way at the end of her journey. When looking into the mirror she sees a different person, her face is swollen by the medications, her skin is grey, making her look older than she is, her shoulders are thin, her abdomen is tense, and her legs are swollen. The resemblance with a pear comes to her resulting in a smile on her face. Her eyes are clear, in them she is still 24, what happened to the rest of her? It's been three years and seven months and it seems like a lifetime. She senses a void inside, causing feelings of loss and alienation living in this unpredictable body (Sekse et al. 2013). This alienation of her body and the taken-for-granted world she has been living in, has come to her in a way she could never imagine.

> Courage is knowing what not to fear (Plato).

Her life has a new perspective, a new horizon, while her surroundings remain as they were. She is ever so alone. Has her journey through the landscape of cancer treatment been worth all of her suffering? The answer remains unanswered to Anne, it's too difficult to answer. The choice of treatment was not a real choice to her at the time, however there are so many things she would like to have known before it all started: the real consequences of treatment, the change of family roles, the social distances, friendships that fade away, the feeling of been inadequate, the loss of home-world, being a burden to both her family, friends and society, being alone. It has been a journey, a single journey, with days, people, feelings and experiences she would never have had if it was not for the cancer, experiences that have made her life richer in many ways. In a sense she has adapted to her environment, "the horizon of our total attitude" (Ludwig 1940), she is in a state of balance; not healthy and not free of disease or disability, feeling illness not being well, not wanting to die, but tired of disease and longing for relief. She is not at ease within, and she is alone.

> If, the woman thinks, she were an airplane that crashed and someone located the little black box, that would be the sentence they found, to hear water, but not to see it … (Øyehaug 2016, 53).[1]

The Future of Ovarian Cancer: Anne's Story in 2040

The history just presented took place in 2020. Today, there are biomarkers of clinical value exist for patients with ovarian cancer, but some promising biomarkers are emerging that hopefully can improve both time of diagnosis and prognosis. Science is moving rapidly but the process of translating scientific findings into clinical practice takes time.

Imagine now that Anne's story is taking place in 2040, would there be any differences? The following projection is based on knowledge that is yet to be implemented in clinical practice.

Jenny performs a gynaecological examination without US but by performing the PapSEEK test. The PapSEEK test may detect both endometroid and ovarian cancer at an early stage, with a specificity of up to 45% in ovarian cancer (Wang et al. 2018). In addition, the test is minimally invasively and can be performed at a routine office visit, as in Anne's case, at her general practitioner. The test comes out positive for ovarian cancer but instead of being referred to the gynaecologist and to a CT scan, Anne is referred to a new form of PET-CT where an ovarian cancer specific tracer is used (Hernot et al. 2019). This examination combined with a new AI

[1]The idea is that we all have a secret truth inside us that is difficult to find and understand. The black box, the tachograph in the plane, contains a description of the plane's movements, you look for it to find answers to what has happened. In the context of Anne's journey, it refers to the fact that she has always felt that somewhere there is a truth that she is unable to get close to.

algorithm predict the biological signature and disease distribution (see also later). In addition, Anne has a laparoscopy performed with a fluorescence camera, and a biopsy resulting in a histological diagnosis: ovarian cancer: FIGO stage IIIC. It's the same diagnosis as in 2020 with a somatic mutation in BRCA1, but instead of waiting four weeks for her diagnosis she's now receives the results in 5 days. They also identify an inherited mutation for hereditary nonpolyposis colorectal cancer (HPNCC). Now all of a sudden this could be her legacy, first perhaps from her mother to her, and now this is something her son will inherit. She will have to inform him, and he has to make the choice of whether or not to have a genetic test. But perhaps even more difficult; he will have to decide whether or not he wants this knowledge of his own genes and the knowledge of the impact it may have on his life.

Anne feels everything is going so fast with tests and examinations, her body senses stress, stress that doesn't go away. The changes in her body have been causing her stress for a long time and she has not yet found her way of coping, unaware of the significant impact on biological stress response (Antoni and Dhabhar 2019). Another test is performed: the two cytokines IL-6 and IL-1β. Overexpressed IL-6 and IL-1β makes her eligible for a clinical trial testing the effect of a psychosocial intervention with cognitive training to reduce stress by coping mechanisms with the aim of reducing biological stress response.

The RMI algorithm and other algorithms used in 2020 have been found to be too complex for clinical practice and not capable of handling the increasingly number of biomarkers, leading the way for artificial intelligence (AI) (Enshaei et al. 2015). Now, the decision of Anne's surgical resection will be based on AI. The AI predicts Anne is likely to end up with suboptimal cytoreduction and rather benefit of neoadjuvant chemotherapy. So instead of primary cytoreduction she starts her treatment with chemotherapy: paclitaxel 175 mg/m^2 body surface and carboplatin AUC=5-6, a platinum-based chemotherapy along with a humanized monoclonal antibody directed against transforming growth factor beta (TGF-b) as the overexpression revealed may lead to chemotherapy resistance and stimulation of tumor blood vessel growth (Arend et al. 2019). The CA125 decreases during chemotherapy, Anne is responding to the treatment with minor side-effects. After her third course of chemotherapy, she is finally eligible for and to have benefit of cytoreduction. Anne has been taking a lot of tests and imaging, most of them she doesn't understand the full meaning of. One of the imaging came positive for a biomarker to improve surgical outcome: the CD24, a potential biomarker for image-guided surgery (Kleinmanns et al. 2020; Wang, Fan, et al. 2018). In 2040, Jenny has a real-time intraoperative guidance increasing Anne's chances for a complete and safe tumor reduction (Hernot et al. 2019). The surgery is followed by maintenance therapy of olaparib and anti-WEE1.

First after 5 years she recurs, the recurrence is identified with the use of liquid biopsies in the for of circulating tumor cells. The mass cytometry profiling shows a profiling that advocates anti-CD73 and anti-PD-L1 treatment. A test for intracellular signaling performed already hours into the 1st treatment cycle indicates that Anne will respond. After being assigned to the treatment for 2 years, she has still relapsed. So still a live, with a disease that might recur again and with her illness. Tired, alone and not at all healthy.

Intermission – The Dichotomy Between Illness and Disease in an Era of Precision Medicine

Personalised medicine, as a result of translating new findings of biomarkers into clinical practice by AI, may improve the prognosis of ovarian cancer, or it may not. It may improve her time of progression-free survival and reduce the agony and violence of cancer treatment, or it may not However, Anne will have a more personalised treatment, giving her less side-effects due to be given treatment specified for her.

As mentioned earlier in this chapter, advanced disease at time of diagnosis is the most important reason to why ovarian cancer is called "the silent killer". Early diagnosis of ovarian cancer is not a quick-fix research area. The journey along the ovarian cancer pathway will be more direct and mobilized with resources needed to conquer the disease, but will the use of more sophisticated biomarkers and the ideal of precision medicine reinforce the dichotomy between illness and disease and improve Anne's journey? Their horizons will still be different, the bridge between them perhaps even further away by their inability to communicate their understanding of illness and disease. The gap between the biophysiological world and the phenomena world will increase. Unless, of course, by 2040 the translation between these two worlds is improving resulting in a new more humanistic meaning of personalised medicine. However, it is likely that the more precise the development in medicine, the greater the distance between illness and disease. Anne's perspective is still the experience of her illness, and perhaps the feeling and experience of her illness will be even more remote from and mismatched with the biological conception of her disease when medical advances increase in sophistication and precision.

The Challenges of Communicating Illness, Disease and Health Between Patients and Clinicians

Ever since Hippocrates in the ancient Greece, physicians have been communicating with patients for reasons of biological, psychosocial and social art. Since the birth of modern medicine, communication with the intent to building a therapeutic relationship have been and still is the heart and art of medicine (Ha and Longnecker 2010). However, communicating the understanding of health, illness and disease, transforming the perspectives and experiences from one person to another in order to share the same understanding and the same horizon, still leaves patients and physicians lost in translation. It seems like as though physicians are having difficulties with building a bridge of understanding towards patients, making it impossible for physicians and patients to see the same horizon leaving a gap between them.

Biomarkers are the future of cancer, at least in the biophysiological world, but also in the phenomena world. The need for science, the need for progress and the desire for a cure is the aim and hope for cancer patients as well as physicians. The glory lies the potential future cure for the disease, because there is no glory in illness, there is no meaning to it and there is no honour in dying of [it]" (Green 2012).

Making Sense of and Finding Meaning Around Health, Illness and Disease

Being diagnosed with ovarian cancer is a traumatic life event. The loss of home-world, the loss of life as Anne knew it before the diagnosis and treatment, demands a new understanding of life, resolving loss and finding a new meaning of life. How Anne meets these demands are crucial for whether or not she will experience good health. The inability to cope may be seen as a sign of disease, with regard to Anne's symptoms that needs intervention like drug treatment or surgery (Bircher 2005).

Anne's meaning of life is fundamentally reshuffled. Beyond the extreme complexity to think about life's meaning in general, how can Anne think about her own new meaning in life with her disease, and communicate that to the physicians? Has it even crossed her mind that this should be worth sharing and discussing? Are physicians equipped to help patients find a new meaning in life with their disease? Moreover, finding meaning in a life with cancer can also trigger guilt: "Nothing is more punitive than to give disease a meaning" and "the more mysterious the disease is made to seem, the more likely we are to supply it with meaning and the greater of moral – if not literal – contagion" (Sontag 1978). How can one give meaning into a disease like cancer? For many people cancer is a disease that equals death; so, what could the meaning be, of having cancer? What must one have done to deserve such a fate? As Sontag argues, it is easier on the patient if he/she manages to not read any guilt-triggering meaning into such diagnosis. Rather, we are born and then we die, and that's it. But beyond those considerations around guilt, some women will find meaning and significance in receiving a cancer diagnose (Davis et al. 1998). They will find meaning in the positive implications of the disease itself, and the experience of loss of home-world. For instance, the small things in life that would have gone unnoticed before, are now perceived as important sources of happiness: the smell of flowers in the spring, the pleasure of experiencing yet another day, to be able to live in the present and not worrying about tomorrow (meaning as significance). Other women will try to find an explanation to why they got cancer. It could be genetic as in BRCA – then it's not your fault, a kind of sense making (meaning as comprehensibility) (Davis et al. 1998). So in many ways, being diagnosed with a disease changes your home-world, and that can constitute a push to find meaning in this new situation. By investigating people's experiences of illness, meaning around illness, and how these experiences of and meaning around illness lead to a disease, it is participating to valuing illness. Looking at meaning also arguably helps bridge illness and disease in more profound and meaningful ways.

The Meaning and Sense of Biomarkers – From a Patient Perspective

Biomedicine is portrayed through its metaphors as a warlike force that seeks to defeat the enemy, the use of military metaphors like cancer being the war to be defeated.

What is the meaning of biomakers? The meaning of biomarkers as they appear in humans: a bloodsample, tissuesample or acites. Howevere they all have a side one can not see at the first glimpse. Even from different persepctives they may still appear the same, but based on experience different functions of biomarkers appears. They are extremely complex and the link between the presence (or absence) of biomarker and type of treatment is far from being straightforward.

No one can claim to have a perfect understanding of what a biomarker is and how exactly it will guide clinical decision-making. So, what is the general understanding among cancer patients of what a biomarker is? Even at Anne's first visit at the hospital, before the diagnosis was established, different tests had been performed and biomarkers had been measured. Are the patients informed about the fingerprints they have left behind in these different tests, and more importantly: are they aware of what it will mean to them?

For many EOC women, receiving the malignant diagnosis is equal to receiving a death sentence: the last stop on their journey. At the beginning of the disease, it is rare that patients are able to fully understand what the use of a biomarker implies for their course of treatment, and their lives in general. This is not facilitated by the fact that the media often portray biomarkers and precision oncology in a very positive and hopeful light, inducing the public to think that cancer cures are just about to be realised (in this book, see the chapter from Stenmark and Nilsen). New cancer drugs are "great hopes" and "revolutionary treatments" – who would not want to have that treatment? To complete this 'ideal' picture of biomarker research and precision oncology, new biomarkers are often depicted as an opportunity for drug development and market growth. When diagnosed with cancer, and having been introduced to biomarkers as hopes and revolutions in the media, patients will claim their right to receive the best, newest treatment and nothing less, regardless of what it might induce in terms of loss for them. Indeed, this treatment might possibly prolong their life, but by how long, and at what cost in terms of quality of life? There is also rarely the explicit consideration that the treatment might not prolong their life at all, or even shorten it.

The question of "what is a good (enough) biomarker" was raised in the book *Cancer Biomarkers: Ethics, Economics and Society* book in a chapter by Blanchard and Wik (2017). What is a good biomarker for HGSC women? Would a diagnostic biomarker in HGSC able to detect 100% sensitivity and specificity, be a perfect biomarker? The uncertainty of biomarkers, like the CA-125, are causing anxiety to EOC patients (Moss et al. 2005). All women of ovarian cancer are familiar with CA-125, it's almost like the CA-125 is the one dependent factor that decides the effect of their treatment, their hope of being cured from cancer relies on a blood sample, on something they can't control or even trust. This biomarker, and others, does not necessarily correlate to the women's health; their experience of wellbeing, or correspond to their clinical state (Strimbu and Tavel 2010).

We performed a focus group interview with women diagnosed and treated for ovarian cancer, asking for their perception on what a good biomarker would be and their perceptions of the concept of biomarkers. The women we interviewed were diagnosed before 2015, and were diagnosed at a late stage of cancer. They had undergone first line therapy with both surgery and chemotherapy. Most of them had received several lines of therapy due to relapse. The overall conception was the lack of information from health professionals of what a biomarker is. None of the participants could remember being introduced to the word biomarker during diagnosis or treatment.

In addition to the difficulty to grasp the complex notion of cancer biomarker, the high hopes and expectations of cancer patients (fuelled by optimistic media discourses like seen above) make it difficult to accept that a biomarker is something complex and imperfect, that is not likely to bring a cure. In the beginning, patients expect to be cured of their cancer diagnosis. They expect to be treated in a unique manner that perfectly matches their diagnostics. They expect to be given the best, newest treatment available and at any cost. If living is the outcome, even for just a few more months, this should be worth it. In addition, patients often expect biomarkers to be safe, as they are described everywhere as the fingerprints of your cancer – there should therefore not be room for error or misinterpretation. They will be disappointed. As said above, there are rarely thorough discussions about the fact that even if you are in a terminal phase of cancer, you have things to lose. Trying any treatment at any cost can actually leave you worse off.

In addition, how should one accept that the same biomarker might grant access to some patients to a treatment, while denying it to others? How can patients accept that they are unique, but in the sense that they are denied the right to try a treatment that they would not benefit from? From a patient's perspective, it may seem unfair, especially in the context of numerous, vocal campaigns claiming the right to try experimental drugs for instance. But if something is to be considered unfair, it would mean that they would have been subjected to injustice, partiality or deception. Patients may have the feeling of being treated unfairly, and this is to be expected when knowing that they are regularly confronted with media discourses about new cancer drugs and new revolutionary cancer treatments. It is perceived as unfair and inexplicable that patients with the same diagnosis and condition are not receiving the same cutting-edge medicine. So, I am unique, but not in a good way, and I am being treated unfairly. We have to bear in mind that many cancer patients, still receiving cancer treatment, were diagnosed before 2015, before President Obama announced the research initiative towards a new era of precision medicine (Collins and Varmus 2015). Many of them have not been informed about biomarkers and how the use of biomarkers is affecting what treatment they are offered.

Biomarkers are measured and identified 'objectively' (at least according to their definition), and they have a key bearing on how we think about our health. How can patients understand and make decisions about their lives when their disease is objectively measured and a biomarker 'decides' the treatment, not considering how patients experience symptoms and the way they are dealing with illness and sickness? Disease seems very far from the lived experiences of illness and sickness in the patient, especially in this context of precision medicine.

Conclusion

The aim of this chapter was to give a glimpse of the challenges in communicating the understanding of health, illness and disease, and eventually decision-making in ovarian cancer from a patient's perspective and a physician's perspective. The two fictional narratives of Anne the patient, and Jenny the gynaecologist was based on literature, focus-group interviews, physician's and the author's own experiences. There is a fine distinction between health, disease, disability, wellbeing and illness, but at the same time it seems difficult to separate them. They belong together as a part of the whole human; alone they would only be fragments of something greater, more complex and multifaceted. In addition, the increasingly precise and targeted advanced in medical research and practice lead to a reframing of these concepts. There is therefore an urgent need for refining the definitions of health and disease to fit with present and future medical culture. The different models of health and disease are still valid, but perhaps it would be useful to consider them side by side, with more holistic perspectives meet more precise and personalised ones. Arguably, the more complex the scientific advances, the more space should be given to patients to voice what health is and what disease is, according to their own experience. What patients want for their life, how they want to manage their disease, and where they find meaning, are crucial aspects in an era of precision oncology.

References

Allemani, C., H.K. Weir, H. Carreira, R. Harewood, D. Spika, X.-S. Wang, F. Bannon, J.V. Ahn, C.J. Johnson, and A. Bonaventure. 2015. Global surveillance of cancer survival 1995–2009: Analysis of individual data for 25 676 887 patients from 279 population-based registries in 67 countries (CONCORD-2). *The Lancet* 385 (9972): 977–1010.

Antoni, M.H., and F.S. Dhabhar. 2019. The impact of psychosocial stress and stress management on immune responses in patients with cancer. *Cancer* 125 (9): 1417–1431.

Arend, R., A. Martinez, T. Szul, and M.J. Birrer. 2019. Biomarkers in ovarian cancer: To be or not to be. *Cancer* 125 (S24): 4563–4572. https://doi.org/10.1002/cncr.32595

Bankhead, C.R., S.T. Kehoe, and J. Austoker. 2005. Symptoms associated with diagnosis of ovarian cancer: A systematic review. *BJOG: An International Journal of Obstetrics & Gynaecology* 112 (7): 857–865.

Bast, R.C., Jr., T.L. Klug, E. St John, E. Jenison, J.M. Niloff, H. Lazarus, R.S. Berkowitz, et al. 1983. A radioimmunoassay using a monoclonal antibody to monitor the course of epithelial ovarian cancer. *The New England Journal of Medicine* 309 (15): 883–887.

Benner, P. 1985. Quality of life: A phenomenological perspective on explanation, prediction, and understanding in nursing science. *ANS. Advances in Nursing Science* 8 (1): 1–14.

Benner, P. 1989. *The primacy of caring: stress and coping in health and illness*. Addison-Wesley.

Berek, J.S., S.T. Kehoe, L. Kumar, and M. Friedlander. 2018a. Cancer of the ovary, fallopian tube, and peritoneum. *International Journal of Gynecology & Obstetrics* 143 (52): 59–78.

———. 2018b. Cancer of the ovary, fallopian tube, and peritoneum. *International Journal of Gynecology & Obstetrics* 143 (S2): 59–78.

Bircher, J. 2005. Towards a dynamic definition of health and disease. *Medicine, Health Care and Philosophy* 8 (3): 335–341.

Blanchard, A., and E. Wik. 2017. What is a good (enough) biomarker? In *Cancer Biomarkers: Ethics, Economics and Society*, ed. A. Blanchard and R. Strand. Norway: Megaloceros Press.

Boorse, C. 1975. On the distinction between disease and illness. *Philosophy & Public Affairs* 5 (1): 49–68.

———. 1977. Health as a theoretical concept. *Philosophy of Science* 44 (4): 542–573.

Bray, F., J. Ferlay, I. Soerjomataram, R.L. Siegel, L.A. Torre, and A. Jemal. 2018. Global cancer statistics 2018: GLOBOCAN estimates of incidence and mortality worldwide for 36 cancers in 185 countries. *CA: A Cancer Journal for Clinicians* 68 (6): 394.

Buamah, P. 2000. Benign conditions associated with raised serum CA-125 concentration. *Journal of Surgical Oncology* 75 (4): 264–265.

Cassell, E.J. 1976. Illness and disease. *Hastings Center Report* 6 (2): 27–37.

———. 1991. *Nature of Suffering and the Goals of Medicine*. Oxford: Oxford University Press.

Chandra, A., C. Pius, M. Nabeel, M. Nair, J.K. Vishwanatha, S. Ahmad, and R. Basha. 2019. Ovarian cancer: Current status and strategies for improving therapeutic outcomes. *Cancer Medicine* 8 (16): 7018–7031.

Chang, S.J., and R.E. Bristow. 2012. Evolution of surgical treatment paradigms for advanced-stage ovarian cancer: Redefining 'optimal' residual disease. *Gynecologic Oncology* 125 (2): 483–492.

Chang, S.J., R.E. Bristow, D.S. Chi, and W.A. Cliby. 2015. Role of aggressive surgical cytoreduction in advanced ovarian cancer. *Journal of Gynecologic Oncology* 26 (4): 336–342.

Collins, F.S., and H. Varmus. 2015. A new initiative on precision medicine. *New England Journal of Medicine* 372 (9): 793–795.

Coticchia, C.M., J. Yang, and M.A. Moses. 2008. Ovarian cancer biomarkers: Current options and future promise. *Journal of the National Comprehensive Cancer Network* 6 (8): 795–802.

Davis, C.G., S. Nolen-Hoeksema, and J. Larson. 1998. Making sense of loss and benefiting from the experience: Two construals of meaning. *Journal of Personality and Social Psychology* 75 (2): 561–574.

De Angelis, R., M. Sant, M.P. Coleman, S. Francisci, P. Baili, D. Pierannunzio, A. Trama, et al. 2014. Cancer survival in Europe 1999–2007 by country and age: Results of EUROCARE-5—A population-based study. *Lancet Oncology* 15 (1): 23–34.

Desai, N.P., and P. Soon-Shiong. 2014. *Combination Therapy Methods for Treating Proliferative Diseases*. Google Patents.

Desai, A., J. Xu, K. Aysola, Y. Qin, C. Okoli, R. Hariprasad, U. Chinemerem, et al. 2014. Epithelial ovarian cancer: An overview. *World Journal of Translational Medicine* 3 (1): 1–8.

Dochez, V., H. Caillon, E. Vaucel, J. Dimet, N. Winer, and G. Ducarme. 2019. Biomarkers and algorithms for diagnosis of ovarian cancer: CA125, HE4, RMI and ROMA, a review. *Journal of Ovarian Research* 12 (1): 28–28.

Elwyn, G., S. Laitner, A. Coulter, E. Walker, P. Watson, and R. Thomson. 2010. Implementing shared decision making in the NHS. *BMJ* 341: 971–975.

Engel, G.L. 1977. The need for a new medical model: A challenge for biomedicine. *Science* 196 (4286): 129–136.

Enshaei, A., C.N. Robson, and R.J. Edmondson. 2015. Artificial intelligence systems as prognostic and predictive tools in ovarian cancer. *Annals of Surgical Oncology* 22 (12): 3970–3975. https://doi.org/10.1245/s10434-015-4475-6

FDA-NIH Biomarker Working Group. 2016. *BEST (Biomarkers, EndpointS, and other Tools) resource*. https://www.ncbi.nlm.nih.gov/books/NBK326791/. Accessed 24 Mar 2021.

Finlayson, C.S. 2017. *The Experience of Being Aware of Disease Status in Women with Recurrent Ovarian Cancer: A Phenomenological Study*. ProQuest Dissertations Publishing.

Fitch, M.I., K. Deane, D. Howell, and R.E. Gray. 2002. Women's experiences with ovarian cancer: Reflections on being diagnosed. *Canadian Oncology Nursing Journal/Revue canadienne de soins infirmiers en oncologie* 12 (3): 152–159.

Foucault, M. 2003. *The Birth of the Clinic: An Archaeology of Medical Perception*, Routledge Classics. 3rd ed. London: Routledge.

Goff, B.A., L.S. Mandel, C.H. Melancon, and H.G. Muntz. 2004. Frequency of symptoms of ovarian cancer in women presenting to primary care clinics. *JAMA* 291 (22): 2705–2712.

Green, J. 2012. *The Fault in Our Stars*. New York: Penguin Books.

Ha, J.F., and N. Longnecker. 2010. Doctor-patient communication: A review. *The Ochsner Journal* 10 (1): 38–43.

Hennessy, B., R.C. Bast, A.M. Gonzalez-Angulo, and G.B. Mills. 2008. Chapter 25 – Early detection of cancer: Molecular screening. In *The Molecular Basis of Cancer*, ed. John Mendelsohn, Peter M. Howley, Mark A. Israel, Joe W. Gray, and Craig B. Thompson, 3rd ed., 335–347. Philadelphia: W.B. Saunders.

Hernot, S., L. van Manen, P. Debie, J.S.D. Mieog, and A.L. Vahrmeijer. 2019. Latest developments in molecular tracers for fluorescence image-guided cancer surgery. *The Lancet Oncology* 20 (7): e354–e367.

Hofmann, B. 2005. Simplified models of the relationship between health and disease. *Theoretical Medicine and Bioethics* 26 (5): 355–377.

Italiano, A. 2011. Prognostic or predictive? It's time to get back to definitions. *Journal of Clinical Oncology* 29 (35): 4718.

Javdekar, R., and N. Maitra. 2015. Risk of Malignancy Index (RMI) in evaluation of adnexal mass. *Journal of Obstetrics and Gynaecology of India* 65 (2): 117–121.

Kleinmanns, K., K. Bischo, S. Anandan, M. Popa, L.A. Akslen, V. Fosse, I.T. Karlsen, B.T. Gjertsen, L. Bjørge, and E. McCormack. 2020. CD24-targeted fluorescence imaging in patient-derived xenograft models of high-grade serous ovarian carcinoma. *The Lancet* 56: 1027823.

Komaroff, A.L. 2019. Advances in understanding the pathophysiology of chronic fatigue syndrome. *JAMA* 322: 499–500.

Kurman, R.J. 2014. *WHO classification of tumours of female reproductive organs*. Lyon: International Agency for Research on Cancer.

Labidi-Galy, S.I., E. Papp, D. Hallberg, N. Niknafs, V. Adleff, M. Noe, R. Bhattacharya, et al. 2017. High grade serous ovarian carcinomas originate in the fallopian tube. *Nature Communications* 8 (1): 1–11.

Ledermann, J.A. 2016. PARP inhibitors in ovarian cancer. *Annals of Oncology* 27 (suppl_1): i40–i44.

Ledermann, J.A., P. Harter, C. Gourley, M. Friedlander, I. Vergote, G. Rustin, C. Scott, et al. 2016. Overall survival in patients with platinum-sensitive recurrent serous ovarian cancer receiving olaparib maintenance monotherapy: An updated analysis from a randomised, placebo-controlled, double-blind, phase 2 trial. *The Lancet Oncology* 17 (11): 1579–1589.

Lisio, M.-A., L. Fu, A. Goyeneche, Z.-H. Gao, and C. Telleria. 2019. High-grade serous ovarian cancer: Basic sciences, clinical and therapeutic standpoints. *International Journal of Molecular Sciences* 20 (4): 952.

Ludwig, L. 1940. The world as a phenomenological problem. *Philosophy and Phenomenological Research* 1 (1): 38–58.

Moore, K., N. Colombo, G. Scambia, B.-G. Kim, A. Oaknin, M. Friedlander, A. Lisyanskaya, et al. 2018. Maintenance Olaparib in patients with newly diagnosed advanced ovarian cancer. *New England Journal of Medicine* 379 (26): 2495–2505.

Mor, G., and A. Alvero. 2013. The duplicitous origin of ovarian cancer. *Rambam Maimonides Medical Journal* 4 (1): e0006.

Morgan, R.J., Jr. 2011. Ovarian cancer guidelines: Treatment progress and controversies. *Journal of the National Comprehensive Cancer Network* 9 (1): 4–5.

Moss, E.L., J. Hollingworth, and T.M. Reynolds. 2005. The role of CA125 in clinical practice. *Journal of Clinical Pathology* 58 (3): 308–312.

Murray, S.A., M. Kendall, K. Boyd, and A. Sheikh. 2005. Illness trajectories and palliative care. *BMJ (Clinical Research Ed.)* 330 (7498): 1007–1011.

Niebrój, L.T. 2006. Defining health/illness: Societal and/or clinical medicine? *Journal of Physiology and Pharmacology* 57 (supp 4): 251–262.

Oken, M.M., R.H. Creech, D.C. Tormey, J. Horton, T.E. Davis, E.T. McFadden, and P.P. Carbone. 1982. Toxicity and response criteria of the Eastern Cooperative Oncology Group. *American Journal of Clinical Oncology* 5 (6): 649–656.

Oldenhuis, C.N., S.F. Oosting, J.A. Gietema, and E.G. de Vries. 2008. Prognostic versus predictive value of biomarkers in oncology. *European Journal of Cancer* 44 (7): 946–953.

Ong, L.M.L., M.R.M. Visser, F.B. Lammes, and J.C.J.M. de Haes. 2000. Doctor–Patient communication and cancer patients' quality of life and satisfaction. *Patient Education and Counseling* 41 (2): 145–156.

Øyehaug, G. 2016. Dreamwriter (Autobiography). *World Literature Today* 90 (6): 50–53.

Paulsen, T., K. Kjaerheim, J. Kaern, S. Tretli, and C. Tropé. 2006. Improved short-term survival for advanced ovarian, tubal, and peritoneal cancer patients operated at teaching hospitals. *International Journal of Gynecological Cancer* 16 (Suppl 1): 11–17.

Pellegrino, E.D. 1981. *A Philosophical Basis of Medical Practice: Toward a Philosophy and Ethic of the Healing Professions*. Oxford: Oxford University Press.

Pool, R., and W. Geissler. 2005. *Medical Anthropology*. Berkshire: Open University Press.

Rustin, G.J., A.E. Nelstrop, P. McClean, M.F. Brady, W.P. McGuire, W.J. Hoskins, H. Mitchell, and H.E. Lambert. 1996. Defining response of ovarian carcinoma to initial chemotherapy according to serum CA 125. *Journal of Clinical Oncology* 14 (5): 1545–1551.

Scully, J.L. 2004. What is a disease? Disease, disability and their definitions. *EMBO Reports* 5 (7): 650–653.

Sekse, R.J.T., E. Gjengedal, and M. Råheim. 2013. Living in a changed female body after gynecological cancer. *Health Care for Women International* 34 (1): 14–33.

Semino, E., Z. Demjén, J. Demmen, V. Koller, S. Payne, A. Hardie, and P. Rayson. 2017. The online use of violence and journey metaphors by patients with cancer, as compared with health professionals: A mixed methods study. *BMJ Supportive & Palliative Care* 7 (1): 60–66.

Simacek, K., P. Raja, E. Chiauzzi, D. Eek, and K. Halling. 2017. What do ovarian cancer patients expect from treatment?: Perspectives from an online patient community. *Cancer Nursing* 40 (5): E17–E27.

Sontag, S. 1978. *Illness as Metaphor AND AIDS and Its Metaphors*. New York: Farrar, Straus and Giroux.

Strimbu, K., and J.A. Tavel. 2010. What are biomarkers? *Current Opinion in HIV and AIDS* 5 (6): 463–466.

Suidan, R.S., P.T. Ramirez, D.M. Sarasohn, J.B. Teitcher, S. Mironov, R.B. Iyer, Q. Zhou, et al. 2014. A multicenter prospective trial evaluating the ability of preoperative computed tomography scan and serum CA-125 to predict suboptimal cytoreduction at primary debulking surgery for advanced ovarian, fallopian tube, and peritoneal cancer. *Gynecologic Oncology* 134 (3): 455–461.

Svenaeus, F. 2005. *Sykdommens mening-og motet med det syke mennesket*. Oslo: Gyldendal akademisk.

Tamm, M.E. 1993. Models of health and disease. *British Journal of Medical Psychology* 66 (3): 213–228.

The Lancet editorial. 2009. What is health? The ability to adapt. *The Lancet* 373 (9666): 781.
Timmermans, M., G.S. Sonke, K.K. Van de Vijver, M.A. van der Aa, and R.F.P.M. Kruitwagen. 2018. No improvement in long-term survival for epithelial ovarian cancer patients: A population-based study between 1989 and 2014 in the Netherlands. *European Journal of Cancer* 88: 31–37.
Toombs, S.K. 1987. The meaning of illness: A phenomenological approach to the patient-physician relationship. *The Journal of Medicine and Philosophy* 12 (3): 219–240.
Ueland, F.R. 2017. A perspective on ovarian cancer biomarkers: Past, present and yet-to-come. *Diagnostics (Basel, Switzerland)* 7 (1): 14.
Vang, R., I.-M. Shih, and R.J. Kurman. 2009. Ovarian low-grade and high-grade serous carcinoma: Pathogenesis, clinicopathologic and molecular biologic features, and diagnostic problems. *Advances in Anatomic Pathology* 16 (5): 267–282.
Wang, Y., L. Li, C. Douville, J.D. Cohen, T.T. Yen, I. Kinde, K. Sundfelt, et al. 2018. Evaluation of liquid from the Papanicolaou test and other liquid biopsies for the detection of endometrial and ovarian cancers. *Science Translational Medicine* 10 (433): eaap8793.
World Health Organization. 1946. *Constitution*. World Health Organization. https://www.who.int/about/who-we-are/constitution. Accessed 31 May 2020.
World Ovarian Cancer Coalition. 2018. *The Every Woman Study*. World Ovarian Cancer Coalition. https://worldovariancancercoalition.org/wp-content/uploads/2018/10/THE-EVERY-WOMAN-STUDY-WOMENS-SURVEY-2018-FULL-RESULTS.pdf.
Wulff, H. 1999. The two cultures of medicine: Objective facts versus subjectivity and values. *Journal of the Royal Society of Medicine* 92 (11): 549–552.

HER2 Revisited: Reflections on the Future of Cancer Biomarker Research

Anne Bremer, Elisabeth Wik, and Lars A. Akslen

Introduction

The HER2 biomarker for breast cancer is an emblematic example of a success story in the field of cancer biomarkers, and it is used to bolster enthusiasm for biomarker research and precision oncology. After the discovery of the ERBB2 (HER2) gene in 1984 by the Weinberg lab, HER2 was eventually demonstrated to be a biomarker of poor breast cancer prognosis. Then, the targeted therapy trastuzumab was developed, aimed for breast cancer patients with HER2 positive tumours, and, in a foresighted move, tested in biomarker-stratified clinical trials. The assessment of HER2 protein expression (by immunohistochemistry) or gene copy number (by *in situ* hybridisation), or a combination of the two, has for many years now been used as the gold standard in the clinic to predict patients' response to anti-HER2 therapy, both in the adjuvant and metastatic settings of breast cancer.

The HER2 success story is often evoked in projecting what precision oncology could materialise into, and as a model for developing other successful biomarkers. To a significant degree, this has meant promoting the advances achieved through HER2, with the effect of creating a 'scientific bandwagon' of efforts to emulate HER2 through 'standardised packages' of theories, methods and technologies.

In this chapter we depart from this tradition, by critical revisiting the HER2 story in a way that highlights not only the enabling conditions and reasons for its success,

A. Bremer (✉)
Centre for Cancer Biomarkers, Centre for the Study of the Sciences and the Humanities, University of Bergen, Bergen, Norway
e-mail: anne.bremer@uib.no

E. Wik · L. A. Akslen
Centre for Cancer Biomarkers CCBIO, Department of Clinical Medicine, University of Bergen, Bergen, Norway
e-mail: elisabeth.wik@uib.no; lars.akslen@uib.no

A. Bremer, R. Strand (eds.), *Precision Oncology and Cancer Biomarkers*, Human Perspectives in Health Sciences and Technology 5,
https://doi.org/10.1007/978-3-030-92612-0_7

but also the complications and challenges that it faced and continues to face along the way; from biological complexity to associated social, political and ethical controversies for instance. In doing so, we aim to show that HER2 confronted many of the limitations that hinder work on other biomarkers and assert that the field can learn as much from these limitations as from HER2's advances. To the extent that HER2 is held up as a standard for all oncology biomarkers, a more nuanced account of its development may provide more realistic expectations of 'good enough' biomarkers. This more nuanced story also helps us interpret what reflection the history of HER2 can lead us to about the possibilities and limits of precision oncology.

In section "The HER2 story", we will review the HER2 history, from its discovery in the lab, to development of diagnostic tools and early clinical trials. In section "Revisiting the story of HER2", we will revisit the story of HER2 by looking at the ethical, legal and social aspects faced by HER2 as a 'standard package' in the HER2/oncogene bandwagon; and discusses one of the key contemporary legacies of the HER2 story, namely: the sociotechnical imaginary of precision oncology. Section "Revisiting the story of HER2" ends with reflections on the need for a greater focus on 'good enough' biomarkers, particularly in a context of precision oncology driven by hyper-precision and the wish for molecular certainty, and it underlines the importance of being open about the low success rate of biomarkers reaching clinical practice, in particular when justifying the risks and opportunity costs of precision oncology. These key reflections are summarised in the concluding section "Conclusion".

The HER2 Story

Discovery and Basic Studies

To understand how this extraordinary achievement for breast cancer patients started, we will shed light on the very first steps of the story, the discovery of the HER2 gene and the accompanying experimental research to understand the function of HER2. But we also need to peek into the time being – the early 80s – trying to understand the field of cancer research at that time.

Theories of cancer being caused by viruses carrying oncogenes, i.e. genes causing conversion of normal cells to cancer cells, to infected human cells have been present for several decades. The first report on gene alterations promoting cancer, resulting from the landmark work of Nobel laureates Bishop and Varmus (Stehelin et al. 1976; Bishop 1983; Varmus 1984), led to a rush in the search for human cancer genes (oncogenes) and their normal counterparts (proto-oncogenes), and how these 'precursor' genes are activated, for example by mutations, amplifications, or gene rearrangements. Many cancer genes were found to be variants of previously identified viral oncogenes.

Studies on a leukaemia virus in chicken led to a major finding in 1984. Several labs came to the conclusion that the *v-erbB* gene is derived from the gene encoding epidermal growth factor receptor, EGFR (Downward et al. 1984; Ullrich et al. 1984). The link between a chicken virus gene and human growth factor signalling was a game changer for cancer research. A model was proposed: "viral oncogenes were misbehaved variants of normal cellular proteins that played fundamental roles in growth factor signalling" (Sawyers 2019).

In the ongoing search for oncogenes, the *HER2/neu* gene was discovered (Schechter et al. 1984). At first, the oncogene *neu* was identified. Other research groups added to this knowledge as they cloned the new gene, naming it *HER2*, Human EGF Receptor 2, due to its resemblance to EGFR (Coussens et al. 1985; King et al. 1985; Semba et al. 1985).

Next step was to identify the functional consequences of alterations in *HER2*, now considered a human oncogene. Connections between other human oncogenes and specific cancer types were identified, like Burkitt's lymphoma and neuroblastoma, but was still uncovered for *HER2* and human cancer.

The HER2 protein was described as a receptor at the cell membrane. One (extracellular) part of the protein was available for blocking by an antibody. The Weinberg group demonstrated reversion of the cancer phenotype when blocking the *neu* gene in cancer cells overexpressing neu (Drebin et al. 1986). The same positive effects were not seen when using the same antibody in cancer cells overexpressing *Ras*, another oncogene. This indicated specificity of the antibody (Sawyers 2019). At the same time, the Ullrich group demonstrated that HER2 took a functional role in transformation of cells from normal to cancer (Hudziak et al. 1987). The functional pathology and treatment potential related to HER2 was now demonstrated.

At this stage, combining the knowledge from the Weinberg and Ullrich labs, the company Genentech initiated a search for relevant HER2 antibodies, proteins developed to bind to the HER2 protein, thereby blocking its cancer promoting effects. They reached the goal of developing an antibody that demonstrated acceptable selectivity towards HER2, leading to growth inhibition of breast cancer cell lines with *HER2* amplification (Hudziak et al. 1989).

When reading stories from the time of these discoveries, we can appreciate nearly an electric sensation in the research races between labs, and the excitement of taking part in what could be a major breakthrough in cancer research (Sawyers 2019).

Biomarker Development

As mentioned, the HER2 biomarker is one of the most successful in contemporary medical oncology (Hunter et al. 2020). There is probably more than one explanation. First, due to the underlying genomic alteration for HER2 overexpression in most cases, *HER2* gene amplification, this marker is much more dichotomous or «on-off» (or two-tiered) than many other tissue-based protein biomarkers, although borderline cases and challenges do indeed exist. In comparison, a completely different biomarker is the Ki67 protein expression for tumour cell proliferation, by many laboratories used in the St. Gallen based surrogate classification of Luminal B breast cancer cases. Ki67 varies continuously from one extreme to the other, without a stepwise pattern, and with much variation in how to stain for Ki67 and how to assess it. Second, in the case of HER2, the biomarker is identical to the target for tailored treatment, and that is the membrane protein HER2, a member of the EGFR family of tyrosine-type of receptors. This 'companion structure' is probably an important but not sufficient explanation. As an example of the opposite, in the case of anti-angiogenesis treatment, the measurement of VEGF protein in tissues or in the blood has not been a success in the prediction of treatment response, VEGF being the target for bevacizumab.

From the discovery phase, and especially after the successful generation of antibodies, a time of intensified *in situ* studies on human cancer tissues were performed, especially looking at expression of HER2 protein in various breast cancers and correlating the findings with clinico-pathologic phenotypes and patient outcome.

In 1987, Slamon and colleagues reported that HER2 was overexpressed in approximately 30% of human breast cancers. In studies of oncogenes, DNA was extracted from primary breast cancer tissues, and Southern blot analyses gave a hit on HER2 (Slamon et al. 1987). In the initial part of the study, a cohort of 103 primary tumours were analysed. HER2 amplification was found in 18% of the cases. Notably, when assigning the tumours to groups of (1) one HER2 copy; (2) 2–5 copies; (3) 5–20 copies; and (4) more than 20 copies, there was no apparent association with the established prognostic variables like tumour size, histologic grade, or oestrogen receptor (ER) status. However, a trend of association was seen between HER2 amplification and the number of lymph nodes with metastasis; HER2 amplification was seen in 32% of the cases with metastasis to more than 3 lymph nodes. Follow-up data like information on recurrent disease and survival was missing for this cohort, and evaluation of the association between HER2 and prognosis could not be made. Based on the trend of association between *HER2* amplification and number of lymph node metastases, the research group interpreted this as a hint that *HER2* amplification could present with prognostic value. To pursue this idea, a cohort of breast cancer patients with lymph node positive tumours was examined. Out of 100 primary tumours (86 with sufficient DNA), 40% showed *HER2* amplification. The finding of an association between *HER2* amplification and increasing number of lymph node metastases was repeated. Additionally, *HER2* amplification was associated with larger tumour size and ER negative tumours and demonstrated strong association with time to relapse and survival. HER2 maintained independent prognostic value when adjusting for established prognostic variables in multivariate survival analysis.

When reading the results from the initial studies (Slamon et al. 1987), we may ask what made Slamon and colleagues pursue their studies on HER2 in primary breast cancer. Many researchers have seen similar results when looking at other biomarkers and decided not to follow that lead any longer. Thus, the difference between statistical and biological significance should always be kept in mind, along with considerations on sample size and statistical power, to avoid type-2 errors or 'error of omission' in the process of interpretation.

Importantly, Slamon and collaborators did not give up their search for a clinically relevant role for *HER2* amplifications, although somewhat weak results from the initial part of the study.[1] Subsequent studies focused on well annotated tumours from patients with long follow-up information, acquired from the collaborator William McGuire, who had established a breast cancer biobank – unique at that time. By this landmark study (Slamon et al. 1987), Slamon and colleagues demonstrated that *HER2* amplified tumours were associated with aggressive tumour features and reduced breast cancer survival, supporting a clinically relevant role for HER2 in breast cancer progression.

Already at this point, discussions on biomarker cut-off came to play. How should HER2 'positivity' be defined? Should the definition follow a strict biological interpretation – indicating that more than two copies align to amplified status? Or would there be need to define amplification according to clinically relevant copy number increase? Slamon and colleagues noted a shorter time to relapse and overall survival in tumours with *HER2* copy number >2, compared to the other – more striking separations between survival curves were seen when the cut-off for *HER2* amplification was defined as >5 *HER2* copies per tumour cell.

In the years to come, and following the initial 'gold rush', the field of precision oncology and cancer biomarker research has expanded significantly with respect to complexity at the methodological and biological levels, as well as in the clinical fields. And to increase this further, HER2 plays important roles in other cancers than in breast tissue. Issues such as definitions of 'positivity' when using immunohistochemistry (IHC) or *in situ* hybridization (ISH), or the combination, have increased with time, for example reflected in the American Society of Clinical Oncology (ASCO) 2018 Guidelines on HER2 testing in breast cancer (Wolff et al. 2018). The topic of tissue heterogeneity and sampling bias, in primary tumours as well as in metastases, and phenotypic development and 'switches' with tumour progression, are just a few questions of concern for practicing pathologists and clinicians. The complexity is still growing, and it is tempting to quote Churchill: '*Now this is not the end. It is not even the beginning of the end. But it is, perhaps, the end of the beginning.*'

Having demonstrated a clinical relevance for HER2, a hunt for a human tolerable antibody started. Techniques of adding a mouse antibody onto a human antibody backbone was described (Jones et al. 1986; Verhoeyen et al. 1988). Shepard and colleagues developed an antibody that selectively killed cells expressing high levels of HER2 (Carter et al. 1992). This had significant implications with regard to toxicity and patient selection, and it was a critical breakthrough in the path to clinical application.

---

[1] In this book, look also at the chapter "Filled with desire, perceive molecules" by Strand and Engen, which discusses the role of persistence in the development of transplantations in AML therapy.

Early Clinical Trials

The development of a humanized antibody, selectively killing cells with high HER2 expression, paved the way for the first clinical trials on trastuzumab (Herceptin®). Experimental models had demonstrated improved effects when combining standard chemotherapy regimens with trastuzumab in HER2 overexpressing cancer cells and xenografts in mice (Pietras et al. 1998; Pegram et al. 1999). Efficacy and safety of trastuzumab given alone to metastatic breast cancer had been demonstrated (Cobleigh et al. 1999), also in combination with chemotherapy, where the results demonstrated no increase in toxicity compared to when chemotherapy was given alone (Pegram et al. 1998).

With this knowledge, Slamon and colleagues set out with the first Phase 3 study on chemotherapy plus trastuzumab to HER2 overexpressing metastatic breast cancer, enrolling patients in the period 1995–1997 (Slamon et al. 2001). Prolonged time to progression was seen in the group receiving chemotherapy plus trastuzumab compared to the group receiving chemotherapy only (median 7.4 vs 4.6 months). Also, response duration and survival time was longer in the group receiving trastuzumab – and this group experienced 20% reduction in the risk of death. At the same time, reports on some severe side effects emerged, demonstrating the often-challenging balancing acts in this field (Slamon et al. 2001).

During the following years, several studies demonstrated clinical benefit from trastuzumab monotherapy in metastatic breast cancer (Baselga et al. 2005; Vogel et al. 2006). Two studies from 2005 demonstrated positive progression and survival effects from trastuzumab given after or concurrently with chemotherapy, in the adjuvant setting (Piccart-Gebhart et al. 2005; Romond et al. 2005), leading to approval of trastuzumab as adjuvant therapy in 2006 (Sawyers 2019).

Some of the important lessons learned from early clinical trials that led to the success of targeting tumour HER2 to a subpopulation of breast cancer patients, were appropriate patient selection, guided by the HER2 tumour status, and accompanied by a robustly validated biomarker test. These early trials would most likely not have demonstrated the progression and survival effects seen, if patients with metastatic breast cancer, across molecular subtypes, were included.

One key to understand the significant and lasting impact of the early trials was the observation that the influence of HER2 blockade was dependent on the stratification using tissue based HER2-status, as there was no overall treatment effect to be observed. Thus, without the 'companion structure' of HER2 tumour status (biomarker) and HER2 blockade (treatment), this field might have been missed initially. However, the momentum continued to increase, with multiple trials of many different anti-HER2 modalities and clinical indications (Wang and Xu 2019).

Revisiting the Story of HER2

We will now revisit the story of HER2 to reflect on the conditions of its inception, some of the reasons of its success, and the challenges met along the way. How do we aim to revisit the emblematic HER2 story? There are different types of scholarship, such as history, philosophy or sociology of medicine, that can do this work of 'revisiting' stories. In this chapter, we revisit the story of HER2 from the related field of Science and Technology Studies (STS), and particularly through the two lenses of ELSI/ELSA[2] and RRI,[3] as they are concerned with what is of interest to us here: how science, technology and society at large are 'co-produced' and change each other's trajectories, and how contributions from the social sciences and the humanities can integrate productively into such processes of co-production.

The concept of ELSI – Ethical, Legal and Social Implications – came about in 1988, at the onset of the Human Genome Project, when ethical, legal and social concerns were raised about the implications of genomic analyses, and notably the risks of using that knowledge to discriminate against people (see for instance McEwen et al. 2014). Shortly after, the field of ELSA – Ethical, Legal and Social Aspects – emerged, and was set up as a research field in its own right, articulated around research programmes and funding schemes (see for instance the European Commission's ELSA programme, 2007). In a similar way to ELSI, the field of ELSA is concerned with how science and technology permeate society and policy (and vice versa), and how emerging science and technologies sometimes leave us with specific problems and issues. Through the ELSI/ELSA lens, we will be able to revisit the different issues that HER2 left us, that have a legal, social, or economic component: ragged edges, cut-offs or questions of fairness, which we can analyse at three levels: (i) basic science; (ii) diagnostics; and (iii) treatment.

In particular, in the latter decade in Europe, the ELSI/ELSA concepts have gradually been supplemented, shifted and to some extent replaced by the emerging concept of RRI – Responsible Research and Innovation. While the ELSI/ELSA approach largely focused on immediate issues of overcoming the ethical, legal and societal obstacles that impede a successful translation and uptake of technology, the continued involvement of STS scholars and other social scientists in life science research led to an increasing appreciation of the need to study, understand and engage with the kind of choices and processes that lead scientists to do research on a particular development or technology. This is what RRI is about: critically looking at the social, political, and scientific interests and choices that shape the trajectory of a particular technology: why are we investing in that research field or technology? Is the research responsible? What kind of intended and unintended effects can we anticipate? Who is concerned and who will this research affect, and how might they become involved in choices that ultimately affect them? Through the RRI lens,

[2] Ethical Legal and Social Implications (ELSI) and Ethical Legal and Social Aspects (ELSA) of new and emerging science and technologies.

[3] Responsible Research and Innovation (RRI).

we will be able to revisit the story of HER2 by seeing it also as an exemplary case that had a broader effect on research and innovation trajectories in the cancer field.

How the HER2 Story Began: Oncogene Research Gathering Steam as a 'Bandwagon'

As described above, HER2 was discovered in the effervescence of the early days of oncogene research. In her seminal paper *The Molecular Biological Bandwagon in Cancer Research: Where Social Worlds Meet* (1988), Fujimura made the parallel between oncogene research and a bandwagon. She explains that oncogene theory and recombinant DNA technologies were 'packaged' in such a way that it became a scientific 'bandwagon[4]'. This scientific bandwagon gained momentum through the neat packaging of a new, and arguably more productive, theoretical model for explaining cancer (and therefore a new definition of cancer) – the oncogene theory – and recombinant DNA technologies for testing the theory (these technologies became standardised, and thus easily transferable). It attracted increasing interest, support and resources from a broad spectrum of actors ranging from public research institutions and laboratories to the private sector and funding agencies. Many groups, relatively close geographically (Cohen and colleagues, Vanderbilt University, USA; the Weinberg Group at MIT; the groups of Ullrich and Coussens at Genentech, USA; Greene and colleagues at Harvard Medical School – MIT; Seeger and colleagues at the UCLA School of Medicine; Minna and Johnson's group at the National Cancer Institute; Slamon and colleagues, University of California, Los Angeles School of Medicine; just to name a few (see the review by Kumar and Badve 2008)), all worked towards understanding how oncogenes worked and could be inhibited. These early oncogene researchers used the oncogene theory-and-technology package to prioritise oncogene research in their cancer research institutions, and in doing so they deeply changed the work organisation in many laboratories, for which the priority became to work on oncogenes, biomarkers or antibodies.

Indeed, in the mid-1980s, "oncogene" became a buzzword that was the centre of attention in oncology research (Hunter and Simon 2007), and the ground-breaking results of Weinberg and colleagues, through experiences with rodent tumours induced by chemical carcinogens, led to an explosion of studies of oncogenes over the following few years (Weiss 2020). Scientific articles related to oncogene research started to flourish in scientific journals, and the media drew attention to these developments as well. In parallel, funding agencies such as the American

[4] Fujimura (1988, p. 261) defines a 'scientific bandwagon' and a 'package' of theory and technology as such: "A scientific bandwagon exists when large numbers of people, laboratories, and organizations commit their resources to one approach to a problem. A package of theory and technology is a clearly defined set of conventions for action that helps reduce reliance on discretion and trial-and-error procedures."

National Institutes of Health channelled increasing funds into oncogene research, going from $5.5 million allocated to 54 projects in 1983, to $103.2 million allocated to 648 projects in 1987. Specific to the HER2 story and further accelerating the pace of the bandwagon, pharmaceutical companies collaborated with academia and regulatory agencies so that developments of the HER2 biomarker and targeted therapies would happen in parallel. In particular, biomarker-stratified trials were useful in efficiently linking the anti-HER2 treatment of trastuzumab and HER2 as an accompanying predictive biomarker.[5]

By 1984, notably with the discovery of HER2 by the Weinberg lab, the "bandwagon sustained its own momentum and researchers climbed on primarily because it was a bandwagon" (Fujimura 1988, p. 262). Fujimura (1988) pins down the success of oncogene research, and the success of HER2, to three main reasons: (i) the oncogene theory-and-technology package provided a new frame to and definition of cancer,[6] and technologies allowed to address relatively 'straightforward' questions within that frame. The apparent simplicity in how oncogenes were seen to work, and the access to well-fitted technologies, made it attractive to engage in large-scale research efforts; (ii) the oncogene package involved novel, pioneering, recombinant DNA techniques which attracted interest from the different actors at play; and (iii) very quickly, researchers working with the oncogene package developed new and valuable knowledge about cancer at the molecular level – an aspect of cancer that had not been the topic of extensive previous investigations. From the story of HER2 described above, we can add four other reasons to further explain the success of HER2: (iv) the dichotomous nature of HER2 – the marker is most often either 'on' or 'off' – facilitated its successful application in the clinic; (v) the persistence of scientists, who despite somewhat inconclusive results from parts of their study, pursued the research on potential clinical application for HER2 amplifications; (vi) in the case of HER2, the biomarker and therapeutic target are the same, which again facilitated the road from the bench to the clinic; and (vii) the alignment between academia, pharmaceutical industries and regulatory agencies, gathered together on the bandwagon was key in the success of HER2.

---

[5] Similarly, at the diagnostic level, the alignment between academia, pharmaceutical companies and regulatory agencies was equally important in linking target drugs to companion diagnostic tests, as shown by the co-development, in 1998, of trastuzumab and the immunohistochemistry assay HercepTest for the detection of HER2 overexpression in breast tumors (Jørgensen and Winther 2010).

[6] The quote in Fujimura's paper reads as follow: "In the molecular biological cancer research bandwagon, cancer was re-packaged as a disease of the cell nucleus and specifically of the DNA. Researchers in other lines of work had previously studied cancer as a disease of the cell, the immune or endocrine system, the entire organism, or the interaction between organism and environment. In the late 1970s and early 1980s, molecular biologists and tumor virologists developed a theory-method package [... which] consisted of a molecular biological *theory* of cancer and a *set of technologies* for testing and exploring the theory. They constructed the oncogene theory so that it mapped onto the intellectual problems of many different scientific social worlds. In addition, by the early 1980s, recombinant DNA technologies were standardized and thus highly transportable between social worlds." (Fujimura 1988, p. 278).

Understanding the Social, Ethical and Economic Implications of the HER2/Oncogene Bandwagon

The analogy between the HER2 story and a scientific bandwagon allows us to look in more details at this bandwagon, and discuss some key social, ethical and economic implications associated with it. In particular, we will look at implications relating to: (i) basic science; (ii) diagnostic; and (iii) treatment.

Basic Science

One significant opportunity of creating an oncogene bandwagon was the capacity to *open up* research and nurture collaboration between different actors, who otherwise were not as aligned as during the HER2 story. We have seen in Section 3.1 how scientific laboratories and institutions, pharmaceutical companies and regulatory agencies together worked around the oncogene *theory-and-technology package*. Two aspects help explain why this collaboration was so successful. First, the recombinant DNA technologies of the package were standardised, and therefore highly 'transportable' to the different actors on the bandwagon. They could all relate to, and work on the 'standard package' without changing it, meaning that there were no major issues of translation, redefinition or reframing of the package: everyone was talking about the same thing, in a clearly defined frame – making collaboration easier. Second, the different actors could see their interests fulfilled through this collaboration, with academia bringing its scientific discovery to the clinic in a uniquely successful and fast way; pharmaceutical companies ensuring that most target patients got a clinical benefit from the treatment while maintaining profits through higher prices; and regulatory agencies in seeing the most effective and safe treatment delivered to each patient (Parker 2018).

However, the bandwagon was also found to be a way of *closing down* research, or in other words, to effectively constrain cancer research trajectories into a domain, where looking at oncogenes was the top priority of cancer research agendas. The phenomenon of the 'bandwagon' finds different expressions in the field of philosophy of science, among which two are interesting for closely looking at this 'closing down' effect. In his paper *History of science and its rational reconstructions*, Lakatos (1970) devised the concept of 'research programmes' as a "sequence of theories within a domain of scientific inquiry". If the move from one theory to the next one is characterised by an advancement of the field, then we are in a 'progressive research programme'. However, if this is not the case, then the programme is 'degenerating', and scientists might ultimately leave the programme to create or engage in a new one. Lakatos' analysis was a response to the famous theory that Thomas Kuhn introduced in his book *The Structure of Scientific Revolutions* (1962), in which he held that research trajectories presented themselves as "paradigms", that is, sets of theories, research methods, assumptions and standards for what constitutes legitimate contributions to a field. He developed a model for science where periods of successful acquisition of new knowledge within the context of the dominant paradigm (periods of 'normal science'), were interrupted by 'scientific revolutions' demanding a 'paradigm shift'. These revolutions would occur after an accumulation of 'anomalies' in the field (typically, repeated failures of the current paradigm to take into account observed phenomena or explain facts). In the paradigm shift, the underlying models, definitions and assumptions of the field are critically questioned, and a new paradigm is established. The new paradigm will ask new questions of old data, giving research a different direction. While every philosopher of science in the 1960s and 1970s admitted that phenomena similar to Fujimura's bandwagons, Kuhn's paradigms and Lakatos' research programmes existed, the main issue of contestation was the rationality of the processes of constraint and change. For Kuhn, such processes were ultimately driven by extra-scientific considerations. For Lakatos, the development of science was inherently a rational and logical process. He conceded, however, that the rationality cannot be determined in real time. Only in historical hindsight can one fully and rightly judge on progressive and degenerative research programmes; on what was a brilliant case of patient persistence and what was an unfortunate lock-in.

Indeed, at some point, the oncogene theory-and-technology package seemed to reach its limits: anomalies were accumulating around the oncogene package, and its paradigm was increasingly criticised for being an "illusion that cancer was as simple as it possibly could be [...and] that a small number of molecular events might explain cancer" (Weinberg 2014, p. 269). In particular, the oncogene package struggled to investigate cancer mechanisms in the face of tumour heterogeneity and complexity (as we discuss in the subsection below). The impressive and outstanding success of HER2 and trastuzumab was exactly that: it stood out, impressed, and remains to date the poster child for biomarkers and targeted therapies. At some point, what was called for was that researchers progressively jumped off the bandwagon of oncogene research to search for new progressive programmes, focusing for instance on immunotherapy (Akkari et al. 2020), and biomarkers of the tumour microenvironment (Laplane et al. 2019).

Diagnostic

There is a certain beauty in the apparent simplicity of HER2. It is one of the rare biomarkers which is most often either 'on' or 'off', and where the biomarker is also the therapeutic target. This explains to a great extent why HER2 has found successful applications in the clinic. However, as seen above, despite the formidable success of HER2, the standard oncogene package started to meet important limits, in particular when facing tumour heterogeneity and overall high complexity of cancer biology, even for HER2. These limits are still visible today at the diagnostic level, where HER2 faces important uncertainties and ethical, social and economic implications, in particular related to: (i) inter- and intra-tumoral heterogeneity which question the reliability and quality of biopsies; (ii) the determination of HER2 positivity and questions of cut-off, with inter- and intra-observer variability. and (iii) the new sequencing and imaging technologies that generate immense amounts of data that need to be governed and made sense of (this latter point is maybe more relevant for new and emerging biomarkers, but might have implications for HER2 as well).

Several models have tried to address the heterogeneity of tumours and the complexity of cancer biology without being capable of singlehandedly grasping it (Boniolo 2017). Basically, each tumour represents an ‘individual organism’, different from all others, and is itself composed of sub-tumours or cellular sub-populations (or clones). This heterogeneity significantly and directly impacts on clinical matters (Boniolo and Campaner 2019) and has important implications for patients who will receive or not receive anti-HER2 treatment based on the HER2 test outcome. Relative to intra-tumour heterogeneity, the heterogeneity within the tumour, Blanchard and Wik (2017) explains that a patient might get a result showing HER2+ or HER2− depending on where in the tumour the biopsy has been taken. Relative to inter-tumour heterogeneity, the heterogeneity between different tumours in one patient, there are some cases where the primary tumour is HER2− and develops into HER2+ metastases, and vice versa. Therefore, as argued by Boniolo and Campaner: “We can no longer speak in terms of, for instance, breast cancer, but properly speaking, we should refer to one of the many possible cancers affecting the breast.” (2019, p. 34) How, under these conditions, can we set up a robust diagnostic algorithm? How many biopsies have to be taken? How many metastatic lesions should be sampled, and how often, and to which costs? This extends to uncertainties relative to how to treat the patient. For instance, Goldhirsch et al. (2009) argue that HER2 overexpression in circulating tumour cells might justify targeted therapy even in the absence of a HER2+ primary tumour; but this remains contested.

This tumour heterogeneity has direct implications on the determination of HER2 positivity and where to place the cut-off. There are indeed some cases where it is not obvious to determine whether HER2 is over-expressed or amplified, or not. As discussed above, two main techniques for determining HER2 positivity are used in clinical practice: (i) immunohistochemistry or (ii) the determination of gene amplification by FISH, CISH or SISH. The threshold for HER2 positivity, and how to ‘correctly’ place the cut-off defining HER2 positive and negative tumours, in rare instances, are still debated (Wolff et al. 2018). Some of the questions without clear answers are referred to in the literature as the ‘cut-off’ problem (Rosoff 2017) and the ‘ragged edge’ problem (Callahan 1990).[7] Although initially perceived as a clean story of HER2 amplification with a clear-cut tissue biomarker, persisting efforts and technology developments have widened our understanding of HER2 biology with corresponding clinical implications. For instance, somatic HER2 mutations might occur in 2–5% of primary breast cancers, mainly among HER2 amplification negative cases (Yi et al. 2020). Observations on Chromosome 17 polysomia or monoso-

[7] In this book, see also the chapter by Leonard Fleck “Just Caring: Precision Health vs. Ethical Ambiguity Can we Afford the Ethical and Economic Costs?”, which discusses the issues of cut-off and ragged edges in a context of rationing cancer health care resources. In particular, Fleck argues that: “A line has to be drawn […] because we have only limited resources for meeting unlimited health care needs. There will always be patients with health needs or health risks just below that line […] The most we can reasonably hope for in this regard is rough justice.”

mia as well as CEP17 centromeric amplifications have made this area even more complicated from a diagnostic point of view.

Correspondingly, on the clinical side, there is not always a clear separation between between responders and non-responders; rather, there is a continuum of responses. For instance, developments of imaging techniques may influence the sensitivity levels of response definitions and detection. Patients who are just below the cut-off will 'fall off' the ragged edge, and not get access to the treatment

Correspondingly, on the clinical side, there is indeed no clear separation between strong responders and non-responders; rather, there is a continuum of responses, and the patients who are just below the cut-off will 'fall off' the ragged edge, and not get access to the treatment (Blanchard 2016; Fleck 2012). Callahan (1990), who first coined the concept of the 'ragged edge', argues that wherever we draw the line, there will always be people just below; and we should therefore try and accept to live with ragged edges: "We can accept [the ragged edge], not because we lack sympathy for those on it, but because we know that, once a ragged edge is defeated, we will then simply move on to still another ragged edge, with new victims – and there will always be new victims. [...] It is a struggle we cannot win. [...] We can ask not how to continually push back all frontiers, smooth out all ragged edges, but how to make life tolerable on the ragged edges; for we will all one day be on such an edge, sooner or later." (Callahan 1990, p. 65).

Finally, a third limit that new and emerging cancer biomarkers face, but that also has implications for the future developments of HER2-linked diagnostics, is related to the new sequencing and imaging technologies which produce huge quantities of data, and explain why, as noted by Boniolo and Campaner (2019), the number of bioinformaticians in the field of oncology has increased exponentially in the last two decades. However, it has proven very challenging to govern these data that are created at an extremely rapid pace, and understand their meaning. These big data, rather than supporting clinical decision-making at the diagnostic level, come to complicate the picture in uncertain ways. The deep-sequencing analyses of tumour DNAs add a new layer of complexity, and the big data that is generated arguably overwhelms our abilities to interpret and make sense of them[8] (Weinberg 2014). The development of large, genomic data sets arguably complicates the patients' choices relative to the use of their genomic information (Mayeur and van Hoof 2021), and it challenges even more their participation in clinical decision-making relative to their treatment, as they might experience an information overload. In addition, the high cost of these technologies, their sophistication and the technical and scientific expertise they demand, challenge their fair and just access, both nationally and globally.[9]

---

[8] In this book, see also the chapter by Roger Strand and Dominique Chu: "Crossing the Styx: If Precision Medicine Were to Become Exact Science", where the authors discuss that in order for precision medicine to become an exact science, everything from cells to patients would have to be conceived as closed systems with deterministic behaviour.

[9] In this book, see also the chapter by Eirik Tranvåg and Roger Strand: "Rationing of personalized cancer drugs: Rethinking the co-production of evidence and priority setting practices" who address the question of fairness in a context of rationing of personalised cancer drugs.

Treatment

After the success of HER2 and adjuvant therapy trastuzumab, other treatment modalities have been explored to propose alternative targeted therapies that could address tumour heterogeneity and the biological complexity represented by redundant activation of signalling pathways downstream of HER2. There is currently a broad selection of anti-HER2 treatments against HER2 tumours of primary resistance, in the form of monoclonal antibodies, antibody-drug conjugates or tyrosine kinase inhibitors for instance. However, it has been difficult to find adequate biomarkers to select between these different modalities. Why has the development slowed down on the biomarker side? Why have the tight collaborations that were happening on the oncogene bandwagon between academia, pharmaceutical industries and regulatory agencies, not continued with the same intensity and simplicity? If we return to Fujimura's notion of oncogene theory-and-technology 'package', we see that at the level of basic science, this package would travel quite seamlessly between the different actors while remaining immutable – the technologies were standardised, and the package was therefore highly transportable without changing its initial shape. At the treatment level, this seems to be a different story. The various actors have confronted the package with the heterogeneity and complexity of cancer, which made it not as transportable and immutable anymore. Multiple treatment modalities are being developed, in trying to address the high complexity of cancer, but the framings, definitions, technologies and interests of academia, pharmaceutical companies and regulatory agencies do not align as well anymore. Issues around data sharing agreements emerge (Antoniou et al. 2019), and the need is growing for pharmaceutical companies to come up with a different business model, as potentially excluding patients from the target treatment population is not compatible with an interest in optimising profits (Parker 2018). Indeed, historically, most pharmaceutical companies have relied on the business model of blockbuster drugs, whereby companies derive their profits on a small number of drugs which can be marketed widely to a broad population (OECD 2011). Shifting to a precision oncology model that relies on targeted therapies for subgroups of patients will reduce the market size for a drug and will off-balance the optimised ratio between profits and development costs. As mentioned in the OECD report from 2011: "Increasingly, to serve the original market, two or three different drugs may be needed, potentially increasing development costs to serve the same market size and accrue the same revenue." (p. 33) A variety of new business models are being adopted by the pharmaceutical industry, tailored to biomarker application in pharmacogenetics and diagnostics. Some of those new business models also use biomarkers to improve the efficacy of existing drugs or work on how to repurpose them. But maintaining profits in a context of increasingly segmented markets, an evolving regulatory environment, and unequal drug developments in terms of speed, costs and efficacy, remains to date a challenge.

Having this array of treatment options give rise to two other ethical, social and economic implications. First, the costs of these treatment options range from $5,134 per cycle for trastuzumab, to $10,290 per cycle for pertuzumab (Hassett et al. 2020).

This raises the question of the fair and just accessibility of such treatments both nationally and globally. Nationally, countries that do not offer public health care schemes might suffer from important discrepancies between the well-insured and non-insured in access to those targeted therapies. In particular, targeted therapies could add billions of dollars per year to the cost of public health care in the USA and in Europe, and these costs will have to be met by the insurance sector (Blanchard 2016). However, Ginsburg and Willard (2009) note that it is not sure whether insurance companies will be able and willing to reimburse these costs. We actually already see some American insurance companies who have begun off-loading the expenses onto patients, leading to 62% of personal bankruptcies being attributed to medical costs, principally cancer (Jackson and Sood 2011). Similarly, at the global level, we experience that personalised cancer therapies increase discrepancies in access to such treatments, not only because of the high costs of treatments which fail to be absorbed by health care systems in developing countries, but also because of the sophistication of the technologies required for the diagnostic part. This means that personalised cancer therapies in general are mostly accessible to the wealthy.

The second implication of having these many options in anti-HER2 treatments, is that it creates an overwhelming choice that comes to complicate the clinical decision-making and the integration of the patient in this decision-making. The different treatment options come with indications about which patients they might be most efficient for (ECOG performance status, size of tumour, earlier treatments, potential side-effects, etc.), but since diagnoses are surrounded by uncertainties, and the expression of HER2 positivity is sometimes unclear, treatment options are chosen on the basis of best, but uncertain, knowledge.

Where Are We Now? The Imaginary of Precision Oncology

We have looked at how the oncogene bandwagon gathered steam and allowed for unique collaborations between academia, pharmaceutical companies and regulatory agencies, thus opening up the field for unprecedented successes such as HER2. We saw how the oncogene bandwagon constrained cancer research trajectories into a domain, where looking at oncogene was the top priority of cancer research agendas. The oncogene 'research programme' met its limits with tumour heterogeneity and the biological complexity, with important social, ethical and economic implications. We saw how some researchers progressively jumped off the bandwagon in the face of these limitations, and the paradigm shifted towards 'progressive' research programmes looking among other at immunotherapy, large-scale (omics) data, biomarkers of the tumour microenvironment, and composite (signature) biomarkers. Having revisited HER2, it is clear that there is a high and ongoing potential for reflexivity and adaptability in the field of cancer research, as a persisting basis for 'scientific revolutions', for researchers to reinvent their field so that it continues to be 'progressive'. In this section we look at one of the key contemporary legacies of the HER2 story, namely: the sociotechnical imaginary of precision oncology.

The HER2 story is still used as evidence to bolster the sociotechnical imaginary of precision oncology. The concept of sociotechnical imaginaries was developed by Jasanoff and Kim in 2013 and defined as "collectively held and performed visions of desirable futures [...] animated by shared understandings of forms of social life and social order attainable through, and supportive of, advances in science and technology" (Jasanoff 2015, p. 25). In other words, sociotechnical imaginaries are visions created and shared by actors in science, industry and politics, of desirable and feasible futures, attainable through science and technology (Strand et al. 2018). They allow for a collective sense-making of social and technological futures, but they are not neutral: they are embedded in social and political negotiations and practices, and "almost always include implicit shared understandings of what is considered to be 'good' or desirable, such as what constitutes "public good" or a "good" society, or how science and technology could meet public needs" (Ballo 2015, p. 12). In that sense, some aspects might be prioritised over others, as some actors have more power than others to participate in the creation of these sociotechnical imaginaries.

Cancer research today is strongly steered by the sociotechnical imaginary of precision oncology. Precision oncology is marketed in policy documents as aiming to achieve "the right therapeutic strategy for the right person at the right time, and/or to determine the predisposition to disease, and/or to deliver timely and targeted prevention" (EC 2015, p. 3). Significant funds are channelled into this ambitious effort. For instance, the EU 7th Framework Programme funded 209 projects on personalised medicine for a total amount of €1.334 billion over the period 2007–2013; and the EU Horizon 2020 funded 167 projects on personalised medicine for a total amount of € 872 million over the 2014–2017 period (EC 2017). In parallel, precision oncology is supported by new technologies relying on big data, such as emerging imaging and new-generation sequencing techniques, often described as being "cost- and time-effective sequencing of tumour DNA, leading to a "genomic era" of cancer research and treatment" (Morganti et al. 2019, p. 9). Significant and important progress has been made in the last few years, with new biomarkers and associated targeted therapies being developed; however, none of these have reached the outstanding success of HER2. These advances and promises are relayed by the media as capable of revolutionising cancer treatment,[10] and are thus accompanied by a strong hope among (future) patients and policymakers alike.

The technoscientific imaginary of precision oncology is producing, among other things, a culture of medicalisation, where the expected medical 'miracles' are no longer perceived as "mirages", but "solidly out there on the horizon[11]" (Callahan

[10] In this book, see the chapter titled "Precision oncology in the news" by Mille Stenmarck and Irmelin Nilsen, where the authors highlight and discuss four unquestioned assumptions in media's articles about precision oncology.

[11] "Hope and reality have fused. Medical miracles are expected by those who will be patients, predicted by those seeking research funds, and profitably marketed by those who manufacture them. [...] The healthier we get, the healthier still we want to become. If we want to live to eighty, why

2003), and it is frequent to see new communities of patients or 'biocollectives[12]' expect and demand a tailored treatment for their diseases[13] (Brekke and Sirnes 2011). There is a fusion of hope and reality, and a confusion between the temporalities of precision oncology, with on one side the 'imaginary' of a desirable future where targeted cancer drugs work without any ambiguity; and on the other side the reality of cancer research today, and the limits it faces with regard to tumour heterogeneity for instance. This confusion fuels the idea that biomarkers are robust enough to (soon) offer solutions and tell us who has, or is at risk of having, a particular type of cancer; who can be treated, with what and when; and how the patient might react to the treatment, including the risk of relapse (Boniolo and Campaner 2019).

We have seen from revisiting the story of HER2 that finding answers to all these questions is rather ambitious, and it would be a shortcut to think that a robust biomarker could help solve all dilemmas related to clinical decision-making, as well as the ethical and social dilemmas related to cut-offs for instance. In a chapter called "What is a good (enough) biomarker?[14]" (2017), Bremer and Wik explored the different dimensions, from the oncology and policy perspectives, according to which a biomarker might be deemed good (enough): analytical validity, clinical validity and clinical utility, as well as improving the health and quality of life of cancer patients and contributing to the sustainability and fairness of healthcare systems. Following that, they discussed the importance of highlighting the opportunity costs of the imaginary of precision oncology. Being steered by the aim of achieving "the right therapeutic strategy for the right person at the right time" (EC 2015), and trying to find 'perfect' biomarkers that can support this endeavour, make us miss important aspects. The key message of Wik and Bremer was that it is impossible for one single biomarker to score high in all of the above-mentioned dimensions. Indeed, choices have to be made when designing a new biomarker, according to its purpose. For instance, while a highly sophisticated and composite biomarker might help better understand the complex cancer biology, it might face important challenges of quality, uncertainty, and difficult implementation in a clinical setting. Similarly, while a simpler biomarker might find broader clinical application and be more widely accessible, the cut-off for patient stratification might be rougher. Therefore, along-

---

not to one hundred? [...] The "mirage of health" – a perfection that never comes – is no longer taken to be a mirage, but solidly out there on the horizon" (Callahan 2003, p. 261).

[12] "New collectives, joined together by shared biomedical traits, now appear in the intersections between science, the economy, and civil society. Patients join forces with biotechnological companies and research groups in order to promote research on "their" disease, and they both influence research agendas and enroll themselves actively as research subjects" (Brekke and Sirnes 2011, p. 349).

[13] In this book, see also the chapters of Hillersdall & Nordahl Svendsen, Strand & Engen, and Stenmarck & Nilsen, which respectively address the tensions between actors' various agendas in first-in-human drug trials; how research endeavours that aim for high levels of precision can shift the focus away from equally important, relevant research; and how precision oncology is framed in the media as a reality that will soon materialise.

[14] The chapter of Wik and Bremer is in the anthology: *Cancer Biomarkers: Ethics, Economics and Society* (2017). Edited by Anne Blanchard and Roger Strand. Norway: Megaloceros Press.

side the search for biomarkers that can do it all, it is also important to highlight the potential for extremely relevant research on how biomarkers can be good enough in certain settings, and how they should be evaluated and implemented. In particular, how the 'quality' of a good (enough) biomarker is evaluated through its 'fitness for purpose' (the biomarker does what it is supposed to do), rather than its capacity to score points in all the above dimensions. This could help curb the spiralling culture of medicalisation, where limits to the realisation of extraordinary treatments are perceived as being only political, not scientific;[15] and where questions of justice and fairness in healthcare distribution are often reduced to a mere problem of lack of accuracy, precision, sensitivity or specificity from the biomarker.

The reality of precision oncology today, is that 99% of published cancer biomarkers fail to enter clinical practice (Kern 2012; see also Ioannidis and Bossuyt 2017 and Ren et al. 2020). Let us look at that 99% through the 'degenerative' and 'progressive' research programmes of Lakatos. At first sight, it would seem rather rational to jump off the precision oncology research programme, to move towards a more 'progressive' type of research. However, Lakatos argues that it is neither irrational nor rational, neither good nor bad, to stay on a degenerating programme: "One may rationally stick to a degenerating programme until it is overtaken by a rival *and even after*. What one must *not* do is to deny its poor public record. […] It is perfectly rational to play a risky game: what is irrational is to deceive oneself about the risk." (pp. 104–105) This means, then, that if precision oncology is regarded as the future of cancer research, we have to accept Kern's claim, be open and honest about the 99% of published biomarkers which don't make it to practice, be ready to justify the risks and opportunity costs of such research efforts, and be transparent about the fact that precision oncology, as shown by the HER2 story, is operating within the limits of tumour heterogeneity and complexity of cancer biology.

Conclusion

In this chapter, we looked at the story of HER2 from its discovery and basic studies, to biomarker development and early clinical trials. We then revisited HER2's story to reflect on the conditions of its inception, some of the reasons for its success, and the challenges met along the way. In particular, we drew a parallel between the story of HER2 and a bandwagon, to see HER2 as a standard theory-and-technology package, that could easily circulate around the network of actors and organisations working on oncogene research, greatly facilitating its development. Nevertheless, revisiting HER2 made clear that despite its extraordinary success, this biomarker operates in a context of high levels of biological complexity, in particular with

[15] "There are no inherent obstacles or pitfalls of science that could stop the realization of revolutionary cures. Therefore, this is not about science; it is all about politics" (Brekke and Sirnes 2011, p. 356).

regard to cancer tumour heterogeneity. HER2 therefore faces legal, social, or economic challenges and dilemmas including ragged edges and where to justly place the cut-off between HER2+ and HER2− patients, questions of fairness in the access of high-priced and sophisticated technologies and therapies, or the difficult partnerships between academia and pharmaceutical companies to bring a scientific discovery to the clinic. Revisiting HER2 also more generally highlighted that the fields of cancer biomarker research and precision oncology, where HER2 belongs, are based on a sometimes confusing blend of hope and reality: hope that targeted therapies will (soon) work for every cancer patient; and the reality of the complexity of cancer biology.

Based on these observations, we reflected upon two aspects relating to the future of cancer biomarker research. First, it is important to not be 'blinded' by the prospects of precision oncology and strive at all costs for hyper-precision and an unachievable molecular certainty, numerical exactness and conceptual rigour. The field of cancer biomarkers could derive much learning from a more pronounced focus on 'good enough' biomarkers: how they can support patients well enough in certain settings, and how they can potentially reconcile cancer as a disease and as an illness, by for instance giving a greater place to the patient's personal experiences of living with cancer. Second, if precision oncology is regarded as the future of cancer research, then we have to accept the uncomfortable claim by Kern (2012) and be honest about the low success rate of 1% of published biomarkers which reach clinical practice. In particular, this means that we should be ready to justify the risks and opportunity costs of precision oncology. As shown by the HER2 story, cancer biomarkers are dealing with intrinsically complex, open and non-deterministic systems from cells to patients, and the field will therefore always operate within the limits of the complexity of cancer biology.

References

Akkari, L., S.F. Bakhoum, S. Krishnaswamy, S. Chen, et al. 2020. The future of cancer research. *Trends in Cancer* 6 (9): 724–729.

Antoniou, M., R. Kolamunnage-Dona, J. Wason, et al. 2019. Biomarker-guided trials: Challenges in practice. *Contemporary Clinical Trials Communications* 16: 100493.

Ballo, I.F. 2015. Imagining energy futures: Sociotechnical imaginaries of the future Smart Grid in Norway. *Energy Research & Social Science* 9 (2015): 9–20.

Baselga, J., X. Carbonell, N.J. Castaneda-Soto, M. Clemens, M. Green, V. Harvey, et al. 2005. Phase II study of efficacy, safety, and pharmacokinetics of trastuzumab monotherapy administered on a 3-weekly schedule. *Journal of Clinical Oncology* 23 (10): 2162–2171.

Bishop, J.M. 1983. Cellular oncogenes and retroviruses. *Annual Review of Biochemistry* 52: 301–354.

Blanchard, A. 2016. Mapping ethical and social aspects of cancer biomarkers. *New Biotechnology* 33 (6): 763–772.

Blanchard, A., and E. Wik. 2017. What is a good (enough) biomarker? In *Cancer Biomarkers: Ethics, Economics and Society*, ed. A. Blanchard and R. Strand. Bergen: Megaloceros Press.

Boniolo, G. 2017. Patchwork narratives for tumour heterogeneity. In *Logic, Methodology and Philosophy of Science – Proceedings of the 15th International Congress*, ed. H. Leitgeb, I. Niiniluoto, E. Sober, and P. Seppälä, 311–324. London: College Publications.

Boniolo, G., and R. Campaner. 2019. Complexity and integration. A philosophical analysis of how cancer complexity can be faced in the era of precision medicine. *European Journal for Philosophy of Science* 9 (3): 1–25.

Brekke, O.A., and T. Sirnes. 2011. Biosociality, biocitizenship and the new regime of hope and despair: Interpreting "Portraits of Hope" and the "Mehmet Case". *New Genetics and Society* 30 (4): 347–374.

Callahan, D. 1990. *What Kind of Life: The Limits of Medical Progress*. Washington, DC: Georgetown University Press.

Cobleigh, M.A., C.L. Vogel, D. Tripathy, N.J. Robert, S. Scholl, L. Fehrenbacher, et al. 1999. Multinational study of the efficacy and safety of humanized anti-HER2 monoclonal antibody in women who have HER2-overexpressing metastatic breast cancer that has progressed after chemotherapy for metastatic disease. *Journal of Clinical Oncology* 17 (9): 2639–2648.

Callahan, D. 2003. *What Price Better Health? Hazards of the Research Imperative*. Berkeley: University of California Press.

Carter, P., L. Presta, C.M. Gorman, J.B. Ridgway, D. Henner, W.L. Wong, et al. 1992. Humanization of an anti-p185HER2 antibody for human cancer therapy. *Proceedings of the National Academy of Sciences of the United States of America* 89 (10): 4285–4289.

Coussens, L., T.L. Yang-Feng, Y.C. Liao, E. Chen, A. Gray, J. McGrath, et al. 1985. Tyrosine kinase receptor with extensive homology to EGF receptor shares chromosomal location with neu oncogene. *Science* 230 (4730): 1132–1139.

Downward, J., Y. Yarden, E. Mayes, G. Scrace, N. Totty, P. Stockwell, et al. 1984. Close similarity of epidermal growth factor receptor and v-erb-B oncogene protein sequences. *Nature* 307 (5951): 521–527.

Drebin, J.A., V.C. Link, R.A. Weinberg, and M.I. Greene. 1986. Inhibition of tumor growth by a monoclonal antibody reactive with an oncogene-encoded tumor antigen. *Proceedings of the National Academy of Sciences of the United States of America* 83 (23): 9129–9133.

EC. 2015. *European Council Conclusions on Personalised Medicine for Patients*. Luxembourg: Publications Office of the European Union, European Council.

———. 2017. *Personalised Medicine Focusing on Citizen's Health*. ICPerMed, European Union.

Fleck, L.M. 2012. Pharmacogenomics and personalized medicine: Wicked problems, ragged edges and ethical precipices. *New Biotechnology* 29 (6): 757–768.

Fujimura, J.H. 1988. The molecular biological bandwagon in cancer research: Where social worlds meet. *Social Problems* 35 (3): 261–283.

Ginsburg, G.S., and H.F. Willard. 2009. Genomic and personalized medicine: Foundations and applications. *Translational Research* 154 (6): 277–287.

Goldhirsch, A., J.N. Ingle, R.D. Gelber, et al. 2009. Thresholds for therapies: Highlights of the St Gallen International Expert Consensus on the primary therapy of early breast cancer 2009. *Annals of Oncology* 20 (8): 1319–1329.

Hassett, M.J., H. Li, H.J. Burstein, and R.S. Punglia. 2020. Neoadjuvant treatment strategies for HER2-positive breast cancer: Cost-effectiveness and quality of life outcomes. *Breast Cancer Research and Treatment* 181: 43–51.

Hudziak, R.M., J. Schlessinger, and A. Ullrich. 1987. Increased expression of the putative growth factor receptor p185HER2 causes transformation and tumorigenesis of NIH 3T3 cells. *Proceedings of the National Academy of Sciences of the United States of America* 84 (20): 7159–7163.

Hudziak, R.M., G.D. Lewis, M. Winget, B.M. Fendly, H.M. Shepard, and A. Ullrich. 1989. p185HER2 monoclonal antibody has antiproliferative effects in vitro and sensitizes human breast tumor cells to tumor necrosis factor. *Molecular and Cellular Biology* 9 (3): 1165–1172.

Hunter, T., and J. Simon. 2007. A not so brief history of the oncogene meeting and its cartoons. *Oncogene* 26 (9): 1260–1267.

Hunter, N.B., M.R. Kilgore, and N.E. Davidson. 2020. The long and winding road for breast cancer biomarkers to reach clinical utility. *Clinical Cancer Research* 26 (21): 5543–5545.

Ioannidis, J.P., and P.M. Bossuyt. 2017. Waste, leaks, and failures in the biomarker pipeline. *Clinical Chemistry* 63 (5): 963–972.

Jackson, D.B., and A.K. Sood. 2011. Personalized cancer medicine – Advances and socio-economic challenges. *Nature Reviews Clinical Oncology* 8 (12): 735–741.

Jasanoff, S. 2015. Future imperfect: Science, technology and the imaginations of modernity. In *Dreamscapes of Modernity: Sociotechnical Imaginaries and the Fabrication of Power*, ed. S. Jasanoff and S.-H. Kim, 1–33. Chicago: Chicago University Press.

Jasanoff, S., and S.H. Kim. 2013. Sociotechnical imaginaries and national energy policies. *Science as Culture* 22 (2): 189–196.

Jones, P.T., P.H. Dear, J. Foote, M.S. Neuberger, and G. Winter. 1986. Replacing the complementarity-determining regions in a human antibody with those from a mouse. *Nature* 321 (6069): 522–525.

Jørgensen, J.T., and H. Winther. 2010. The development of the HercepTest–from bench to bedside. In *Molecular Diagnostics–The Key Driver of Personalized Cancer Medicine*, ed. J.T. Jørgensen and H. Winther, 43–60. Singapore: Pan Stanford Publishing.

Kern, S.E. 2012. Why your new cancer biomarker may never work: Recurrent patterns and remarkable diversity in biomarker failures. *Cancer Research* 72 (23): 6097–6101.

King, C.R., M.H. Kraus, and S.A. Aaronson. 1985. Amplification of a novel v-erbB-related gene in a human mammary carcinoma. *Science* 229 (4717): 974–976.

Kuhn, T. 1962. *The Structure of Scientific Revolutions*. Chicago: The University of Chicago Press.

Kumar, G.L., and S. Badve. 2008. Milestones in the discovery of HER2 proto-oncogene and trastuzumab (Herceptin™). *Connections* 13: 9–14.

Lakatos, I. 1970. History of science and its rational reconstructions. In *PSA: Proceedings of the Biennial Meeting of the Philosophy of Science Association*, vol. 1970, 91–136. D. Reidel Publishing.

Laplane, L., D. Duluc, A. Bikfalvi, et al. 2019. Beyond the tumour microenvironment. *International journal of cancer* 145 (10): 2611–2618.

Mayeur, C., and W. van Hoof. 2021. Citizens' conceptions of the genome: Related values and practical implications in a citizen forum on the use of genomic information. *Health Expectations* 00: 1–10.

McEwen, J.E., J.T. Boyer, K.Y. Sun, K.H. Rothenberg, N.C. Lockhart, and M.S. Guyer. 2014. The ethical, legal, and social implications program of the National Human Genome Research Institute: Reflections on an ongoing experiment. *Annual Review of Genomics and Human Genetics 15*: 481.

Morganti, S., P. Tarantino, E. Ferraro, et al. 2019. Next generation sequencing (NGS): A revolutionary Technology in Pharmacogenomics and Personalized Medicine in cancer. In *Translational Research and Onco-Omics Applications in the Era of Cancer Personal Genomics*, ed. E. Ruiz-Garcia and H. Astudillo-de la Vega, 9–30. Cham: Springer.

OECD. 2011. *Policy Issues for the Development and Use of Biomarkers in Health*. Paris: OECD.

Parker, D. 2018. The evolving role of biomarkers in oncology clinical trial design. *Clinical Leader Guest Column; accessed January 2022; https://www.clinicalleader.com/doc/the-evolving-role-of-biomarkers-in-oncology-clinical-trial-design-0001*

Pegram, M.D., A. Lipton, D.F. Hayes, B.L. Weber, J.M. Baselga, D. Tripathy, et al. 1998. Phase II study of receptor-enhanced chemosensitivity using recombinant humanized anti-p185HER2/neu monoclonal antibody plus cisplatin in patients with HER2/neu-overexpressing metastatic breast cancer refractory to chemotherapy treatment. *Journal of Clinical Oncology* 16 (8): 2659–2671.

Pegram, M., S. Hsu, G. Lewis, R. Pietras, M. Beryt, M. Sliwkowski, et al. 1999. Inhibitory effects of combinations of HER-2/neu antibody and chemotherapeutic agents used for treatment of human breast cancers. *Oncogene* 18 (13): 2241–2251.

Piccart-Gebhart, M.J., M. Procter, B. Leyland-Jones, A. Goldhirsch, M. Untch, I. Smith, et al. 2005. Trastuzumab after adjuvant chemotherapy in HER2-positive breast cancer. *New England Journal of Medicine* 353 (16): 1659–1672.

Pietras, R.J., M.D. Pegram, R.S. Finn, D.A. Maneval, and D.J. Slamon. 1998. Remission of human breast cancer xenografts on therapy with humanized monoclonal antibody to HER-2 receptor and DNA-reactive drugs. *Oncogene* 17 (17): 2235–2249.

Ren, A.H., C.A. Fiala, E.P. Diamandis, and V. Kulasingam. 2020. Pitfalls in cancer biomarker discovery and validation with emphasis on circulating tumor DNA. *Cancer Epidemiology and Prevention Biomarkers* 29 (12): 2568–2574.

Romond, E.H., E.A. Perez, J. Bryant, V.J. Suman, C.E. Geyer Jr., N.E. Davidson, et al. 2005. Trastuzumab plus adjuvant chemotherapy for operable HER2-positive breast cancer. *New England Journal of Medicine* 353 (16): 1673–1684.

Rosoff, P.M. 2017. *Drawing the Line: Healthcare Rationing and the Cutoff Problem*. New York: Oxford University Press.

Sawyers, C.L. 2019. Herceptin: A first assault on oncogenes that launched a revolution. *Cell* 179 (1): 8–12.

Schechter, A.L., D.F. Stern, L. Vaidyanathan, S.J. Decker, J.A. Drebin, M.I. Greene, et al. 1984. The neu oncogene: An erb-B-related gene encoding a 185,000-Mr tumour antigen. *Nature* 312 (5994): 513–516.

Semba, K., N. Kamata, K. Toyoshima, and T. Yamamoto. 1985. A v-erbB-related protooncogene, c-erbB-2, is distinct from the c-erbB-1/epidermal growth factor-receptor gene and is amplified in a human salivary gland adenocarcinoma. *Proceedings of the National Academy of Sciences of the United States of America* 82 (19): 6497–6501.

Slamon, D.J., G.M. Clark, S.G. Wong, W.J. Levin, A. Ullrich, and W.L. McGuire. 1987. Human breast cancer: Correlation of relapse and survival with amplification of the HER-2/neu oncogene. *Science* 235 (4785): 177–182.

Slamon, D.J., B. Leyland-Jones, S. Shak, H. Fuchs, V. Paton, A. Bajamonde, et al. 2001. Use of chemotherapy plus a monoclonal antibody against HER2 for metastatic breast cancer that overexpresses HER2. *New England Journal of Medicine* 344 (11): 783–792.

Stehelin, D., H.E. Varmus, J.M. Bishop, and P.K. Vogt. 1976. DNA related to the transforming gene(s) of avian sarcoma viruses is present in normal avian DNA. *Nature* 260 (5547): 170–173.

Strand, R., A. Saltelli, M. Giampietro, K. Rommetveit, and S. Funtowicz. 2018. New narratives for innovation. *Journal of Cleaner Production* 197 (2018): 1849–1853.

Ullrich, A., L. Coussens, J.S. Hayflick, T.J. Dull, A. Gray, A.W. Tam, et al. 1984. Human epidermal growth factor receptor cDNA sequence and aberrant expression of the amplified gene in A431 epidermoid carcinoma cells. *Nature* 309 (5967): 418–425.

Varmus, H.E. 1984. The molecular genetics of cellular oncogenes. *Annual Review of Genetics* 18: 553–612.

Verhoeyen, M., C. Milstein, and G. Winter. 1988. Reshaping human antibodies: Grafting an antilysozyme activity. *Science* 239 (4847): 1534–1536.

Vogel, C.L., M.A. Cobleigh, D. Tripathy, J.C. Gutheil, L.N. Harris, L. Fehrenbacher, et al. 2006. Efficacy and safety of trastuzumab as a single agent in first-line treatment of HER2-overexpressing metastatic breast cancer. *Journal of Clinical Oncology* 20 (3): 719–726.

Wang, J., and B. Xu. 2019. Targeted therapeutic options and future perspectives for Her2-positive breast cancer. *Signal Transduction and Targeted Therapy* 4: 34.

Weinberg, R.A. 2014. Coming full circle – From endless complexity to simplicity and back again. *Cell* 157 (1): 267–271.

Weiss, R.A. 2020. A perspective on the early days of RAS research. *Cancer and Metastasis Reviews* 39 (4): 1023–1028.

Wolff, A.C., M.E.H. Hammond, K.H. Allison, B.E. Harvey, P.B. Mangu, J.M.S. Bartlett, M. Bilous, et al. 2018. Human epidermal growth factor receptor 2 testing in breast cancer: American society of clinical oncology/college of American pathologists clinical practice guideline focused update. *Journal of Clinical Oncology* 36 (20): 2105–2122.

Yi, Z., G. Rong, Y. Guan, J. Li, L. Chang, H. Li, B. Liu, et al. 2020. Molecular landscape and efficacy of Her2-targeted therapy in patients with Her2-mutated metastatic breast cancer. *NPJ Breast Cancer* 6: 59.

The Dynamics of the Labelling Game: An Essay On FLT3 Mutated Acute Myeloid Leukaemia

Caroline Engen

Cancer is widely recognised as a human malady that embodies all elements of the triad of *disease*, *illness* and *sickness*; causing physiological malfunction, severe subjective symptoms and distress, and associated with certain specific social claims and obligations. The word alone spurs discomfort and fear, and is often shadowed by associations of struggle, pain, suffering and death (Vrinten et al. 2017). Historically these associations are both well placed and justified. The term cancer, originating from the Greek word "karkinos", meaning crab, well captures with what malicious and ruthless force a malignant tumour can invade and spread its claws in a destructive and merciless manner throughout the human body. Until very recently the only form of cancer recognized in clinical medicine resembled this very image. Paired with mutilating surgery and toxic medical treatment terror has been considered a very suiting response to cancer. Currently, however, there are indications that the content and organisation of the category of cancer as a pathological process is rapidly evolving. We are presently in an era where a new form of medical practice is taking shape and gaining ground; precision medicine. Founded on the historical relationship between improved pathophysiological understanding and the subsequent development of efficient medical solutions this emerging medical approach proposes to change the way we approximate cancer as a disease process on several dimensions. Based on increased precision and resolution the hypothesis is that individual classification of cancer at a molecular level, rather than macroscopically and morphologically, as is the tradition, will promote improved medical management and result in reduced cancer related morbidity and mortality (Ashley 2016). As the study and clinical management of cancer steadily distances itself from the form and shape cancer takes on when it manifests itself clinically new questions and

C. Engen (✉)
Centre for Cancer Biomarkers CCBIO, Department of Clinical Medicine, University of Bergen, Bergen, Norway
e-mail: Caroline.Engen@uib.no

A. Bremer, R. Strand (eds.), *Precision Oncology and Cancer Biomarkers*, Human Perspectives in Health Sciences and Technology 5,
https://doi.org/10.1007/978-3-030-92612-0_8

challenges are, however, bound to emerge. In this essay I would like to reflect upon some of the novel surfaces of friction that are currently materializing as the boundaries of cancer are set in motion by the shift in dimensionality that precision medicine promotes; moving from the macroscopic to the molecular level of understanding cancer, and from the study of humans to the study single individuals. To illustrate the complexity and intricacy of some of the issues that arise I will base the discourse on two interconnected narratives exemplifying the transition; the historical storyline about the coming of a molecularly defined subgroup of acute myeloid leukaemia (AML), and the story of the biological evolution from a normal hematopoietic progenitor cell to an AML cell. Based on these two stories I will proceed to discuss how the molecular level of resolution currently challenges not only how we academically and scientifically go about defining cancer the disease, but ultimately how this change in resolution also implies vast clinical and social consequences, correspondingly influencing both the illness and sickness aspects of cancer. As scholars of medical philosophy has not yet resolved in agreement on the central epistemological and ontological questions about the nature of disease and disease classification I will not dare to take a definite position. I rather try to reflect on the phenomena of cancer from the position of the simple clinician and translational cancer researcher that I am; as a product of biology, but always framed and managed by humans, and therefore essentially shaded by both aspects of underlying nature and normativity.

Leukaemia was first described as a distinct clinical entity in the midst of the nineteenth century. Initially the disease was recognised based on symptoms, distinct clinical observations and autopsy findings. Based on this level of resolution the first division of leukaemia into two subgroups shortly followed; an acute, storming and rapidly lethal form of the disease, contrasted to a more indolent chronic form of leukaemia. Only when microscopic assessment of the blood was available by the midst of the century did one approximate the nature of leukaemia as a neoplastic condition originating in the hematopoietic system and as the microscopic method became more refined the understanding of leukaemia gradually improved (Kampen 2012). Based on morphological distinctions leukaemia was subsequently subjected to an additional stratification; leukaemia originating from cells committed to the myeloid lineage of the hematopoietic system opposed to leukaemia deriving from the lymphatic cell lineages. Founded on morphological and cytochemical attributes further sub-classifications followed, and by 1976 three groups of acute lymphoblastic leukaemia and six distinct morphological subgroups of acute myeloid leukaemia were proposed, illustrating the gradual compartmentalization of the disease (Bennett et al. 1976). Contemporarily, the clonal origin of leukaemia was gradually uncovered, and the relationship between the disease phenotype, the cell of origin and genetic damage was gradually revealed (Nowell 1976). This resulted in a steady shift from a descriptive morphologically based system for organising the different sorts of leukaemia to a more pathophysiological and functionally founded organisational structure, frequently based on the molecular level of genetic aberrations. Initially cytogenetic features were the main focus, but as methods for molecular genetics became more refined and available also mutations have gradually found their place in the organisational system (Bennett 2000; Vardiman et al. 2009; Arber et al. 2016).

A molecular feature that has been devoted immense attention in AML from the very beginning of this shift towards molecular pathophysiological characterisation of cancer is the protein FLT3. The gene coding for this protein was identified as recurrently mutated in AML already in 1996, and the presence of mutations within this gene has consequently been linked to both phenotypical and functional properties. The presence of FLT3 mutations has been demonstrated to be statistically associated with features like age and gender, and clinical characteristics like leucocytosis and high bone marrow blast counts at time of diagnosis. The group has further been correlated with cytomorphological features, cytogenetics and molecular genetics, as well as with more functional properties like expression of certain immunophenotypical markers and intracellular signalling patterns. Ultimately the mutation has been repetitively coupled with patient outcome, predicting high likelihood of poor response to treatment, high relapse rate and inferior overall survival. Based on these findings and bound together by this shared pathophysiological feature the group has gradually been thought of as a confined subgroup of AML, and has consequently been treated as a distinct biological entity in preclinical and clinical research as well as in clinical practice (Lagunas-Rangel and Chavez-Valencia 2017).

As mentioned in the introduction, the leading clinical motivation for molecular characterisation of cancer and the ultimate validation of such an endeavour is closely linked to the utility of the approach. Within the field of haemato-oncology this might be best illustrated by the discovery of the Philadelphia chromosome, and the BCR-ABL oncoprotein in chronic myeloid leukaemia (CML). The molecular framing of this disease resulted in the subsequent development of pharmaceutical agents like Imatinib (Gleevec), specifically targeting the oncoprotein, resulting in remarkable therapeutic advances for this patient group (Deininger et al. 2005). Descriptively and functionally FLT3 mutated AML shares resemblance to BCR-ABL positive CML. As myeloid malignancies AML and CML share very similar cells of origin, and they are both seemingly driven by genetic aberrations resulting in an oncoprotein in the form of a constitutively active tyrosine kinase, resulting in a comparable proliferative advantage. Encouraged by the success of BCR-ABL targeted therapy FLT3 very early singled out as an attractive therapeutic target in AML. During the last two decades considerable effort has been devoted to developing therapeutic agents that would selectively benefit this patient group through specific inhibition of FLT3. 15 years after the initiation of the first clinical trials exploring the benefit of such FLT3-targeted therapy only very recently a few trials have demonstrated modest clinical benefit (Stone et al. 2017; Cortes et al. 2019; Perl et al. 2019). The fact that significant clinical improvements for this patient group still seems far out of reach is discouraging, and many possible explanations have been put forward attempting to explain the limited responses (Engen et al. 2014). At the core however, the lack of substantial achievements may possibly indicate fundamental limitations in the current labelling of FLT3 mutated AML. Accumulating data derived from the past 20 years is gradually exposing the heterogeneous nature within this confined group. The summation of temporal, spatial, multidimensional and high-resolution analysis of this group has revealed vast inter- and intra-individual heterogeneity. The gene can be damaged in multiple ways; most frequent by internal tandem duplications of varying lengths and motifs, followed by point mutations and occasional insertions or deletions. Uniparental disomy of the part of the chromosome entailing the mutated gene is also a common aberration, resulting in homozygous mutations and loss of the wild type allele. All the variants are validated as functionally significant, although with varying implications. The fraction of the leukemic disease that entails a FLT3 mutation is in addition highly variable and dynamic, ranging from a diminishing portion of the leukaemia cells to defining the entire tumour cell population. A large portion of the patients actually has several sub-clones, characterised by different distinct FLT3 alterations. The mutations can even occur or disappear through a single clinical disease course. Several of these features have further been linked to disease specific outcome; the length of the mutation, portion of FLT3 mutated cells, the insertion site of the length mutation, and the amino acid sequence of the duplicated region are just some of the features various investigators have attributed functional significance with prognostic implications (Lagunas-Rangel and Chavez-Valencia 2017).

While the inter-individual heterogeneity of FLT3 mutations makes the group hard to study, it might essentially be the intra-individual heterogeneity and kinetics that generates the greatest challenges in the practical assessment and categorisation of this patient group. What this intra-tumour variability implies, however, is that alterations in the FLT3 gene most frequently represent a late event in the process of leukemogenesis, and this is essentially where the great value in the data positions. When the individual findings are aligned and considered together they add up like pieces in a puzzle, revealing truths about the second story I now want to tell – the tale of the origin and evolution of leukaemia; the transition from a healthy cooperative and obedient hematopoietic progenitor cell to an insensitive and anarchistic leukaemia cell, and the story starts long before the FLT3 mutation enters the narrative.

Cancer is often described as characterised by their monoclonal origin, and this implies also for FLT3 mutated AML. The leading hypothesis, supported by strong emerging evidence, suggests that preceding the FLT3 mutation, maybe by as much as decades in some patients, an initial molecular alteration occurred in a long-lived hematopoietic stem or progenitor cell. This alteration resulted in functional changes in the cell, providing a survival advantage, securing that as time passed by the decedents of this very first cell survived and/or multiplied at a higher rate than the other progenitor cells. This ultimately generated a pool of cells characterised by this common alteration, confining a condition we now call clonal haematopoiesis of indeterminate potential (CHIP) (Genovese et al. 2014). As occurrence of genetic damage and epigenetic modifications are frequent events, at a point in time an additional alteration occurred, providing an added advantage to one of the cells, and then this cell again fostered a group of decedents with shared properties. This process continued in a branched manner, gradually generating a vastly heterogeneous pool of cells with varying degree of deviant behaviour, although with certain remaining similarities. At a certain point one of these cells gained a property that influenced also the interaction with surrounding cells and tissue, and a situation where normal hematopoietic function was affected arose, resulting in what we might define as a pre-leukemic condition (Shlush et al. 2014). Eventually the sufficient damage befell one pre-leukemic cell resulting in the development of the full blown leukemic phenotype. Within the group we currently characterise as FLT3 mutated AML potentially, and in some cases probably, it is the acquisition of exactly the FLT3 mutation that generates the leukemic phenotype and initiates the clinical presentation of the disease. The augmented proliferative drive the damaged FLT3 protein adds to the already damaged cell could be the sufficient addition of disruption needed for the cell to divide at such a rate that it and its descendants manages to overtake the bone marrow environment, effectively causing disruption of normal hematopoietic functioning and resulting in the presentation of symptoms, and clinical findings.

The two stories presented, the first about the formation of FLT3 mutated AML as a diagnostic unit and the second about the evolution of FLT3 mutated AML as a biological entity, can be condensed to illustrate two central challenges; (1) The soundness of FLT3 as a marker for the transition from a non-disease to a disease and (2) the legitimacy of FLT3 as marker of a specific confined disease. The generalisation of these two questions transfer us towards a major challenge in precision oncology; if and how a defined molecular feature or event can reliably and truthfully delineate something healthy from something diseased, or serve as robust foundation for dividing and stratifying specific diseases. So after the exercise of contracting down from the level of illness and clinical symptoms to the resolution of single molecules in leukaemia I would like to reverse and expand the perspective, attempting to demonstrate the relevance of this example when considering the challenges within the overarching category of cancer.

Of both intellectual and clinical interest is the question of whether the acquisition of a specific molecular alteration can delineate the point in carcinogenesis when a cell or a group of cells transit from being normal or healthy to becoming a disease or a condition that can be defined as a pathological process. Returning to the example of leukemogenesis and whether the attainment of a FLT3 mutation can serve as marker for the conversion from non-leukaemia to leukaemia it seems from a naturalistic perspective plausible that the FLT3 mutation may well denote a cell fraction characterised by the fully developed leukemic phenotype, at least on a single cell level. The clinical relevant disease however is frequently demonstrated to entail additional complexity. Experience treating FLT3 mutated leukaemia has shown that although the addition of a FLT3 mutation may be the sufficient element founding a confined unit of cells responsible for the promotion of the clinically dе́but of leukaemia, it is clearly not necessary, as there are other mechanisms that can produce a very similar phenotype. After successful induction therapy in a FLT3 mutated patient the leukaemia can recur as a FLT3 wild type disease, or even being characterised by an unrelated novel FLT3 mutation. The existence of AMLs characterised by only a small fraction of FLT3 mutated cells indicates the same conclusion, implying the presence of alternate leukemic drivers in the remaining FLT3 wild type blast population. The presence of several distinct FLT3 mutations concurrently within the same AML provides further proof; as such mutational patterns are shown to define different distinct cellular populations with discrete functional properties. This heterogeneous pattern is certainly not valid just in FLT3 mutated AML but also in other cancers where it has been shown that although an individual cancer is usually characterised by a common driver mutation, distinct cellular subsets are characterised by additional but discrete driver mutations, and that the composition is dynamic. The spatial heterogeneity is particularly striking with evidence suggesting that metastatic disease often derives from ancestral cells, preceding the dominating genotype of the primary tumour (Gerlinger et al. 2012; Johnson et al. 2014; Gibson et al. 2016). The parallel co-occurrence of several distinct disease characterising driver mutations challenge the core of precision oncology and raises the question of how these situations should be interpreted and managed.

The conclusion one might derive from this is that if FLT3 denotes the transition on a single cell level, then most patients with AML have many different and distinct parallel leukaemias. In addition most AML patients probably have several additional conditions, including several distinct pools of clonal hematopoietic cells and multiple pools of pre-leukemic cells. The same situation would consequently apply for most cancers at the time when they revile their malignant nature. Precision-wise one may therefore consider the situation as managing several distinct diseases and conditions within single individuals, some times not even separated by time, but simply by molecular characteristics resulting in distinct functional properties. The fundamental challenge here is that dependent on the strength of our magnifying glass all cancer cells could in nature be unique, both in descriptive and in functional terms. When considering the process of carcinogenesis in mechanistic terms, as the result of numerous sequential molecular alterations, happening one at a time in parallel in every single cell; one acetylation here, one posttranslational modification there, one double stranded DNA brake here, and one phosphorylation way over there, it is evident that no cancer cells are identical in molecular terms. Supportive evidence of the functional consequences of this is derives from studies of relapsed AML patients, after treatment with selective FLT3 inhibitors. These therapeutic agents pose a very specific selection pressure on the leukemic cells, and the recurring result is emergence of a polyclonal pattern of novel FLT3 mutations conferring resistance to tyrosine kinase inhibitor treatment. This indicates a narrow selection of cells, otherwise not detectable, that likely under no other condition would be selected for or would stand out as functionally different to the bulk disease, except exactly under these very specific conditions (Man et al. 2012; Smith et al. 2012, 2017; Baker et al. 2013).

The thought experiment above leads to the possible conclusion that every individual cancer cell may ultimately represent a discrete functional unit (at least potentially – dependent on applied selection pressure) and may thereby even be considered individual cancers in biological terms. However, at this level of resolution the clear-cut patterns and delineations of what can be considered normal or deviant becomes difficult to grasp, and the close relationship between the level of dimensionality and what can be considered normal or anomalous becomes very clear. Recalling the transitions of dimensionality illustrated in the stories told about FLT3 mutated AML this becomes very clear. In developed countries cancer is a leading cause of death and a very common disease affecting up to one third of the population in a lifetime perspective (McGuire 2016). Within the category of cancer haematological malignancies are however rather rare, and the lifetime likelihood of being diagnosed with AML is very low. Incidence is related to age, and presentation of AML early in life is extremely uncommon. Even more improbable is the development of FLT3 mutated AML. Increasing the sensitivity when assessing FLT3 mutated AML patients I have tried to demonstrate that molecularly they are all unique; in fact all of their cancer cells might ultimately represent distinct biological entities, as derived from the conclusion above. Reversing the reasoning the same challenge emerges; damaged hematopoietic progenitor cells with future leukemic potential are most likely almost ubiquities and if we just use a sensitive method enough we would probably show that we all possess such cells occasionally (Young et al. 2016). Enrichment of such cells as to fit the current definition of CHIP is however slightly more rare (Genovese et al. 2014), and the presence of pre-leukemic cells posing a danger to normal hematopoietic homeostasis is likely much more infrequent. Overt leukaemia remains very rare. The same reasoning applies when considering changes in function – are we to consider fluctuations in in function on the level of the individual, of the organ, of the tissue or the individual cells? The chosen dimension of investigation seems to strongly influence the statistics and thereby conclusions produced, and the question of what the appropriate level of inquiry should be seems of outmost importance. How should we understand cancer and the dangers of cancer under such varying perspectives? To take this analysis to its outer limits, what it implies is that (1) we all have cancer if we just look closely enough, and (2) all these cancers will be singular diseases that on a molecular level only can be grouped together by the use of simplifications.

From a clinical perspective this conclusion may not intuitively seem very useful, so moving slightly from the discourse of what can be considered real to what is practically relevant within the realm of applied medicine some significant practical questions emerge. From a clinical and therapeutic perspective it is not just about what is and what is not, but maybe more importantly about what could and should be done. The principal goal of precision oncology is after all to improve patient care and outcome. Molecular markers are merely considered the tools required for achieving this. Although surrounded by significant hope and hype, the unsatisfactory results in the clinical management of FLT3 mutated AML (Prasad and Gale 2016) unfortunately mirrors the current status and accomplishments of precision oncology far better than the triumphant story of tyrosine kinase inhibitors and their achievements in the treatment of CML (Prasad et al. 2016). As I previously challenged the validity and utility of labelling of FLT3 mutated AML as a discrete diagnostic subgroup based on the lack of clinical improvements, one might question if a similar challenge may essentially be an underlying variable currently restricting the potential of precision oncology at large.

To start at the very practical oriented end the viability of precision medicine as project, both scientifically and clinically, is supported and dependent on powerful and high-resolution technological solutions, allowing uncoupling of the subjective and clinical findings of cancer traditionally causing illness, by focusing the gaze on ever- more narrowly defined composites and interactions of the biological processes involved in the development of cancer. An apparent challenge is nevertheless that the precision of the lines we draw is rigorously limited to the qualities of our scientific assays. In fact, examining the historical narrative of the compartimalisation of leukaemia it is from beginning to end largely framed and defined by available technology, ranging from the early development of the microscope to the recent technological advances of single cell next generation sequencing. Considering FLT3 mutated AML in specific it is clear that technological limitations shade the true incidence and prevalence of FLT3 mutations in AML. The characterisation of this gene in AML is routinely performed by polymerase chain reaction amplification, fragment analysis and Sanger sequencing. The conditions of the assay, ranging from the amount of DNA put into the reaction to the *in silico* interpretation of the data ultimately influences the sensitivity of this analysis. The ratio between FLT3 mutated and non-mutated patients in AML it is therefor reflecting the chosen settings of the scientific analysis rather than the "true" naturally occurring ratio. While most articles discussing FLT3 mutated AML states that the mutation occurs in approximately one third of AML patients assessments by more sensitive methods have revealed that the portion of FLT3 mutated patients is substantially larger (Ottone et al. 2013). This ultimately limits the validity of the compartimalisation of subgroups of cancer based on the presence or absence of most molecular traits.

Moving from the use of molecular markers as delineation of important and relevant disease fractions to the use of molecular markers as indicators of disease from a clinical perspective, at least within the frames of precision medicine, the point demarking the transition from a state of subclinical disease to debut of clinical disease may essentially not be the point of greatest interest. While only the last molecular steps in the process of carcinogenesis may ultimately be good markers and predictors for clinically relevant disease, they may conversely be of little use as predictive markers, they may be unfit as efficient therapeutic targets and they may serve as poor markers for therapy surveillance and disease recurrence. This is at least indicated from the experience of the study and treatment of FLT3 mutated AML.

Predicting the onset of cancer is a major focus in precision medicine; with the goal of preventing the development of clinical relevant disease. This in essence means that one is actively searching for occult pathological conditions not yet promoting symptoms in individuals that consider them selves as healthy. The fact that FLT3 mutations are late events in leukemogenesis makes them poor markers for prediction of future disease. It is rather the mutations characteristic of clonal haematopoiesis and pre-leukaemia that are feasible for early detection. Searching, identification and management or premalignant conditions does however come with some major pitfalls. What the story of the biological evolution of FLT3 mutated AML has demonstrated is that the earliest steps of leukemogenesis is only weakly related to the presentation of overt leukaemia in the future. Applying highly sensitive methods mutations defining CHIP, and accordingly also recurrently mutated in AML, has been demonstrated almost to be ubiquitously present in adult individuals (Young et al. 2016). We know that very few of these individuals will ever progress to develop any sort of clinical relevant haematological malignancy. Screening for and treatment of CHIP and pre-leukaemia therefor presupposes a willingness to withstand a substantial increase of total individuals diagnosed and treated for something that if left untouched would never develop into clinically relevant disease.

Moving from prediction to action the experience from targeted treatment, including FLT3 targeted therapy, indicates that therapeutic success, as in achievement of clinically meaningful responses, may necessitate a broader aim than targeting what is understood as the latest acquired molecular drivers of the disease. The entire pool of cells descending from the initial monoclonal origin may possibly be vital parts of the biological entity of the disease, and cancer cure may in biological terms signify that not only the various disease fractions but also the premalignant fractions of the disease must be managed to secure the prevention of disease recurrence. Many have suggested that earlier molecular events may serve as much more attractive markers, not only in prediction and prevention of cancer, but also as therapeutic targets and in surveillance of disease recurrence. Combination treatment or sequential therapeutic regiments are the alternate strategy, but how many targets that need to be considered to achieve therapeutic efficacy in relation to how many targets that can be managed without accumulating unacceptable toxicity is still unanswered questions. Importantly, however, we know from the study of minimal measurable disease in AML that remaining fractions of both clonal haematopoiesis and pre-leukaemia is frequent, and that although associated with variable risk of recurrence it is by far deterministic (Hirsch et al. 2017), bringing us back to the ambiguous nature of pathological conditions that are not causing clinical implications when assessed. Again, increasing therapeutic intensity based on traces of remaining cells characterised by their heritage as descendants of the first monoclonal cell of origin may signify significant overtreatment. A crucial question in the undertaking of predictive oncology – both predicting disease development and disease recurrence, is if the development of cancer causing clinical relevant disease will ever be possible to reliably foresee. Considering the level of randomness that seemingly is involved in the evolution of cancer there is a real possibility that the prediction of cancer forever will be associated with estimating risk rather that certainty. A deciding element in the way we end up managing early stage cancer and evidence of remaining non-disease promoting cells after therapy may therefor reside far beyond the limits of biology, but rather reflect our willingness and ability to manage risk.

In this essay I have attempted to illuminate some of the challenges that arise in the process of transforming the boundaries of the category of cancer in two central dimensions – from groups to individuals and from the macroscopic level to the molecular level. As the potency of our magnifying glass increase, and the shift towards the highest imaginable level of resolution advances the distinct diagnostic boxes are becoming increasingly refined and precise. The consequence seems to be that the validity of what is "normal" from a quantitative and statistical perspective seems to gradually dissolve. The heterogeneity of cancer is no longer limited to distinctions between cancers of the blood compared to cancers of the gut, or to distinctions between cancers driven by specific mutations. The variation is made visible, tangible and relevant down to the level of not only individuals but of single cancer cells. Cancer is, thus, not a static biological being; it is in essence evolution and thereby dynamic in character. The same kinetics, however, seem also to apply for cancer when considering its broader character; as the totality of cancer as a disease, illness and sickness. As a malignant tumor is known to spread throughout the human body cancer has similarly gained a pervasive grip on contemporary western society, where it plays a central role in the structuring of both healthcare politics, healthcare services, and the private healthcare and pharmaceutical industry. There is a multitude of individual stakeholders in this process, not only suffering, but also profiting on cancer, and ultimately where the lines are drawn around the category of cancer has potential pervasive cultural, political and economical consequences. The central question of this essay sounds: when the macroscopic boundaries of cancer the illness and disease dissolve and loose their relevance, where are the new borders to be drawn? The complexity of any relevant answer should probably entail elements not only on where but also how, why, and by whom these decisions are to be made. Independent of the origin of the applied boundaries of cancer however; as a naturally defined pathological process, confined by available technology, drawn as a result of social constructivism and normative choices, or all of the above, the category remains powerful not simply as a creature of nature but through its collective position in society. The expansion, reduction or stratification of the concept therefore has widespread effects outside the realm of biology. Already the meaning of cancer has changed in such a way that it does not always equal the devastating, destructive and mortal human malignancy that we often associate with the word, and the traditionally firm relationship between cancer as a disease, illness and sickness is consequently fading. If the goal of precision medicine is to be achieved – to reduce cancer related suffering, then as the biologically founded disease fraction of the category cancer expand, cancer as an illness and sickness must synchronously decrease. Ultimately the high level of resolution in the study and understanding of cancer challenges not only how we traditionally go about classifying cancer, but even more notably how we consider it as a distinct and definite human malady and health threat. Together these elements indicate that the current transferral towards precision oncology is not only a practical and clinical endeavour but represents both epistemological and ontological challenges that need to be addressed.

References

Arber, D.A., A. Orazi, R. Hasserjian, J. Thiele, M.J. Borowitz, M.M. Le Beau, C.D. Bloomfield, M. Cazzola, and J.W. Vardiman. 2016. The 2016 revision to the World Health Organization classification of myeloid neoplasms and acute leukemia. *Blood* 127 (20): 2391–2405.

Ashley, E.A. 2016. Towards precision medicine. *Nature Reviews Genetics* 17 (9): 507–522.

Baker, S.D., E.I. Zimmerman, Y.D. Wang, S. Orwick, D.S. Zatechka, J. Buaboonnam, G.A. Neale, et al. 2013. Emergence of polyclonal FLT3 tyrosine kinase domain mutations during sequential therapy with sorafenib and sunitinib in FLT3-ITD-positive acute myeloid leukemia. *Clinical Cancer Research* 19 (20): 5758–5768.

Bennett, J.M. 2000. World Health Organization classification of the acute leukemias and myelodysplastic syndrome. *International Journal of Hematology* 72 (2): 131–133.

Bennett, J.M., D. Catovsky, M.T. Daniel, G. Flandrin, D.A. Galton, H.R. Gralnick, and C. Sultan. 1976. Proposals for the classification of the acute leukaemias. French-American-British (FAB) co-operative group. *British Journal of Haematology* 33 (4): 451–458.

Cortes, J.E., S. Khaled, G. Martinelli, A.E. Perl, S. Ganguly, N. Russell, A. Kramer, et al. 2019. Quizartinib versus salvage chemotherapy in relapsed or refractory FLT3-ITD acute myeloid leukaemia (QuANTUM-R): A multicentre, randomised, controlled, open-label, phase 3 trial. *Lancet Oncology* 20 (7): 984–997.

Deininger, M., E. Buchdunger, and B.J. Druker. 2005. The development of imatinib as a therapeutic agent for chronic myeloid leukemia. *Blood* 105 (7): 2640–2653.

Engen, C.B., L. Wergeland, J. Skavland, and B.T. Gjertsen. 2014. Targeted therapy of FLT3 in treatment of AML-current status and future directions. *Journal of Clinical Medicine* 3 (4): 1466–1489.

Genovese, G., A.K. Kahler, R.E. Handsaker, J. Lindberg, S.A. Rose, S.F. Bakhoum, K. Chambert, et al. 2014. Clonal hematopoiesis and blood-cancer risk inferred from blood DNA sequence. *The New England Journal of Medicine* 371 (26): 2477–2487.

Gerlinger, M., A.J. Rowan, S. Horswell, J. Larkin, D. Endesfelder, E. Gronroos, P. Martinez, et al. 2012. Intratumor heterogeneity and branched evolution revealed by multiregion sequencing. *The New England Journal of Medicine* 366 (10): 883–892.

Gibson, W.J., E.A. Hoivik, M.K. Halle, A. Taylor-Weiner, A.D. Cherniack, A. Berg, F. Holst, et al. 2016. The genomic landscape and evolution of endometrial carcinoma progression and abdominopelvic metastasis. *Nature Genetics* 48 (8): 848–855.

Hirsch, P., R. Tang, N. Abermil, P. Flandrin, H. Moatti, F. Favale, L. Suner, et al. 2017. Precision and prognostic value of clone-specific minimal residual disease in acute myeloid leukemia. *Haematologica* 102 (7): 1227–1237.

Johnson, B.E., T. Mazor, C. Hong, M. Barnes, K. Aihara, C.Y. McLean, S.D. Fouse, et al. 2014. Mutational analysis reveals the origin and therapy-driven evolution of recurrent glioma. *Science* 343 (6167): 189–193.

Kampen, K.R. 2012. The discovery and early understanding of leukemia. *Leukemia Research* 36 (1): 6–13.

Lagunas-Rangel, F.A., and V. Chavez-Valencia. 2017. FLT3-ITD and its current role in acute myeloid leukaemia. *Medical Oncology* 34 (6): 114.

Man, C.H., T.K. Fung, C. Ho, H.H. Han, H.C. Chow, A.C. Ma, W.W. Choi, et al. 2012. Sorafenib treatment of FLT3-ITD(+) acute myeloid leukemia: Favorable initial outcome and mechanisms of subsequent nonresponsiveness associated with the emergence of a D835 mutation. *Blood* 119 (22): 5133–5143.

McGuire, S. 2016. World Cancer Report 2014. Geneva, Switzerland: World Health Organization, International Agency for Research on Cancer, WHO Press, 2015. *Advance in Nutrition* 7 (2): 418–419.

Nowell, P.C. 1976. The clonal evolution of tumor cell populations. *Science* 194 (4260): 23–28.

Ottone, T., S. Zaza, M. Divona, S.K. Hasan, S. Lavorgna, S. Laterza, L. Cicconi, et al. 2013. Identification of emerging FLT3 ITD-positive clones during clinical remission and kinetics of disease relapse in acute myeloid leukaemia with mutated nucleophosmin. *British Journal of Haematology* 161 (4): 533–540.

Perl, A.E., G. Martinelli, J.E. Cortes, A. Neubauer, E. Berman, S. Paolini, P. Montesinos, et al. 2019. Gilteritinib or chemotherapy for relapsed or refractory FLT3-mutated AML. *The New England Journal of Medicine* 381 (18): 1728–1740.

Prasad, V., and R.P. Gale. 2016. Precision medicine in acute myeloid leukemia: Hope, hype or both? *Leukemia Research* 48: 73–77.

Prasad, V., T. Fojo, and M. Brada. 2016. Precision oncology: Origins, optimism, and potential. *Lancet Oncology* 17 (2): e81–e86.

Shlush, L.I., S. Zandi, A. Mitchell, W.C. Chen, J.M. Brandwein, V. Gupta, J.A. Kennedy, et al. 2014. Identification of pre-leukaemic haematopoietic stem cells in acute leukaemia. *Nature* 506 (7488): 328–333.

Smith, C.C., Q. Wang, C.S. Chin, S. Salerno, L.E. Damon, M.J. Levis, A.E. Perl, et al. 2012. Validation of ITD mutations in FLT3 as a therapeutic target in human acute myeloid leukaemia. *Nature* 485 (7397): 260–263.

Smith, C.C., A. Paguirigan, G.R. Jeschke, K.C. Lin, E. Massi, T. Tarver, C.-S. Chin, et al. 2017. Heterogeneous resistance to quizartinib in acute myeloid leukemia revealed by single-cell analysis. *Blood* 130 (1): 48–58.

Stone, R.M., S.J. Mandrekar, B.L. Sanford, K. Laumann, S. Geyer, C.D. Bloomfield, C. Thiede, et al. 2017. Midostaurin plus chemotherapy for acute myeloid leukemia with a FLT3 mutation. *The New England Journal of Medicine* 377 (5): 454–464.

Vardiman, J.W., J. Thiele, D.A. Arber, R.D. Brunning, M.J. Borowitz, A. Porwit, N.L. Harris, et al. 2009. The 2008 revision of the World Health Organization (WHO) classification of myeloid neoplasms and acute leukemia: Rationale and important changes. *Blood* 114 (5): 937–951.

Vrinten, C., L.M. McGregor, M. Heinrich, C. von Wagner, J. Waller, J. Wardle, and G.B. Black. 2017. What do people fear about cancer? A systematic review and meta-synthesis of cancer fears in the general population. *Psychooncology* 26 (8): 1070–1079.

Young, A.L., G.A. Challen, B.M. Birmann, and T.E. Druley. 2016. Clonal haematopoiesis harbouring AML-associated mutations is ubiquitous in healthy adults. *Nature Communications* 7: 12484.

Crossing the Styx: If Precision Medicine Were to Become Exact Science

Roger Strand and Dominique Chu

> None the less, he said, he meant to peg away until every peasant on the estate should, as he walked behind the plough, indulge in a regular course of reading Franklin's Notes on Electricity, Virgil's Georgics, or some work on the chemical properties of soil. (Nikolai Gogol: Dead Souls, Part II, Chapter III)

Introduction

This chapter features in a book that discusses the issues at stake and the matters of concern with precision oncology. Our contribution discusses how precision oncology, or rather a particular interpretation of it, is imagined to become an exact science. We furthermore address a matter of concern in that regard, namely the implications such a realization would have for the further development of life science and ultimately, medical practice. In a somewhat dramatic fashion, we liken that matter of concern to the crossing of Styx, which was one of the rivers in Greek mythology that separated our world from Hades. Over the following pages we shall develop our argument for believing so. Essentially, exact science demands tractable scientific problems. In order for the problems to be tractable, they have to concern explicitly and rigorously defined systems. Currently, medical attempts at representing, understanding and intervening direct themselves at wholes and parts of

R. Strand (✉)
Centre for Cancer Biomarkers, Centre for the Study of the Sciences and the Humanities, University of Bergen, Bergen, Norway
e-mail: roger.strand@uib.no

D. Chu
School of Computing, University of Kent, Canterbury, UK
e-mail: D.F.Chu@kent.ac.uk

A. Bremer, R. Strand (eds.), *Precision Oncology and Cancer Biomarkers*, Human Perspectives in Health Sciences and Technology 5,
https://doi.org/10.1007/978-3-030-92612-0_9

organisms that do not satisfy the demands from exact science. So, in order for medicine to become exact science its subject matter has to change.

We emphasize that we address only one particular interpretation of "precision". Since the term "precision medicine" was launched in 2011 as a policy initiative (NRC 2011), it has been used ambiguously to refer to, on the one hand, how individual genetic and molecular information is already included in research and clinical practice, and on the other, a more or less Utopian sociotechnical imaginary of perfect and precise personalisation of medicine (Blasimme 2017, see also Engen, same volume Chap. ''Introduction to the Imaginary of Precision Oncology"). In a similar vein, the Springer/Nature journal *npj Precision Oncology* defines:

> … precision oncology as cancer diagnosis, prognosis, prevention and/or treatment tailored specifically to the individual patient based on the genetic and/or molecular profile of the patient. High-impact articles that entail relevant studies using panomics, molecular, cellular and/or targeted approaches in the cancer research field are considered for publication.[1]

It could be argued that cancers have been diagnosed and treated for decades in this way, taking into account the molecular characteristics of the patients, or using "cellular approaches". The "precision" in this particular definition is in other words implied, either by assuming that the use of certain contemporary and timely laboratory methods by itself qualifies as precise, or by alluding to an ideal of precision that is assumed or realised by the type of science to which one aspires, namely *exact science*. In this chapter, we shall explore the latter alternative: That what above is being called "genetic and/or molecular", is conceived as part of a broader development within life science towards what sometimes is called *systems biology*, characterised by bioinformatic methods, large quantitative data sets, numerical precision and at least the ambition of mathematical rigour. The vision may also include the use of machine learning and artificial intelligence; however, that will not be our main focus.

Mathematical rigour will have to involve models and formal reasoning. Precise and accurate measurement by itself does not make for exact science; Lord Rutherford would have to admit that it would remain a particular form of stamp collecting. The component that is missing in the *npj Precision Oncology* definition above, is that of *computational modelling*. The systems biology imaginary centres around the potential of computational modelling to change fundamentally biology into an exact science, on a par with physics and chemistry. In this way, the argument goes, biology may finally also provide quantitative knowledge that will enable it to predict, control and engineer life.

There is of course disagreement about the plausibility of the vision of an exact biology, both on principled and practical terms. Philosophically, anti-reductionist arguments against the plausibility have had the upper hand, while reductionist imaginaries have prevailed in research policy (for a discussion of that apparent paradox, see Strand 2022). In this chapter, we shall pursue a different question: If computational models indeed come to prevail and become the norm for good life science

[1] https://www.nature.com/npjprecisiononcology/about/aims

research, what would be the consequences? This question appears to have received little attention so far: Anti-reductionist critics have found it irrelevant since they do not believe in the vision anyway, while proponents appear to be convinced that the consequences will be uniformly beneficial.

The authors of this chapter have followed the development of computational methods into life science for more than two decades. In our notes, we wrote 10 years ago:

> There is a sense now that mathematical rigour will eventually be an important part of reasoning in the biosciences. At the moment this is not yet the case. Most articles published in biology journals rely on experimental results and verbal reasoning based on these results. Mechanisms are described using diagrams and descriptions in plain English.

At the time of writing, 2021, this seems still to largely hold true, even for journals such as the mentioned *npj Precision Oncology*. In this respect the biosciences are different from sciences such as physics and chemistry where valid reasoning includes formal reasoning methods, such as mathematical proofs, computational simulations or calculations. In physics, knowledge is typically encoded in systems of equations (differential equations) or other types of formal models.

In what follows we explore two questions. Firstly, we want to analyse the nature of such a transition from verbal to formal models. In a possible future where every biological discovery has to be supported by a corresponding formal model, is it conceivable that formal models would replace more traditional types of biological knowledge, consisting of verbal models supported by experiments? Or would formal and verbal models co-exist? Based on a typology of currently used formal models in the biosciences, what can be said about the scope and purpose of such models? The second question we wish to investigate relates to the implications of the use of new methods. Assuming the use of computational methods in biology, life science and medicine continues, how might that transform the questions that are asked within the science, and equally important, how might it change the perception and use of the resultant knowledge and its accompanying power?

Types of Models

There is a variety of formal models in biology. For what follows it is useful to categorise these. The categories are reasonable in the view of the authors and broadly reflect the types of models that can be found in the wider biological literature. However, the authors admit that they use a broad brush, glossing over many details and that different categorisations could be equally reasonable. That said, our categorisation will not be essential for the main point of this contribution, but merely an orientation to help the reader think about modelling on biology.

The first and oldest class of models are *small scale mathematical models* of biological systems. These are typically sets of differential equations or other explicit mathematical relations. Mathematical modelling in this sense is not a new phenomenon at all in biology. Theoretical biology is an established field, but mostly with limited impact on mainstream biology with few exceptions. Within the theory of evolution mathematical modelling has gained some traction with real biologists. Apart from that, mathematical modelling in biology often clarifies mechanisms in biology. Examples include modelling separate pathways or parts of pathways, uptake dynamics in bacteria, small-scale reaction diffusion systems and the like. The models can be very simple and may relate well known physical effects to biological systems. One of the simplest examples is the Cherry-Adler model (Cherry and Adler 2000) which shows under which conditions two genes that repress one another display oscillatory behaviours. More complex models may investigate the details of regulatory pathways, establish the robustness of signalling pathways, or establish fundamental limits on biological sensing. (McGratch et al. 2017; Govern and Rein ten Wolde 2014; Halasz et al. 2016; Eduati et al. 2020; Adlung et al. 2017)

While small-scale models are typically formulated with mathematical rigour and afford analytic solutions they nearly always depend on radical idealisation and skim over much biological detail. This is perhaps also the reason why they tend to have limited impact within mainstream bioscience. *Computer simulations* are an alternative method of biological modelling. Computational simulations may be based on systems of equations, but could also implement biochemical reaction systems or even entire ecological systems. The key difference between computational and pure mathematical solutions is that the former are numerical in nature, rather than producing analytical solutions. This allows much more complicated models and by the same token, reduces the need for idealisation. As a result, these computational models are much closer to the experiment and can even make numerical predictions of specific systems, that is, provided the data used to build and parametrize the model are of sufficient quality. Examples are simulations of the entire translation system in yeast (Chu et al. 2012, 2014), or simulations of entire brains (Markram et al. 2015).

With modern computers and simulations technologies fairly large systems can be simulated within reasonable time. When it comes to systems consisting of a large number of particles computational cost can still limit a modelling project. It is conceivable that one may overcome many of those problems in the future when hardware becomes even cheaper and modelling technologies advance beyond their current state. There is still another problem of large scale computational models: Parameter values. Once one knows (or believes to know) the structure of a biological system, i.e., which protein interacts with which protein, which genes control one another, what pathways look like and the like, it is quite straightforward to encode it. Yet, determining the quantitative details of interactions is chronically difficult. The relevant empirical data is often hard and expensive to get by. Even if it is available, it is inherently uncertain and often species- and context-specific. Missing parameters are a major challenge for everyone who tries to model biological systems. Again, it is conceivable that in the future it becomes easy to measure these parameters and this problem goes away. Yet, at the moment we are not at this stage and the lack of quantitative information is a major impediment for the development of system wide models.

While such computational models may be quite large-scale in the sense that they represent a large number of interactions, they are focussed on a particular research question. Typically, they will be published together with a particular set of experiments. As such, these computational models are not unlike the mathematical models in that their development requires the ingenuity of the modeller to select relevant features of the system to be modelled and exclude irrelevant ones. This process of selection should not be misunderstood as a necessary evil of modelling. Instead, understanding what is and what is not relevant for a particular purpose is an essential part of specifying the model and the system, and the one that connects scientific values and practice to social values and practice. We shall return to this important point below.

The final type of model we wish to describe are what we call *system wide models* (SWM). "System wide" here usually means that everything known about a particular type of interaction is included in the model. This could mean that the model includes all metabolic reactions, or all genes and how they regulate one another. In this sense, system wide models are repositories of all available knowledge of a particular domain (e.g., the proteome) for a particular organism. They are very different from the computational models that clearly focus on a particular research question in that making these models does not require the ingenuity of the modeller, but could be and regularly *is* automated. Human input is usually still required for sanity checks and overall quality control, yet the main work of the model construction is performed by computer programs that query databases or perform automated literature searches and text mining to establish models.

There are many different approaches to SWMs. One important methodology is Flux-Balance Analyses. These are models that concentrate on an assumed steady-state state of a system. Their all-encompassing scope requires substantial idealisations, including assuming a steady-state and mass action kinetics only. Yet, with those assumptions in place all available information about the metabolism of a cell can be encoded in a (large) model and examined using appropriate software tools.

There are a number of other tools and models within the wider field of computational biology that can reasonably be considered as computational models. Databases, network representations ("omics") and the like are also SWMs, in the sense that they are representations of organisms. Yet, they are not runnable.

A final type of model used in biology are informal or verbal models. Unlike any of the above types, verbal models are to some extent personal to the individual scientists and consist of the particular understanding that researchers have of a process. So, for example, a biologist may understand how translation works. She will then also be able, to some extent, to communicate this understanding to her students using everyday language or diagrams and cartoon models.

In some sense, this informal understanding is the real understanding of a system. Experts have it. On the other hand though, verbal reasoning and understanding is inherently imprecise and susceptible to logical, quantitative and other errors. For example, it is very difficult, even for an expert, to understand the behaviour of even moderately large gene networks by informal reasoning alone.

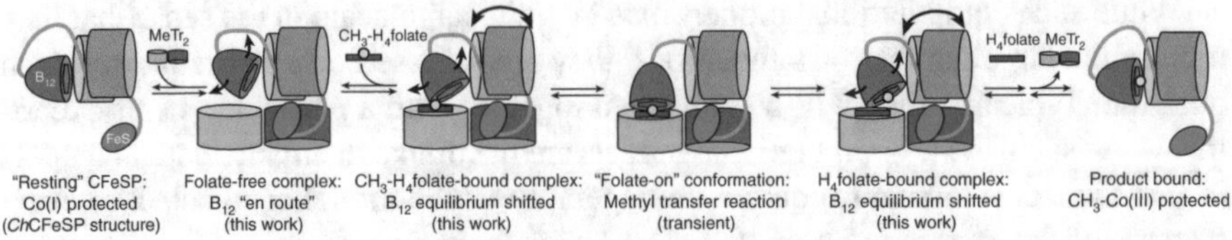

Fig. 1 An example of a cartoon model (Kung et al. 2012). Reprinted by permission from Springer Nature: Nature, © 2012

Kung et al. (2012), call their Figure 4a "cartoon model". It is reproduced below (Fig. 1).

This drawing displays several key features: There are schematic, drawn elements that intend to correspond to a material constituent (typically a biomolecule or complex). Next, there are arrows that can be thought of as chemical reaction arrows and that enable representation of change in the dynamical system. In the case of verbal models, drawings are replaced by names and descriptions, and arrows are replaced by sentences that describe possible events. We will therefore treat verbal models as equivalent with cartoon models.

An informal model such as the one above has indeed a type of entailment structure (denoted by arrows by drawings of different system states in this particular example). However, the informal model cannot be "run" in the sense of being calculated or implemented in a computer. If one looks at Kung's model above, one can deduce by one's own inferential powers that there are some constraints in state space and the passages that can take place within it. In this sense it provides predictions, albeit qualitative and often quite imprecise and probabilistic ones. Such predictions are applied e.g., in medicine, when a candidate drug is chosen for a particular therapy because it is known that the patient could benefit from its effects *if* they occur. Predictions are also useful in the research process itself because they enter the cycle of speculating and hypothesizing about unknown biological mechanisms.

Where cartoon models meet their limits, mathematical or computer models can be used as a virtual laboratory to explore and understand the mechanisms. Not only will the computation be able to track a much larger number of interactions than what a human can do by informal deductive reasoning alone. More importantly, in order to formalise the model into a form that can be runnable by an algorithm, the modeller will request a lot of information that is not included in the informal model, about the kinetics of interactions, allowed concentration ranges for chemical components, etc. The choice to construct a formal model causes a need for more precise information which, together with computational precision, paves the way for improved prediction.

Understanding the Purpose of Formal Models

Unification

Traditionally, models in science are thought of as catering for two main needs: Explanation and prediction. Precisely what explanation means and under which condition one can be satisfied that a model is predictive are difficult topics in their own right. A thorough treatment of those issues goes well beyond the scope of the present contribution. Hence, instead of giving detailed explanations we are satisfied here with highlighting essential aspects connected to models.

In processes in which a scientific field develops into increased quantification, physics is often referred to as an exemplary science. Notably, in physics, explanation and prediction nearly always go hand in hand. A model in this case may be a relation that has been derived from some known physical equations. When adjusted to specific circumstances, it can then predict the result of an experiment. Whether or not this prediction is correct can and usually is tested by designing an experiment that implements the basic assumptions of the model. This is a well-known modus operandi of physics.

Models in physics can be explanatory in that they relate phenomena that are apparently different to one another. Typically, this happens by reducing them to a common underlying theory. So, for example, the theory of electricity and magnetism can be understood by reducing them to the underlying theory of electromagnetism, and in essence to Maxwell's equations. Unification is a powerful and intellectually very satisfying mode of explanation.

Can models in biology fulfil a similar role? As far as prediction is concerned, some computational biologists will certainly recognise the cycle of prediction and experimental corroboration. There is nothing more satisfying for a modeller than to see an experiment reproducing what she predicted will be the case.

The similarity with physics, while apparent at first, is superficial though. The nature of the prediction is of a very different kind in biology than in physics. Experimental physicists tend to design their experiments in relation to an existing theoretical prediction. So, for example, if somebody predicted (credibly) that Higgs bosons exist, then experimentalists will try to find ways to confirm/reject this theory in practice. Note that the prediction of the existence of Higgs bosons is itself an application of physical insights tightly coupled with mathematical reason. In essence, the prediction of Higgs boson has been stipulated as a consequence of existing physical theories. In this sense, the experiment is subservient to the theory.

No such subservience exists in biology. Computational biologists do not normally derive their models from existing biological law-like theories. Instead, they encode existing biological understanding of particular structures and processes into a computer program which they then run. In this sense, computational models are more akin to special purpose reasoning tools, rather than case specific consequences of a general theory.

For example, if a modeller wishes to predict the speed of ribosomes motion over the mRNA, then she needs to first read through the experimental literature to find out how ribosomes move over the transcripts. This will involve an understanding of the degeneracy of the genetic code, of tRNAs and how they bind to the ribosome. Nearly everything that goes into the model will have ultimately been discovered by somebody through a large number of pain-staking experiments. In this sense, in computational biology the theory is subservient to the experiment.

Consequently, the role of models and their relation to experiments are different in biology than in physics. The experiment confirms the usefulness of a model in a specific context rather than corroborating a law-like theory about the world in general. Models can have an explanatory function in biology, but explanations do not derive from unification. They are of a different sort. Accordingly, unification is not a likely scope and purpose of the future biosciences even if formal models come to prevail.

Prediction

Rather than unification, the more frequent vision of a future quantitative biology, in particular in research policy, is that formal models will provide precise predictions of the behaviour of biological systems, including patients who then could benefit from truly precise medicine.

It seems indeed reasonable to expect improved predictive abilities in the biosciences. We shall explain why by highlighting important features of verbal and cartoon models. However, we shall also argue why the ambition of an all-encompassing exact biology might be unfeasible, in short, because there is an inevitable trade-off between precision and the complexity of the biological phenomena to be modelled. A more likely knowledge base for high-precision biological engineering is the type of synthetic biology that is based in relatively simple systems composed by artificial molecular species (e.g., biobricks) designed to have few interactions with native biological compounds. However, biological mechanisms are rarely simple in the sense that there are only a few relevant components. Instead, biological processes are regulated by a large number of elements. Consider the following example as found in Strath et al. (2009) (Fig. 2).

Here, the model elements include biomolecules, biological processes and cellular compartments, while arrows denote chemical reactions, physical movements and regulatory signals. We are not presented with an array of instances from state space as in Fig. 1. Instead, the figure invites questions such as “What would happen if the extracellular level of TGFβ increases?” and one could in principle try to “run” the model by human thinking, following the arrows in one’s mind. In practice, however, this procedure would not yield robust and precise results. The number and nonlinearity of interactions and the lack of quantitative information about their magnitudes and kinetics, render the answer indeterminate. This model provides very little precise information about the dynamics of the system that it intends to model; rather, it is an inventory of material and functional constituents and their interactions.

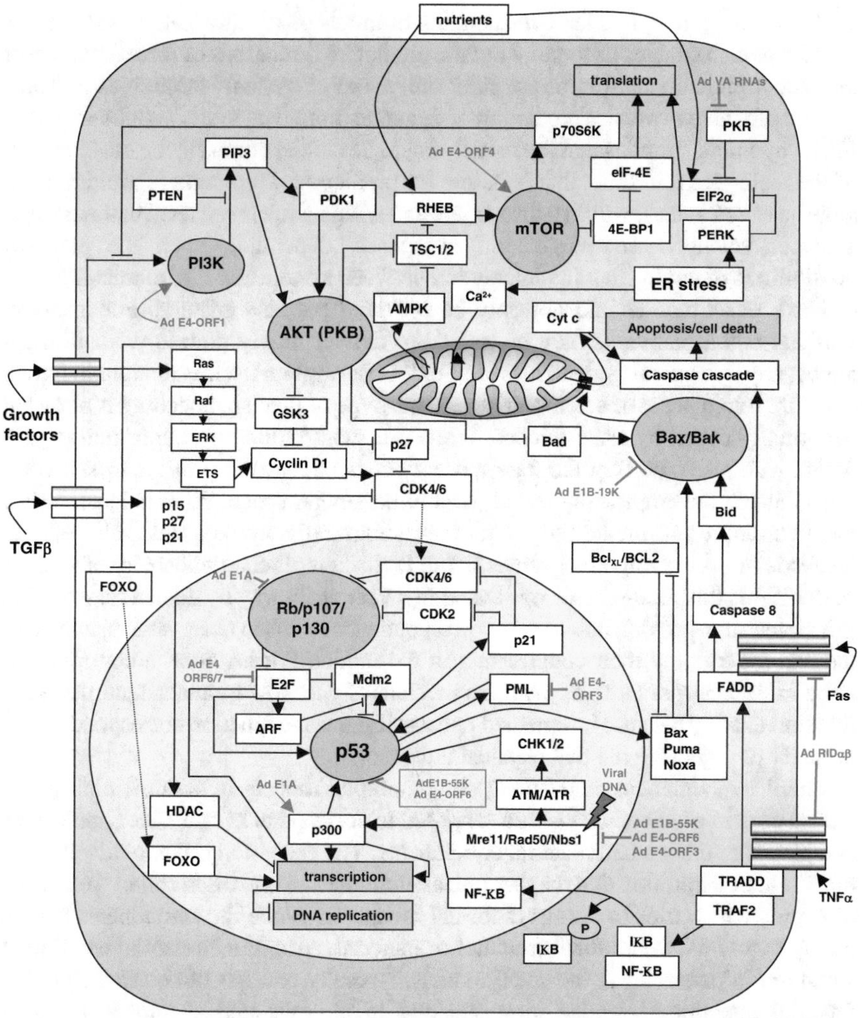

Fig. 2 An example of a complicated cartoon model (Strath et al. 2009). Reprinted under Creative Commons license CC-BY 2.0

Where cartoon models meet their limits, mathematical or computer models can be used as a virtual laboratory to explore and understand the mechanisms. Not only will the computation be able to track of a much larger number of interactions than what a human can do by informal deductive reasoning alone. More importantly, in order to formalise the model into a form that can be runnable by an algorithm, the modeller will request a lot of information that is not included in the informal model, about the kinetics of interactions, allowed concentration ranges for chemical components, etc. The choice to construct a formal model causes a need for more precise information which, together with computational precision, paves the way for improved prediction.

At the same time, it is the consideration of precision of information and computation that also enables us to point out the predictive limitations of formal models of biological phenomena. Let us consider the vision of systems biology as an exact science that allows precise prediction of dynamic biological systems (Kitano 2002). First, "dynamic" implies that the model should be able to represent the development of the real system in time, that is, allow for prediction. Predictions should in principle be exact, at least within a time frame and within a tolerance level that is acceptable for scientific and engineering purposes, depending upon the practical possibilities of controlling the interaction between system and environment.

Next, exactitude should not only be provided but also ensured. For instance, artificial neural networks may be trained to deliver highly precise predictions of almost any dynamical system; however, unless the system itself is thought to have a structure reminiscent of a neural network, this type of precise modelling is not what is normally meant by exact science. While one would trust predictions that remain within well-tested parts of the state space, there is no reason a priori to trust predictions outside the empirically tested parts of parameter space. What is lacking, is a justified claim that model and system are structurally similar, and that model is "realistic" in some important sense. In the logical empiricist philosophy of science of the 1960s this requirement was stated in a very strict way by demanding "bridge principles" that provide one-to-one correspondence between observable elements of the real system and their counterparts in the model. For instance, a controversial issue in the interpretation of quantum mechanics was how to understand the wave function exactly because it remained unclear if the wave function corresponded to a property of objects in the real physical universe.

Small mathematical models and computational models in systems biology do indeed consist of elements that are supposed to correspond to molecules, molecular complexes or other small material components. The exception is the relatively marginal research tradition that the theoretical biologist Robert Rosen called "relational biology", which tries to construct formal models in which the components can be purely functional rather than structural or material. Also, the inferential entailment structure, in particular in the small models, typically consists of deterministic differential equations intended to correspond to physical and chemical processes involving and governing the state functions of the material constituents. Equations may represent chemical reactions, transport, diffusion, enzymatic catalytic activity, etc. Such models may in principle be exact and provide precise predictions.

In practice, however, the precision is challenged on three fronts. First, the model will typically include equations with parameters, the values of which must be estimated either by experiment or, if this is not possible, also in part by reverse engineering approaches. Secondly, it is impossible to model a biological phenomenon without simplifying it. Already a single cell includes too many chemical and physical interactions and too much spatial detail for a model, and only some of these can be included. A higher organism includes a high number of tissues and a very high number of cells, none exactly identical; it interacts with a changing environment; and through reproduction it takes part in the evolutionary process. Very little of this, if anything, can be included in a computational model that aims to faithfully represent chemical and physical interactions, and even less in an analytical model. In practice, there is a trade-off between biological relevance and what can be achieved by a reasonable modelling effort. Thirdly, while it is certainly likely that computer power will continue to increase and that future computational models may be larger, this does not by itself necessarily increase precision. Large models introduce computational complexity. A model may be expanded so that it demands twice the computational expense to run it; however, a proper sensitivity analysis or at least an exploration of the model's behaviour across its parameter space may demand much more than a doubling of computational expense. We believe that many modelling practitioners may verify the practical difficulties of tuning models into a biologically relevant behaviour: even with the "correct equations", the everyday experience of the modeller is that most of the time the model runs produce noise and useless results. In practice, a larger model does not necessarily mean less noise; the opposite may be the case.

Robust predictions are therefore only to be expected in cases where there is good strategy for how to idealize and simplify the system into a model, either by drawing clear boundaries around the physical system, or by including only specific phenomena inside it, or, as in the case of flux balance analysis, state clear assumptions about the processes to be studied (in that case, model metabolic processes in terms of steady-state kinetics of perfectly available chemicals in solution). Predictive success will consequently depend on how reasonable these assumptions are in the specific case, that is, on the knowledge that the system already is approximately simple in a way that corresponds to the simplified model. Another way of saying this is that the model may yield robust predictions if we have a robust understanding of the biological system and know what questions we reasonably can ask about it. This is not in any way unique to biology. For instance, in the aftermath of the financial crisis, economists have had to explain that their models could not predict the system breakdowns that actually occurred because the possibility was not included in the design.

System Wide Models: Descriptive Models and Repositories of Information

It may be objected that already contemporary systems biology indeed counts with a number of what we above call SWMs, that is, large, system-wide models (SWMs), and that unlike the models discussed in the previous section, they do not need the ingenuity of the modeller to the same extent as simulation models do. While SWMs normally are computerized, they often are not executable. Instead, these models are mostly collections of data that is somehow collected, either by systematic search of the literature, a large number of individual submissions by experimentalists, or by a specific large scale experiments. Such SWMs are often collections of components, functional annotations of components (e.g., "this gene is involved in metabolism"), interaction tables of molecules, expression data, quantitative data (e.g., reaction rates) and the like. There are now numerous databases, most of which are accessible to the wider community, which store SWMs; one well known example is the KEGG database of pathways. (KEGG: Kyoto Encyclopedia of Genes and Genomes).

SWMs are frequently automatically created, although sometimes curated as well. Compared to nearly any other type of model they are less coupled to a specific context or purpose. Still, in order to be of use in practical science, the information contained within SWMs must somehow be related to a question. There are many conceivable ways in which this could happen. One way is that the data are processed, for example by statistical analysis. A next step to this is to use the raw information in SWMs to generate new knowledge using the help of statistical or machine learning/AI methods. For example, raw sequencing data could be used to generate to find promotor sites or introns. The result of those exercises is most often a derivative SWM.

SWMs lack many attributes that are usually associated with models. They are not built with respect to a theoretical framework, they are not unifying, and they are not predictive. As such, they are more akin a stamp collection than a theory. Nevertheless, SWMs allow scientists to reason about systems, for example by systematically comparing the differences between diseased tissue and healthy tissue or by extracting communalities between taxonomies of species. SWMs are accordingly models proper in the sense of being tools to help reasoning.

The existence of such models has led to a new type of "hypothesis-free" biological research which is mostly concerned with finding new ways to administer the new wealth of data, to find more efficient ways to cross-link available information in useful ways (Kuperstein et al. 2015).

A type of SWMs that has gained particular importance in systems biology are *Flux-Balance-Analyses* (FBA) models. FBA models, which may be an intermediate case between computational models and SWMs, can be "run," in the sense that one can set up optimisation algorithms to find particular solutions that are consistent with given constraints. Yet, the model itself is not dynamic, but consists of a description of the topology of a metabolic network plus constraints. Together with an assumption of optimality FBA models can then be used to compute fluxes through the network. This can then be used in order to infer missing components or to analyse differences in the biochemistry of various cell types.

FBA models are SWMs in that they can be constructed semi-automatically from information stored on databases and can map entire metabolic networks of organisms. All those usages of SWMs do not generate direct understanding of particular systems, but rather more information, sometimes only aggregate views of detailed data that was already present. The important conclusion to draw from this, is that SWMs may play an important role in future biosciences, but that they are not end products in themselves. They have to be used by other modelling practices to produce prediction and understanding, be they verbal or formal mathematical or computational models.

Explanation

Arguably, as of now the most important function of biological models is that of explanation. Explanation in biology frequently means uncovering "mechanisms." So, for example, transcription factors and their binding dynamics to the operator site together with an understanding of the action of RNA polymerase can explain how the activity of one gene can be regulated by that of another genes.

It is acceptable within biological science to describe such mechanisms qualitatively. That is, a number of experimental results together with a coherent verbal story is sufficient to satisfy editors in prominent journals that the relevant mechanism is interesting and can explain some biological phenomenon. Hence, verbal models are explanatory as long as it is supplemented by experimental evidence. So, for example, one could show that Gene A ceases to be regulated when Gene B is deleted. Furthermore, one could provide specific assays that demonstrate that within a certain area upstream of Gene A mutations can cancel the regulatory action of Gene B. Finally, one could directly demonstrate that the product of Gene B binds with high affinity to the sequence motif found upstream of Gene A.

This sort of evidence is mostly qualitative, elucidating the basic structure of the biological system. It provides little information about how strong the regulation is beyond very general qualifiers (i.e. the regulation is "very strong" or is "weak" etc...). For small systems such qualitative and semi-quantitative descriptions can be sufficient to create a good understanding of the system. However, they could be insufficient if the system is of moderate complexity. Non-linearities, for example, make it chronically difficult to understand how a system behaves, even if it consists of a few components only. Therefore, as long as they are used by themselves, verbal models are limited in their explanatory powers in biology.

Combining verbal models with computational reasoning can enhance explanations, in that it can add quantitative detail to mechanisms. Rather than saying that Gene A is regulated by Gene B, a quantitative model could add some understanding of how fast the regulation is and how the overall function of the regulation is achieved by this regulation. An example of this is the case of methylation in the case of the regulation of the fim switch in E.coli as described in (Chu et al. 2008). In this article the authors describe how the metabolism of sialic acid is turned on upon take-up of this nutrient. However, the known mechanistic model of the pathway activation conflicted with separate known information about the toxicity of sialic acid. Upon closer inspection of available experimental data, there appeared to be a contradiction. However, using a dynamical computational model, the authors could show that the apparent contradiction could be resolved if the different timescales of activation of the pathways are taken into account. This reasoning depends crucially on the separation of timescales in the regulatory dynamics. This sort of effect is impossible to describe by verbal reasoning. However, it should be noted that once the case is made formally by simulation, the formal argument can then be reincorporated into a verbal model and described using plain language. It is not necessary to re-run the formal model each time one talks about the system.

The quantitative understanding of the system can often be transformed into a suitable verbal model plus a reference to some formal model that demonstrates the claimed effect. In fact, if the formal model is to be explanatory at all, then it *has to be* translated into a verbal model in order to be useful. Computations by themselves only yields numbers. These numbers can be used for the purpose of prediction but do not convey understanding. Only when meaning is attached to them by interpretation, and they are related to a network of knowledge can they lead to understanding.

A Possible Tension Between the Requirements of Formal Models and the Need for Conceptual Flexibility and Ambiguity in Discovery

In the logical empiricist philosophy of science of the mid-twentieth century, the distinction between the context of discovery and the context of justification was devised in order to clarify the epistemic status of philosophy of science itself. Processes of discovery were considered to be informal, creative and a proper object of psychological research. The task of justification of scientific knowledge, on the other hand, was seen to be one of rational reconstruction, that is, the application of logic to demonstrate the valid relationship between scientific knowledge and its objects. Later developments in the philosophy of science have shown that the distinction between the context of discovery and the context of justification in itself is a simplification. If we accept the distinction as a first approximation, however, we can note how modelling in science serves two functions that at first sight appear quite different.

Often, the so-called modelling relation (Rosen 1985) is taken to describe the relationship between model and system (Fig. 3).

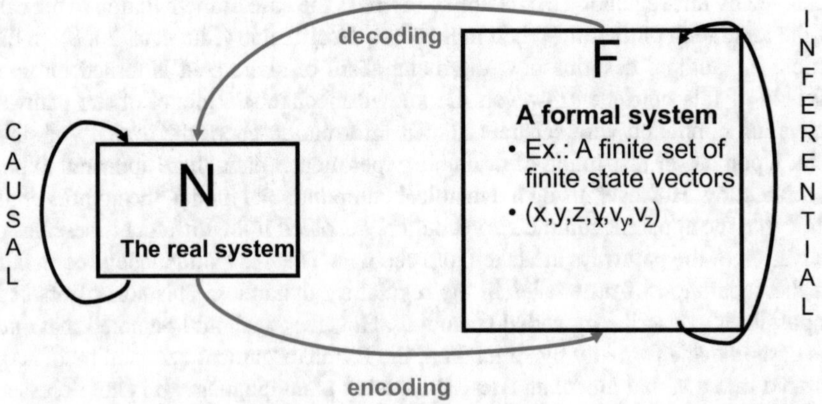

Fig. 3 Rosen's modelling relation

This can be understood as a schema for rational reconstruction, that is for pursuing the question of the validity of the scientific model. For instance, one could argue that the model is valid if F and N are isomorphic and the encoding and decoding mappings are isomorphisms. This would indeed amount to the strictest possible logical formulation of the vision of exact science. On the other hand, in science as practice – in the context of discovery, as it were – there is usually no external position from which one can directly inspect the properties of F and N and compare their structure. Rather, the objective or research is to describe and understand a partially unknown natural phenomenon and the model F is a tool in the pursuit of that objective. With the philosopher Immanuel Kant we can say that there is no way of directly knowing the thing in itself. Our cognition takes place with the help of and through our cognitive apparatus, in which concepts and models play a main role. What we know about the real natural phenomenon is what is hypothesised by our best (formal or informal) model F. At an early stage in discovery in the biosciences as currently practiced, the best model will almost invariably be an informal, verbal one.

The argument has been made that in the biosciences, it can be helpful for discovery if one's best model – that is, one's preconceived ideas of the system – is indeed tentative, flexible and ambiguous. This is due to the nature of living systems. While many scientific experiments have the simple objective of measuring properties of known entities, molecular biologists, biochemists and microbiologists routinely do experiments with more unknowns. For instance, one may suspect the existence of a certain biological activity, signal or pathway, and tries out a number of experimental systems in order to see if the suspected phenomenon can be observed in a stable and reproducible manner. Furthermore, one would typically like to ascribe the function to a specific biomolecule or complex, but the phenomenon might not exist in a purified solution and might only be observed in an intact or close-to-intact biological structure as a living cell (Strand et al. 1996).

Rheinberger (1997) described the process of discovery in the life sciences as a dynamic interplay between modifying the experimental system and modifying one's description of it. In the critical phases of discovery, the interpretations of the experiments might change on a daily basis. Concepts and even research questions may be changed, refined and rejected on the path towards a stable and interesting signal from the experimental system. If the process is successful, gradually an "epistemic thing" emerges, which is neither a consolidated phenomenon nor a clear concept yet. As the experiments become more reproducible and stable, the phenomenon is at some point said to exist, and its identity is given by its epistemic counterpart, that is, the model as of the time of consolidation. During such processes, conceptual flexibility and ambiguity seem to be an advantage; one could speculate if it is a prerequisite. If so, that would make a case for the usefulness of informal models as tools for discovery.

What we do know from the history of science, is that verbal models and cartoon models have played an important role in biological discovery. This does not preclude the possibility that computational models could serve as tools for discovery. We would like to speculate, however, that they would be different tools. For instance, the use of computational models in discovery would easily encourage questions about specific interactions, the value of kinetic parameters, etc. In this way, they might influence the life sciences to adopt a style of experimentation that is more similar to physics, chemistry and macroscopic biology. Whether this is a good approach depends on the body of available knowledge of a given biological system. If one is convinced that the inventory of constituents largely is known, it may be a sound approach to focus on quantification of their properties. If not, the general inclusion of computational modelling may actually slow down discovery, in particular if there is a division of labour where the experimentalists struggle to understand the content of the model and the modellers have little practical knowledge of biology. Cartoon models/verbal models facilitate thinking of the type "What is the function of this biomolecule/signal/pathway?" Sometimes this approach is too simplistic; however, it remains unclear what kind of biology one would get if such questions get dismissed altogether as too fuzzy and informal.

The Modelling Process: The Challenges of Radical Openness and Contextuality

Above we described the challenges to predictive precision that are encountered in modelling practice. We shall now turn to the possible implications of the responses that these challenges foster in terms of how the biosciences may develop.

Let us return to Fig. 3. The choice of a model (or a model design) implies a positive statement about the natural phenomenon under inquiry, a statement that ideally can be verified, corrected or refuted by experiment and observation. However, insofar as the natural phenomenon has not been completely identified (and accordingly being under inquiry), the choice of model (or model design) also acts to frame the phenomenon, that is, to delimit the research object. Certain elements and aspects of the real world will fit into the frame provided by the model, while others remain invisible, not measured or otherwise outside the scope of the model. While this fact is generally appreciated with respect to the particular choice of elements, we wish to draw attention to the most general implication of modelling, which is that the natural phenomenon will be framed in terms of a natural system (Chu et al. 2003). Chu (2011) has described how this constraint translates into two practical challenges that may occur in the modelling process: radical openness and contextuality.

Radical openness is a feature of the phenomenon under inquiry that is observed in the inability to successfully delimit the model/system. In order to improve the predictive power of the model, the model may be expanded by including elements from the environment that strongly interact with system; but this new definition lead to the identification of other strong interactions with the environment, and so on.

Contextuality is similar to radical openness, only that the problem of delimiting the model/system resides inside it, in the indefinite richness in the number and nature of properties of model elements and interactions between them. The model may seem predictive and exact but suddenly fails because it did not take into account properties and interactions that had not been considered or measured before.

Both challenges result from a richness in properties and interactions in the natural world. Exact science has three strategies to meet these challenges:

1. Exact science tries to avoid or minimise radical openness by searching for parts of the natural world that either appear relatively isolated from the environment ("looking for Nature's seams") or that can be isolated in the laboratory (or produced by technology).
2. Physics and chemistry try to avoid contextuality by constructing complete physicalist models of all properties of the elementary material constituents and developing a unified theory of all forces that act upon them. This amounts to a reductionist programme and it explains the importance of the assumption of the perfect identity of elementary particles in the same quantum state, and the unacceptability of "hidden variables".
3. Finally, economics as a non-physicalist exact science tries to identify independent layers of law-like behaviour (of rational agents and market transactions) that are robust against contextuality.

Biology in its full scope studies a lot of different phenomena: structures such as biomolecules, cells, organisms, species and ecosystems and processes such as metabolism, reproduction, animal behaviour, and morphogenetic and phylogenetic development, to mention but a few. If emphasis is put on prediction and precision, the biosciences will have to employ all three of the strategies mentioned above. Already contemporary life science on the molecular and cellular/sub-cellular level is oriented towards (informal) models, and it focuses on "systems-like" biological phenomena: structures such as organisms, their spatial compartments and material constituents, and processes that involve material constituents. Framing phenomena of life as systems produces a bias towards constancy rather than change, and similarity rather than variation. We expect this bias to become stronger if computational models become the norm also because prediction will become a more central value and hence radical openness can be less tolerated. We would expect computational systems biology to consolidate the emphasis on material structure and single organisms in contemporary life science. Paradoxically, the strengthening of computational models and systems biology, which often is presented as a non-reductionist programme, may indeed in this sense lead to less appreciation of biological complexity.

A Drive Towards Computational Models

The emergence of journals such as PLoS Computational Biology and research initiatives such as the Centre for Digital Life Norway reflects a shift in the biological paradigm. We suspect that an important trigger for this shift is the development of high-throughput experimental methods. Storing the results of modern experimental techniques (i.e., microarray, ChiP-seq, etc..) requires sophisticated computational solutions. Even more so, intricate algorithms are required to understand the meaning of those. These new experimental techniques have made the use of computational tools indispensable and in the slipstream of this a new interest in modelling has evolved.

We believe that computational modelling is not merely riding a bandwagon of computational methods establishing themselves in unrelated areas. Instead, computational modelling becomes necessary also in traditional biosciences in order to manipulate, interpret and understand data and results. The results of this research usually lead to descriptive models but not by themselves to detailed understanding of mechanisms. For example, a sequenced genome only describes the DNA, but does not allow the user to understand the functional significance of the sequence. One use further computational processing to predict, for example, transcription factor binding sites and promotor sites. Yet, even with the best algorithms this only leads to a list of candidate locations with a given functional relevance. This is interesting knowledge that is routinely produced and deposited in databases. Yet, by itself it does not produce actual biological knowledge. Apart from the uncertainty that is attached to this data, a list of binding sites says very little about what is actually going on in the organism. The entries in the databases only become actually useful when they feed into the work of the biological investigator who confirms predicted knowledge experimentally and weaves the facts into a coherent mechanistic understanding of a concrete system.

High throughput biology is a just one symptom of a rapid method development that has led to an astonishing ability to manipulate and measure biosystems. These methods by themselves force the biomedical researcher into more complex verbal models that combine quantitative and qualitative information, often including nonlinearities. Very quickly, verbal reasoning is not powerful enough anymore to integrate this information. This is only exacerbated by the large amount of information available from SWMs resulting in a deluge of information that needs to be integrated. In order to have an understanding of biological systems at the level of detail that is implied by the available data it is necessary to use computational reasoning as an enhancement to verbal argument.

We suspect that this trend will intensify, and with it, the need for computational processing. At the same time, it must be remembered that computational models do not per se provide understanding, but are reasoning tools to aid the intuitive understanding of biological systems. By themselves these models cannot provide any understanding, but they need to be related to verbal models and transformed into verbal models, at least by our present concept of what it means to understand something.

In summary: As long as biomedical research concentrated on a few genes at a time and their local effects, there was no need to outsource reasoning to a machine. Progress in biotechnology led to a refinement of measurement techniques. This in turn allowed high throughput technologies which necessitates the use of computers to administer and analyse the data deluge.

Crossing the Styx

Science, technology and society are co-produced in entangled processes that include sociotechnical imagination. In our case, the development of high-throughput laboratory methods, the increased focus on computational modelling and the emergence of the sociotechnical imaginary of precision medicine are three processes that are causally entangled and of course also form part a larger causal complex that includes the political economy of medical practice and research.

We have now reviewed certain features of types of models and modelling practices. If precision medicine were to become an exact science, it would imply that computational modelling would take a prominent if not wholly dominant place. It remains to discuss what the implications of such a development might have on medical technology and practice.

We find it useful to reflect on how technology normally works. The usual engineering solution to the problem of radical openness and contextuality referred to above is to construct simple and easy to predict systems rather than applying models to highly complicated systems. While many natural systems display nonlinearities and in general behaviour that is difficult to model, mechanical systems may have linear behaviour that can be modelled with precision exactly because they are designed to. For instance, railways are designed to have small and predictable friction between rails and train wheels. Well-designed mechanical systems can be predicted and controlled with extreme precision not because the universe is governed by simple and linear laws, but because nonlinear behaviour is deliberately excluded and prevented by skilful design of the system. This is how it is possible to send successfully spacecrafts to other planets or to develop and distribute vaccines (Latour).

Medical science and technology have ample examples of highly simplifying strategies, ranging from the "cut, burn and poison" of cancer medicine to lobotomy, electroshock therapy and various psychopharmaceuticals that aim to reduce suffering by reducing brain complexity. However, because the underlying body of knowledge is not exact, these technologies are more likely to fail, especially if one's purpose is to restore and protect biological complexity. If one is satisfied with killing the patient, these technologies can all be applied without failure.

The imaginary of personalized and precision medicine contains the purpose to maintain or restore the subtle and delicate homeostasis of human health in the presence of multicausal networks that drive the individual towards illness and possibly death. It wants to achieve this by tailoring the treatment, that is, finding the right drug and dose to the right patient at the right time. This ideal is not new; it is the heritage from patient-centred clinical practice which by means of consultation and communication with the patient sought to tailor the doctor's intervention. However, patient-centred practice, in all its imprecision, builds on hermeneutic knowledge of the single individual; what used to be called idiographic rather than nomothetic knowledge in philosophy of science, or simply experiential as opposed to evidence-based knowledge. The design problem in precision medicine is that it wants to achieve tailoring (which is sometimes possible but fallibly so by means of inexact science) by relying on exact science, which does not translate into technologies that tailor solutions around natural complexity. Exact science translates into technology that changes the system so that it keeps within the boundaries and parameter space of the model. The ambiguities in the use of the concept of precision medicine, and notably precision oncology, bear witness of this design problem. Certain diagnostic and therapeutic practices are already called precision oncology; it is just that the therapy outcomes are not precise.

Science is one of the most powerful institutions in modern society. It offers not only the knowledge base for the development of technology but also a large part of the knowledge with which modern human beings understand the world and themselves. It provides not only facts and explanations, but also indirectly guides us in the choice of questions and perspectives.

A transition from cartoon- and verbal model-based life science to a precision medicine based on systems biology and mathematical computational models will undoubtedly lead to new and improved knowledge of innumerous biological phenomena. At the same time, we have argued that it may direct the research focus even more towards controllable and predictable phenomena in relatively closed systems with regular behaviour. This is because such systems provide tractable problems for computational models. The resulting body of biological knowledge may reinforce modern human beings' understanding of life as essentially predictable, understandable and controllable, and which therefore is provides a suitable substrate for industrial and economic exploitation. In this sense, precision medicine and systems biology present themselves as a business case for the bioeconomy. Still, as of today, precision medicine is inexact and largely retains a concept of and an interest in biological understanding close to its inexact past. A next logical step, however, could be the gradual dominance of computational modelling, which would imply an even stronger shift towards instrumentality, reductionism and the view that life is predictable and controllable.

We have called this paper "Crossing the Styx". The shift to a biology dominated by computational models might be likened to crossing a river from which there is little possibility of return. In Greek mythology, Styx was one of the rivers that separated the World of the Living from Hades, the World of Death. Still, curiously, the souls continued a kind of life in Hades; but it was a different life. Often it was assumed to be an inferior life – quite the opposite of the optimistic visions and not the least all the sales talk that surrounds precision medicine, for which also practising scientists are responsible. We leave the evaluative aspect with the reader, well aware that the metaphor in itself may seem provocative. Still, we see two senses in which one could follow a quite different and broader debate than the one pursued in this paper.

The first is to what extent precision medicine could become an exact science by redefining its purpose and thereby solving the problems of radical openness and contextuality. We already noted how this problem was solved by military research that focuses on destroying life rather maintaining health. Killing people was successfully translated into true engineering problems. A less radical alternative is to make the criteria of medical success as simple as possible, for instance in terms of standardized clinical outcomes such as survival or progression-free survival, perhaps under hospitalized and strongly medicated conditions. For instance, a precision oncology that merely focuses on short-term delay of death due to cancer, has a better chance of successful translation into precise clinical practice than if quality of life is considered as part of the problem and accordingly part of the system. A step in the direction towards this rather extreme scenario would be the scientific dominance of computational modelling, SWMs and the disappearance of classic informal and verbal models and hence human understanding as a scientific product. It would all be data, models and clinical outcomes.

Some would argue from a cultural, perhaps humanistic point of view, that such a development of medical science and practice is undesirable and should be avoided; hence implying that it can be avoided. Still, it would have to be admitted that it could be seen as just a next step of what the philosopher Jürgen Habermas called the colonization of the lifeworld by technology, and other scholars have called medicalization processes throughout the twentieth century and into the twenty-first. If one takes the metaphor of Styx, one might be tempted to ask if human civilisation in this respect is itself becoming senescent and replacing its human faculties by formal reasoning and machines. True to the theoretical concepts of co-production and sociotechnical imagination, however, one could argue that there is no necessity in this development. Future science can become different. For instance, cancer medicine can become more tailored by resisting the Utopian ideal of exact science and rather combining high-throughput methods and other biomedical developments with the patient-centred focus of the art of medicine, staying with living, as it were.

References

Adlung, L., S. Kar, M.-C. Wagner, B. She, S. Chakraborty, J. Bao, S. Lattermann, et al. 2017. Protein abundance of AKT and ERK pathway components governs cell type-specific regulation of proliferation. *Molecular Systems Biology* 13 (1): 1–25.

Blasimme, A. 2017. Health research meets big data: The science and politics of precision medicine. In *Cancer Biomarkers: Ethics, Economics and Society*, ed. A. Blanchard and R. Strand, 95–109. Bergen: Megaloceros Press.

Cherry, J.L., and F.R. Adler. 2000. How to make a biological switch. *Journal of Theoretical Biology* 203 (2): 117–133.

Chu, D. 2011. Complexity: against systems. *Theory in Biosciences* 130: 229–245. https://doi.org/10.1007/s12064-011-0121-4

Chu, D., R. Strand, and R. Fjelland. 2003. Theories of complexity. *Complexity* 8 (3): 19–30.

Chu, D., J. Roobol, and I.C. Blomfield. 2008. A theoretical interpretation of the transient sialic acid toxicity of a *nanR* mutant of *Escherichia coli*. *Journal of Molecular Biology* 375 (3): 875–889.

Chu, D., N. Zabet, and T. von der Haar. 2012. A novel and versatile computational tool to model translation. *Bioinformatics* 28 (2): 292–293.

Chu, D., E. Kazana, N. Bellanger, T. Singh, M.F. Tuite, and T. von der Haar. 2014. Translation elongation can control translation initiation on eukaryotic mRNAs. *The EMBO Journal* 33 (1): 21–34.

Eduati, F., P. Jaaks, J. Wappler, T. Cramer, C.A. Merten, M.J. Garnett, and J. Saez-Rodriguez. 2020. Patient-specific logic models of signaling pathways from screenings on cancer biopsies to prioritize personalized combination therapies. *Molecular Systems Biology* 16 (2): 1–13.

Govern, C.C., and P. Rein ten Wolde. 2014. Optimal resource allocation in cellular sensing systems. *Proceedings of the National Academy of Sciences of the United States of America* 111 (49): 17486–17491.

Halasz, M., B.N. Kholodenko, W. Kolch, and T. Santra. 2016. Integrating network reconstruction with mechanistic modeling to predict cancer therapies. *Science Signaling* 9 (455): 1–16.

KEGG: Kyoto Encyclopedia of Genes and Genomes. https://www.genome.jp/kegg/.

Kitano, H. 2002. Computational systems biology. *Nature* 420: 206–210.

Kung, Y., N. Ando, T. Doukov, L.C. Blasiak, G. Bender, J. Seravalli, S.W. Ragsdale, and C.L. Drennan. 2012. Visualizing molecular juggling within a B12-dependent methyltransferase complex. *Nature* 484: 265–269.

Kuperstein, I., E. Bonnet, H.-A. Nguyen, D. Cohen, E. Viara, L. Grieco, S. Fourquet, et al. 2015. Atlas of Cancer Signalling Network: A systems biology resource for integrative analysis of cancer data with Google Maps. *Oncogenesis* 4: e160.

Markram, H., E. Muller, S. Ramaswamy, M.W. Reimann, M. Abdellah, C.A. Sanchez, A. Ailamaki, et al. 2015. Reconstruction and simulation of neocortical microcircuitry. *Cell* 163 (2): 456–492.

McGratch, T., N.S. Jones, P. Rein ten Wolde, and T.E. Oulridge. 2017. Biochemical machines for the interconversion of mutual information and work. *Physical Review Letters* 118 (2): 028101.

National Research Council (US). 2011. *Toward Precision Medicine: Building a Knowledge Network for Biomedical Research and a New Taxonomy of Disease*. Washington (DC): National Academies Press.

Rheinberger, H.J. 1997. *Toward a History of Epistemic Things: Synthesizing Proteins in the Test Tube.* Stanford: Stanford University Press.

Rosen, R. 1985. *Anticipatory Systems: Philosophical, Mathematical, and Methodological Foundations*. New York: Springer.

Strand, R. 2022. The impact of fantasy. In *Personalized Medicine in the Making Philosophical Perspectives from Biology to Healthcare*, ed. C. Beneduce and M. Bertolaso. Springer International Publishing.

Strand, R., R. Fjelland, and T. Flatmark. 1996. In vivo interpretation of in vitro effect studies with a detailed analysis of the method of in vitro transcription in isolated cell nuclei. *Acta Biotheoretica* 44 (1): 1–21.

Strath, J., L.J. Georgopoulos, P. Kellam, and G.E. Blair. 2009. Identification of genes differentially expressed as result of adenovirus type 5- and adenovirus type 12-transformation. *BMC Genomics* 10: 67.

Publication Bias in Precision Oncology and Cancer Biomarker Research; Challenges and Possible Implications

Maria Lie Lotsberg and Stacey Ann D'mello Peters

Introduction

Multiple sources of bias are found in health research. Two of the most discussed biases include selection bias and confirmation bias. Selection bias happens when participants or other types of research material such as biobank material, cell lines or mouse strains, are not randomly selected for a study. The best way to select for people or animals in a research study is through randomisation where everyone within the group that are investigated are selected randomly to attend the study, however, this is not always possible (Medical Research Council n.d.). Second, confirmation bias appears when researchers, intentionally or unintentionally, look for information or patterns in their data that confirm the ideas or opinions that they already have (Medical Research Council n.d.). Other types of biases which are especially relevant for clinical trial studies or observational studies include (i) channelling bias, where patients within the study are not randomly selected for a given study subgroup (Lobo et al. 2006), for example if prognostic biomarkers or degree of illness affects which cohort the patients are placed in; (ii) performance bias, for example in clinical trials involving surgery which can have technical variability between surgeons; (iii) interviewer bias, which refers to a systematic difference between how information is gained or interpreted in the different groups, for example if the interviewer ask different questions or formulate them differently when interviewing the different groups; (iv) recall bias, where respondents or patients

M. L. Lotsberg (✉)
Centre for Cancer Biomarkers CCBIO, Department of Biomedicine, University of Bergen, Bergen, Norway
e-mail: Maria.Lie@uib.no

S. A. D'mello Peters
Department of Biomedicine, University of Bergen, Bergen, Norway
e-mail: Stacey.Dmello@uib.no

A. Bremer, R. Strand (eds.), *Precision Oncology and Cancer Biomarkers*, Human Perspectives in Health Sciences and Technology 5,
https://doi.org/10.1007/978-3-030-92612-0_10

need to remember what has happened in the (more or less distant) past; (v) observation bias, where participants alter their behaviour in the knowledge that they are studied; (vi) chronology bias, when historic controls are used as a comparison group for patients undergoing a therapeutic intervention; and (vii) transfer bias, where patients drop off from the study, and it has to be considered whether these patients are fundamentally different from those who remained in the study (Pannucci and Wilkins 2010). Bias also happens after a study is completed – we have for instance publication bias, which will be the focus of this chapter, meaning that the published results are not a representative selection of all results within a study, and not all studies are published causing unfavourable results to be less reported.

Meaningful research outcomes have been defined as findings that advance their respective field of research and have a practically useful effect on society (Helmer et al. 2020), but this requires research to be shared with peers, decision makers and citizens. The main format of research dissemination is through publications in peer reviewed scientific journals. However, other formats of information sharing also play an important role, like oral or poster presentations, newspaper articles, books, web pages, research archives, informal discussions, and all kinds of forums where research results and related information are being communicated. Research results that are not shared with anyone will not be of any value to anyone other than the researchers who performed the study. This is why disseminating and communicating research results, and more generally, opening up research to other's scrutiny, is an integral part of the role of a researcher.

Publication bias occurs in scientific research if the outcome of a study influences whether or not the results will be published in a scientific journal, presented at conferences or otherwise distributed and made available for society as a whole (Song et al. 2010). Publication bias could thus be defined as "the selective publishing of research based on the nature and direction of the findings" (Marks-Anglin and Chen 2020). When studies with significant or favourable results are more likely to be published than those with non-significant, unexpected or unfavourable findings, it skews the balance in the pool of available research results, thus causing a bias in favour of so-called 'positive' results (Song et al. 2010; Marks-Anglin and Chen 2020). Factors that determine the selection of results include experimental outcome and how the results sit in light of the original hypothesis and previously published work. For example, an experimental result is often considered 'positive' when a difference that is statistically significant is observed. In other cases, 'representative' results may be considered in the light of the original hypothesis and selected in a manner that excludes contradicting outcomes, best suits the hypothesis and fits into the logical flow of the paper, in order to maximise the probability that the results are accepted for publication in peer reviewed journals.

This chapter aims to explore publication bias in the context of precision oncology and cancer biomarker research; why it exists, implications it has for researchers, patients, and society, as well as reflecting on the deeper roots of the problem. Section "Evidence of publication bias in medical research" provides evidence of publication bias in medical research in general, and how this applies to cancer biomarker research specifically. Section "The impact of publication bias on the validity of the scientific literature and contribution to the reproducibility crisis" explains the different types of publication bias based on whether or not the statistical hypothesis is true or false and how this has an impact on the validity of the scientific literature, and the contribution of publication bias to the reproducibility crisis. Section "Discussion: Publication bias in precision oncology and cancer biomarker research; implications and reflections on the deeper roots of the problem" discusses possible implications of these biases for patients, researchers and the scientific society and the general public, and offers reflections on how to minimise the occurrence of publication bias.

Evidence of Publication Bias in Medical Research

The term 'publication bias' started appearing sporadically in the literature in the 1980s but the number of publications have increased remarkably over the years, as illustrated in Fig. 1 (Marks-Anglin and Chen 2020; Simes 1986; Easterbrook 1987; Boisen 1979; Begg 1985). However, although the term publication bias did not appear in the literature until 1979, the concept itself was discussed much earlier (Marks-Anglin and Chen 2020; Editors 1909). In 1959, statistician Theodore Sterling and colleagues presented evidence that published results are not representative of all scientific studies (Sterling 1959; Sterling et al. 1995). Sterling found that as much as 97% of the papers published in some of the major journals in the field of psychology had statistically significant findings for their major scientific hypothesis, highly indicative of publication bias in the field (Sterling 1959; Sterling et al. 1995).

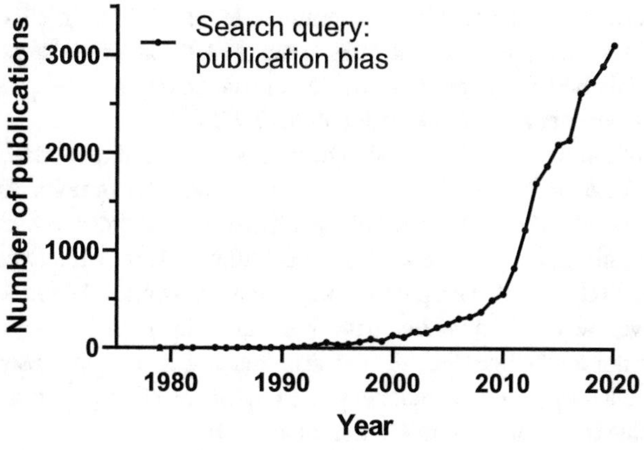

Fig. 1 Number of publications available at PubMed (https://pubmed.ncbi.nlm.nih.gov/) for the search query «publication bias» between 1979 and 2020. (The figure was created by the authors in GraphPad Prism v8.3.0)

In a retrospective study published by Easterbrook and colleagues in *The Lancet* in 1991, the authors followed 487 research projects and found that studies with statistically significant results were more likely to be published than studies that were statistically nonsignificant (Easterbrook et al. 1991). In addition, research projects with significant results also led to a greater number of publications and presentations, and the results were published in journals with higher impact factors (Easterbrook et al. 1991). It has also been found a greater tendency towards publication bias in observational or laboratory experimental studies compared to studies of randomised clinical trials (Easterbrook et al. 1991). Easterbrook and colleagues further claimed that "the most serious potential consequences of publication bias would be an overestimate of treatment effects or risk-factor associations in published work, leading to inappropriate decisions about patient management or health policy" (Easterbrook et al. 1991).

Multiple studies have later investigated publication bias in the scientific literature in a systematic way (Dickersin and Min 1993; Franco et al. 2014; Driessen et al. 2015; Vera-Badillo et al. 2016; Scherer et al. 2018). These studies have looked at projects receiving ethical approvals, external funding, reports to authorities or conference abstracts, and studied the correlations between the amount and type of scientific publications and whether or not the study gained positive or significant results (Marks-Anglin and Chen 2020). It was clear from many of these studies that publication bias indeed occurs and positive results are more likely to be published than negative results (Dickersin and Min 1993; Driessen et al. 2015; Vera-Badillo et al. 2016; Scherer et al. 2018). Reports have also shown that as many as 50% of studies may not be published in any given area of research and that it is more than twice as likely that null results will not be published or communicated (Shields 2000). In addition, these studies also demonstrated that other types of publication related biases exist including time-lag bias where favourable results are published within shorter time (Ioannidis 1998; Shields 2000), citation bias meaning that favourable results are more cited (Nieminen et al. 2007; Shields 2000), and sponsorship-bias in the way that studies sponsored by industrial funding are less likely to be published compared to government funded research (Marks-Anglin and Chen 2020; Scherer et al. 2018; Lexchin et al. 2003).

Evidence of publication bias has also been reported for clinical trial publications (Simes 1986; Vera-Badillo et al. 2016; Bardy 1998). As an example, Simes and colleagues reported in 1986 that when only published results from clinical trials were considered, combinational chemotherapeutic regimes were statistically preferable compared to single agent therapy in ovarian cancer (Simes 1986; Sterling et al. 1995). However, when all registered trials were included in their analysis, the statistically significant advantage disappeared. Another important bias observed in clinical trials in oncology is the under-reported toxicity which is essential for the approval of new treatments (Vera-Badillo et al. 2016).

Publication Bias in Precision Oncology and Cancer Biomarker Research

Cancer therapy has greatly developed over the years, and as new and more targeted therapies become available, cancer therapy is moving from standard treatment regimens to a more personalised and tailored therapy, also referred to as precision oncology. However, despite the development of multiple different molecularly targeted therapies, most patients with advanced cancer will not experience durable clinical response from targeted therapies (Marquart et al. 2018). As cancer therapy becomes more personalised, there is therefore a constant need for novel predictive biomarkers to guide tailored therapy based on the patients and the tumours' unique characteristics, as the treatment is no longer solely based on the tumour type. The BEST (Biomarkers, EndpointS and other Tools) glossary defines a biomarker as "a characteristic that is measured as an indicator of normal biological processes, pathogenic processes, or responses to an exposure or intervention, including therapeutic interventions" (FDA-NIH Biomarker Working Group 2016). In the field of oncology, biomarkers have multiple applications including diagnosis and subtyping of cancer (diagnostic biomarker), and they can also be used to estimate prognosis (prognostic biomarker), predict treatment effect (predictive biomarker), or to monitor the treatment effect or cancer recurrence over time by longitudinal sampling (FDA-NIH Biomarker Working Group 2016; Wu and Qu 2015).

Cancer biomarker research ranges from experimental studies to clinical applications and involves various types of studies including cell culture and animal models, research on humans or human material (including databases and clinical trial studies) or even computational modelling. Publication bias in precision oncology can occur at any stage of the process ranging from the early discovery to the clinical validation of new biomarkers. Within a complex biological system as tumours are, there is a high degree of both intra-tumour and inter-tumour heterogeneity which also changes over time and affects drug responses. Cancer biomarkers include a wide range of molecules, including DNA, mRNA, enzymes, metabolites, transcription factors, and cell surface receptors (Wu and Qu 2015), and many of these are continuous variables (e.g. protein expression) or exists only in a certain proportion of the cells or tissues analysed (e.g. frequency of DNA mutations). Since biomarker definitions are often based on measurements of a continuous variable, such as protein expression or proportion of biomarker positive cells, there is therefore not a clear cut-off between a biomarker positive and biomarker negative sample. In addition, the analysed material is typically only taken from a small part of the tumour and this subsampling might not be representative for the whole primary tumour and potential metastases that could have a very different biology than the primary tumour it is derived from. Biomarkers are frequently used to guide therapy, and even the therapeutic outcome for the patient, typically measured as responders or non-responders in accordance with a biomarker, does not have a clear cut-off as some patients can have a partial response.

The story of HER2 in breast cancer is often referred to as an example of a successful biomarker story. Multiple copies of the gene encoding for the HER2 receptor causes cancer cells to be more responsive to growth signals making the tumour more aggressive, and HER2 gene amplification is associated with worse prognosis than HER2 negative breast cancer when kept untreated (Lakhtakia and Burney 2015). However, research during the 1990s led to development of the monoclonal antibody trastuzumab (brand name: Herceptin) which specifically targets the HER2 receptor and significantly improves the outcome of HER2 positive breast cancer patients (Slamon et al. 2001; Lakhtakia and Burney 2015). This biomarker could thus be considered as "ideal" in the sense that it has entered clinical use as a biomarker that effectively identifies a subgroup of breast cancer patients that are more likely to benefit from HER2 targeted treatment, thus contributing to a more tailored and effective breast cancer treatment, which also saves the biomarker negative subgroup from potential side effects and toxicity from a treatment that is less likely to have an effect. However, although HER2 is frequently used as a textbook example of the "ideal" cancer biomarker, it is still not perfect in determining treatment response and some patients develop therapy resistance. Thus, it could be argued that it is more accurate to talk about biomarkers as "good enough" rather then "ideal", leaving more room for accepting the complexity and uncertainty of biological systems that biomarkers are based upon (Blanchard and Wik 2017). It has also been detected HER2 positive metastases in patients with HER2 negative primary breast cancer and vice versa (Xiao et al. 2011; Ulaner et al. 2016), which illustrates the complexity of the tumour biology and highlights some of the challenges with applying a cancer biomarker in clinical practise.

The example of the HER2 biomarker illustrates that even for the most promising biomarkers, the outcome of a biomarker test is not absolute in predicting patient response to a therapy, it can only place the tumour or the patient in a group that has statistically higher or lower chances of some degree of response (Fleck 2017). When biomarkers are included in clinical practice and thus accepted as valid predictors of biological or clinical outcome, this creates a 'skew' in the available literature caused by publication bias, and it will have an impact on the set threshold for biomarker positive or negative samples or when deciding which biomarker defined subgroups of patients that will receive a given therapy. It is also likely that the more complex a field of research is, the more it will be influenced by publication bias. Indeed, the outcome from studies of complex, uncertain, and non-linear systems will have more variations, and therefore more room for subjective selection of results prior to publication. Publication bias will then in turn create an illusion from the literature that the biology behind the findings is less complex and more certain than it actually is (Blanchard 2016). In a field such as precision oncology where the researchers are aiming for perfection, this could also increase the risk of publication bias as there might be less room for publication of negative results.

The Impact of Publication Bias on the Validity of the Scientific Literature and Contribution to the Reproducibility Crisis

The fact that many scientific studies are difficult or even impossible to replicate or reproduce has become so evident that the term 'reproducibility crisis' is used to describe this phenomenon (Miyakawa 2020; Twa 2019). A major contributor to this crisis is believed to be publication bias caused by the fact that statistically insignificant results are rarely published or discussed in scientific publications (Marks-Anglin and Chen 2020). In an online survey performed by *Nature* answered by more than 1500 participating scientists, 70% of the researchers answered that they had tried and failed to reproduce others' experiments (Baker 2016). When the researchers were asked what led to these problems of reproducibility, more than 60% mentioned the strong pressure to publish, or 'publish or perish' culture, and selective reporting of results (Baker 2016).

In principle, there are two different scenarios of publication bias based on statistical significance depending on whether the statistical hypothesis is true or not: the *false hypothesis bias* and the *true hypothesis bias* (Sterling et al. 1995). Typically, the statistical null hypothesis (H_0) is defined as 'no differences between the groups'. To illustrate these two types of biases, we define the two groups as two different biomarker based subgroups named X and Y. The statistical null hypothesis will then be defined as no differences in treatment effect between the groups, meaning that the treatment effect is similar in group X and Y. A study is then performed to see if there is evidence to disprove H_0, and if it is, the study concludes by rejecting H_0 and accepting that there are differences between the groups (often referred to as accepting H_1). Importantly, the statistical null hypothesis should not be confused with the scientific hypothesis, which typically will be to disprove the statistical null hypothesis. In our example, the statistical null hypothesis is defined as no differences between group X and Y, while the scientific hypothesis will be that there are in fact differences in the treatment effect between the two biomarker defined subgroups.

In all statistical testing, the null hypothesis can either be wrongly rejected (Type I error) or the test can fail to reject a false null hypothesis (Type II error). The experiment or study performed is defined as significant based on a significance level *alpha*, often set to 0.05 (5%). Given that H_0 actually is true and the significance level is set to 0.05, there will then be a 5% chance that the null hypothesis will be (wrongly) rejected, also referred to as a type I error. This will be the case every time this same experiment is performed, and if either the same researchers perform this experiment many times or the same experiment is repeated by different investigators (that might not even be aware of each other's research), the chances that a type I error will occur by chance in at least one of the performed studies will accumulate over time and number of experiments/studies performed. Then, if only the one or the few studies that showed statistically significant results are published, while the majority of the studies that showed insignificant results are ignored in the sense that they are not published or otherwise made available, this will lead to a situation where the published results are not at all representative for all the experiments that have been performed. Multiple repetitions of the same experiment will thus accumulate the chance of a wrongly rejected true statistical null hypothesis, also referred to as *the true hypothesis bias*. In addition, simply increasing the number of replicates, applying another type of statistical test, increasing the statistical power by comparing only selected subgroups, or removing so-called 'outliers', may be the difference between a significant and non-significant result, and are other examples of how publication bias can skew the statistics in favour of increasing the chances of type I errors.

The second scenario is that the statistical null hypothesis is false. In our example this means that there actually is a difference between group X and Y. In this case, publication bias will cause a bias against elimination of type II errors, meaning a bias in favour of eliminating false negative results, referred to as the *false hypothesis bias*. Although this will cause a bias in the direction of the correct conclusion, it will still have implications since it will skew the results presented in the literature that will make it look like the differences are bigger, or at least more significantly different, than they actually are. This could have implications for example when the benefit and side-effects or toxicity of a treatment are considered against each other or when evaluating the validity of a biomarker. Therefore, no matter if the statistical null hypothesis is true or false, publication bias will make the probabilities of statistical type I and type II errors different for the reader than for the initial researchers that performed the study (Sterling et al. 1995), and it will "skew" the available literature by increasing the chances of Type I errors and decreasing the chances of Type II errors.

Publication bias also serves as a threat to the validity of meta-analysis (Marks-Anglin and Chen 2020). Meta-analysis is a method that combines the results from multiple similar studies and aims to make it possible to draw conclusions with a higher degree of certainty. Meta-analyses are frequently used in oncology, for example when evaluating how good a new treatment regime is compared to standard treatment, or it can also be used to evaluate the validity of a cancer biomarker. Meta-analyses are based on the assumption that the meta-study summarises all relevant studies, or at least a representative selection (Sterling 1959). However, publication bias will have an impact on the conclusion of a meta-analysis if it only includes published results. It should also be mentioned in this setting that other types of biases including citation bias and time-lag bias could skew the results of meta-analysis (Marks-Anglin and Chen 2020). However, although the meta-analysis could have a wrong conclusion based on a biased 'selection' of only published data, meta-analyses still tend to be trustworthy and are especially convincing since they cover multiple studies. However, this discussion on publication bias invites us to handle those meta-analyses with a critical eye.

Discussion: Publication Bias in Precision Oncology and Cancer Biomarker Research; Implications and Reflections on the Deeper Roots of the Problem

Implications for Patients, Decision-Making in Clinical Practice and Socio-economic Aspects

In the case of medical research in general and also precision oncology and cancer biomarker research specifically, publication bias will have many possible implications that eventually affect the patients, for instance through consequences for medical practice and evidence-based medicine (Marks-Anglin and Chen 2020). More specifically, policy and decision-making processes rely on the scientific literature, and publication bias can therefore result in inappropriate decisions about health policy and patient management (Marks-Anglin and Chen 2020; Easterbrook et al. 1991). Worst case scenario, publication bias in the field of oncology can cause inappropriate estimation of the balance between treatment effects and toxicity, resulting in inappropriate treatment of cancer patients. Ideally, a cancer biomarker should be reliable, cost-effective and powerful in detecting and monitoring cancer risk, cancer detection and tumour classification so that improved medical decisions can be made and the patients will receive the most appropriate therapy (Wu and Qu 2015; Blanchard and Wik 2017). Biomarkers are thus important for subtyping patients into groups, for example when a new treatment regime is considered for use in clinical practice. Publication bias in the field has therefore a direct impact on this decision-making.

There is not always an obvious cut-off for biomarkers, both in relation to what is a positive or negative sample and whether the defined subgroup will benefit from a given therapy. Publication bias will in this setting skew the literature which could affect where these cut-offs are set and further which patients that are given the therapy. One such example of a biomarker where there is no obvious cut-off or implementation of the biomarker is the use of the protein expression of programmed cell death ligand 1 (PD-L1) to predict response to immune checkpoint inhibitors (ICB). Although PD-L1 expression is established as a biomarker to predict response to ICB, its clinical utility as a biomarker remains to be further investigated. Clinical trials are not consistent in their conclusions of weather PD-L1 predicts response to ICB, and the biomarker defined cut-offs are varying as much as from >1% to >50% of PD-L1 positive cells necessary to define the tumour as PD-L1 positive (Yi et al. 2018). When the results are less clear like in the example of PD-L1, it is also likely that publication bias (if it exists) could have greater implications than if the results are clearer, as only a small bias in the literature then can make a big difference when decisions about patient treatment are made.

A cut-off can either be selected prior to the study based on previous knowledge or experience or by applying a statistical method to the data to estimate new cut-off values (Woo and Kim 2020). When applying a statistical method there are two different approaches either based on the biomarker distribution itself or a selection can be made based on the association between biomarker and outcome (Woo and Kim 2020). A popular method to predict biomarker based patient outcome is to select a cut-off value that minimises the p-value when comparing the outcome in different groups (Woo and Kim 2020; Polley and Dignam 2021). However, this strategy of minimising the p-value results in highly unstable p-values and increases the chances of significant findings when the biomarker is not associated with outcome (Polley and Dignam 2021). It can thus be argued that this method directly causes publication bias or misinterpretation of results since the cut-off is selected in a way that corresponds to the most statistical significant difference in the data set and the significance of the chosen cut-off value therefore tends to be overestimated causing an increased rate of false positive errors. Although methods have been developed to reduce this effect of false positives, the lack of reproducible biomarker cut-offs is still a challenge that might have hindered the adaptation of biomarkers into clinical practice (Polley and Dignam 2021).

A type of bias that is related to publication bias is the overinterpretation of results, in the sense that the meaningfulness of the result can be embellished with overly optimistic terms (Fong and Wilhite 2017) or the speculation for its application in the clinic may be exaggerated to maximise acceptance for publication. In a systematic review of ovarian cancer biomarkers, Ghannad and colleagues found that interpretation bias is abundant in evaluation of cancer biomarker studies and that it is a practice of making study findings appear more favourable than what could be justified from the results (Ghannad et al. 2019). The authors further claim that this misinterpretation or overinterpretation may lead to an unbalanced and unjustified optimism in the performance of potential biomarkers, and the published literature might suggest stronger evidence than what is justified. The most frequent misinterpretations found in their study include claiming other purposes of the biomarker that were not investigated, mismatch between the aim and the conclusion and incorrect presentation of the results (Ghannad et al. 2019). In particular, the most frequent mismatch in the results was the selective reporting of the most positive or statistically significant results in the abstract. This illustrates again that the 'ideal' of precision oncology, aiming for perfect biomarkers to support perfect clinical decision-making and highly tailored treatments to individual patients, puts a high pressure on researchers to put forward positive results that support this ideal. We can see how difficult it is, then, to totally avoid the practice of publication bias: the ideal of precision oncology demands 'perfect' results and biomarkers, but since the biology around cancer biomarkers is so complex, these results can only be achieved through a biased analysis, interpretation, and presentation of results.

Publication bias in precision oncology and cancer biomarker research also has broader socio-economic aspects. It is known that, in addition to the devastating effects that cancer has on patients and their families, the economic consequences of cancer are enormous (Wu and Qu 2015). Cancer-related economic costs include the direct health care resources and the cost of expensive cancer therapies, and it also includes loss of human capital due to early mortality or inability to work because of the disease (Wu and Qu 2015). When a new drug or treatment regime is evaluated in a subgroup of patients based upon a set of biomarkers, the health benefit are evaluated against the cost, and publication bias could skew this equation in favour of increased benefit of a treatment, which potentially could lead to approval of a treatment that otherwise would not have been approved for the given clinical application, thereby affecting health care resource allocations. A lot of resources are used in biomarker research, and despite that, the harsh reality is that less than 1% of published biomarkers end up entering clinical practice (Kern 2012). There are many possible explanations for this including the complexity of malignant tumours, but publication bias is also one out of many contributors to the fact that only few of the potential biomarkers end up reaching a clinical application. For example, some studies might find that a particular protein X is indicative of response to a treatment in their study while other studies may not find that this trend exists in their cohort. If the former studies are more likely to be accepted for publication while the latter will not even be considered submitted, this could cause a biased availability in the literature of the evidence for using X as a biomarker. This will in turn have an impact on designing new research projects evaluating biomarker X, and publication bias in the follow up studies, will further escalate the problem, and is likely to affect the number of biomarkers that in the end will end up in the clinic. Biomarker research is often funded by the government or funding agencies, and as a general rule the goal of all government funded research should be to benefit the community. If research results are not published nor otherwise made available, the knowledge gained from this research cannot be used for the benefit of society and it could therefore be argued that these resources could be better spent somewhere else. It could also be argued that it is unethical to perform research without publishing it, both in respect to the funding agencies, but also in respect to the participants including patients or volunteers who contribute to the research material, and also with respect to the society in general since tax money is used to fund government funded biomarker research.

Implications for Researchers and the Scientific Community

As seen in section "The impact of publication bias on the validity of the scientific literature and contribution to the reproducibility crisis", publication bias also has important implications for the researchers and the validity of the scientific literature. Justifying the design of new research projects relies heavily on previously published studies and literature, and when these are not representative, researchers run the risk of basing further studies on false premises. To obtain positive results is especially important for PhD candidates (and other early-stage researchers) that are early in their career. They have a limited time to do their research, but they are faced with the pressure of 'publish or perish' and are expected to have publications preferably in high impact journals, so that they can contribute with new and valuable knowledge to their field in order to graduate. Publication bias leading to the reproducibility crisis causes a situation where new PhD projects could be based on a skewed or even wrong literature, which is increasing the chance for the candidates to have difficulties publishing their findings – soon enough, they might realise that their project is actually a dead end. In addition, supervisors might encourage PhD candidates to test hypotheses that are considered as dead ends from the beginning, and cases has also been discussed at forums (including https://academia.stackexchange.com) where multiple candidates within a research group are set to do very similar projects, and whoever finishes the task first will get their name on the publication (Lowe 2019). Approaches like this further increase the rate of publication bias in the field, and the pressure to publish is likely to be one of the reasons for the mental health challenges in science. In a study by Levecque and colleagues, an increased prevalence of mental health problems for PhD candidates was observed, compared to the highly educated general population, and a third of PhD candidates in the study was at risk of a psychiatric disorder (Levecque et al. 2017; Pain 2017). Encouraging research into dead ends also causes a waste of time, money and research effort since multiple researchers perform the same studies potentially without being aware of each other's null results. Statistically insignificant results are therefore significant in their own right because they provide valuable information to scientists designing new studies which will ultimately save researchers time and resources that could be more efficiently spent.

Negative results that are either insignificant or disprove the original scientific hypothesis could be at least as important as the positive results. For example, one of the now world's best-selling breast cancer drugs, Tamoxifen, first synthesized in 1962 as a contraceptive pill in the pharmaceutical laboratories of ICI (now part of AstraZeneca) was not patented because it stimulated, rather than suppressed, ovulation. The project was nearly stopped but was reportedly saved partly because team leader, Arthur Walpole, threatened to resign, and pressed on with a project to develop tamoxifen for the treatment of breast cancer. It was initially used as a palliative treatment for advanced breast cancer but later became a best-selling medicine in the 1980s, when clinical trials showed that it was also useful as an adjuvant to surgery and chemotherapy in the early stages of the disease and even later, trials showed that it could prevent occurrence or re-occurrence of disease in at high-risk individuals. Tamoxifen therefore became the first preventive for any cancer, helping to establish the broader principles of chemoprevention, and further extending the market similar drugs (Quirke 2017).

A primary goal of research is to test hypotheses, but the researchers are in no control of whether this process will lead to a 'positive' finding. If you are unlucky and end up with only null results in your project, or you are not able to replicate the 'common knowledge' in your field (which might be wrongly represented in the literature because of publication bias), then we know that these results are more difficult to publish. Difficulties by publishing contradictive findings could thus result in the literature only supporting a certain hypothesis or established scientific opinion in the field although there are a lot of unpublished data supporting the opposite hypothesis (Prinz et al. 2011). This in turn will have consequences for the researcher's career as scientists are generally judged and ranked by their number of publications, impact factors and citations, when applying for an academic position or project funding. The lack of control over experimental outcomes generates unbiased results, and so it seems ironic that this core aspect of research that is beyond the researcher's control plays such an influential role in whether they have a 'successful' future. It is also ironic that this dilemma exists in the field of science where logic and fairness are the pillars of its foundation. Hence, it is inevitable that these factors are likely to become major influencers driving motivation, overshadowing consideration of patient/societal benefit. Researchers may also become mentally and physically stressed because their careers and livelihood can be dependent on the attainment of positive results.

Reflecting on the Roots of the Problem of Publication Bias

In order to reflect on ways to avoid, as much as possible, the problem of publication bias, we first need to discuss what is causing the problem. There are at least two possible explanations of why we have publication bias: (i) researchers might decide not to submit 'negative' results because they are in a system where negative results might jeopardise their career or their opportunity for future funding; (ii) the journals are more likely to reject a manuscript where the results are 'negative' because the ideal of precision medicine does not leave much space for negative results. The reality is probably a combination of the two. Not all studies performed are even prepared as manuscripts to be submitted to a scientific journal, and not all results from a particular study are included in the final manuscript. Further, the submitted manuscript could be rejected in the peer-review process. It is likely that publication bias occurs in all these steps, and for every step, the likelihood of proceeding to the next step of this process is higher for positive results.

The publishing process is highly competitive, and to publish in a high-quality journal you are required to have good quality data, but is it enough to have high quality research or do you also have to have the 'right' results? For example, one of the criteria of publishing in the journal *Nature* is that the papers "are of outstanding scientific importance" [https://www.nature.com/nature/for-authors/editorial-criteria-and-processes]. More proof is also generally needed to go against the established knowledge than to publish something that already has great support in the literature. Therefore, misleading knowledge or false positive results could remain 'common knowledge', especially in fields such as cancer research, that are based on a highly complex and heterogeneous biology.

Authorship bias involves misattribution in publications and could indirectly be related to publication bias. It is not unusual to add individuals who contribute nothing to the research effort research papers or grant proposals. In some cases, editors pressurize authors to add citations that are not relevant to their work. Adding highly recognized author names to manuscripts has become a common practice. Junior academics are more likely to add individuals in positions of authority or mentors to papers. A study showed that 60% added an individual because they thought the added scholar's reputation increased their chances of a positive review (Fong and

Wilhite 2017). This type of bias will thus have a lot of the same implications as publication bias if adding a name to the paper increases the chances of making it through the peer-review process and adds to the number of citations after publication, since the paper is not considered solely by the quality of the research but also by the names on the author list.

In some cases, researchers might feel that they are forced to biasedly select their data in order to get the data published. This could for example be selection of results that fits a logical flow of events, or they can select only successful replicates of an experiment, redo the statistics, include only selected subgroups in the analyses to get significant results and so on (Fig. 2). Other types of selection bias introduced by the researcher might be selecting only the findings that are statistically significant or fit the hypothesis. A more crucially dangerous selection is to intentionally exclude replicates without any logical reason other than to present the replicates of the

Fig. 2 Example of how the mindset of researchers can contribute to publication bias. (The figure was created by the authors with BioRender.com)

experiment that was expected or 'successful'. It is not practical to publish all research that has ever been performed, and therefore some degree of selection is required. Many journals also have strict word count limitations, leaving no room for all results in the manuscript, and the scientist is therefore forced to prioritise the most 'important' results. It is therefore not evident exactly where to draw the line between what is an acceptable selection of data or if certain types of data selection could be considered as data falsification and thereby fall under the definition of scientific misconduct, and there have even been some historical cases reported when researchers have gone so far as to intentionally fabricate their results (Else 2019; Stebbing and Sanders 2018; Müller et al. 2014; "Beautification and fraud" 2006; Fanelli 2009). Pressure to increase the number of publications coupled with the increased difficulty of publishing, can motivate academics to violate research norms even though majority of academics disapprove of this, others suggest that it is just the way the game is played. Examples include falsifying data, falsifying results, opportunistically interpreting statistics, and fake peer-review. A study reported that 1.97% academics admit to falsifying data, although this is likely understated (Fong and Wilhite 2017).

Shields reported in 2000 that one of the most typical factors influencing publication bias is investigators who do not submit their research for publication due to a lack of enthusiasm and the consequential drive to publish only the statistically significant studies, or the educated assumption that null outcomes are given low publication priority. He speculates whether the publication of null studies is more commonly driven by junior investigators who must publish to become known, or busier senior investigators who are less intrigued by null findings. Most importantly, he concludes that also journals contribute to publication bias when they refuse to publish null studies (Shields 2000).

In order to reduce publication bias, some journals like Cancer Epidemiology, Biomarkers and Prevention have begun to publish null results in specified formats where the articles are brief enough to encourage researchers to submit their findings but also sufficiently robust to ensure that the strengths and limitations of the study are discussed in light of other studies in the field (Shields 2000). One attempt to reduce publication bias is to have separate journals that specialise in publishing null results that only base their peer-review process on the quality of the research and have no requirements for the outcome of the study. Another alternative could be to have requirements for scientific journals to report a balance of significant and insignificant findings. A third strategy could be not to make the problem of publication bias in peer-reviewed journals disappear, but rather by minimising its impact by making research results available elsewhere, for example through publicly available databases or archives such as bioarchive (BioRxiv.org) and medarchive (MedRxiv.org). The limitation of such archives is that there is no control over what is published since the manuscripts are uploaded without going through a peer-review process, and it will therefore be up to the reader to evaluate the quality of the research. This has strong limitations as it is possible to cite or refer to such articles and the plausibility of unchecked citations can easily become overlooked.

Registration of clinical trials prior to the results is another attempt to reduce publication bias and ensure that the results become publicly available regardless of the results. Registries of clinical trials have therefore been created to increase transparency and reproducibility, and these registries have also been used to study publication bias and its impact on meta-analysis (Marks-Anglin and Chen 2020). Multiple funding agencies have therefore encouraged or made it mandatory with trial registration including the US Food and Drug Administration (FDA) (FDA n.d.), the International Committee of Medical Journal Editors (ICMJE) (De Angelis et al. 2004), and the National Institutes of Health (NIH) (Zarin et al. 2016).

Clinical trial registration has been implemented to reduce publication bias, but this only partly solves the problem and is not sufficient to eliminate publication bias completely. While these registries have relatively good coverage today, this was not the case previously and only 20 years from now the grey literature was barely available (Marks-Anglin and Chen 2020). Meta-analyses tend to cover studies spanning decades of work and there will still be a bias in these meta-analyses although recent results from unpublished trials are included in the analysis. In addition, despite the registries there can still be publication bias within which results that are published or otherwise reported from the trials. In the case of cancer biomarker studies, these will often not be directly included as a part of the clinical trial design. Biomarker studies could for example be retrospective studies investigating potential biomarkers of clinical trials already performed. Indeed, biomarker studies are often either retrospective studies of clinical trials or even pre-clinical *in vitro* or animal studies. When a biomarker study is investigating many potential biomarkers, it is likely to think that the candidates that show significant results are much more likely to be included in the reports/publications than those that did not show significant differences in patient outcome.

In this section, we have seen that to address publication bias, we need to go to the root of the problem and question the mindset of both the researchers, journals/editors and the general community perception, and abolish the stigma that null results are less meaningful than positive results. The impression that there is a direct relationship between statistical significance and scientific importance is not always true. It can thus be argued that popularisation of reporting P values starting in the early twentieth century has led to an overuse of statistical testing (Marks-Anglin and Chen 2020). The competition amongst researchers that are valued based on their number of publications, citations, and impact factors, combined with a constant race for funding or extended contracts, is a system which creates a risk of favouring or 'selecting' the researchers that are most biased in their presentation of results to continue and even propel their careers as opposed to those who are more open minded and honest about their research results. More focus on the problem of publication bias, increased awareness and methods developed to understand and address the problem, and to study the extent of publication bias is important in this context.

Conclusion

Publication bias within the field of precision oncology and cancer biomarker research, meaning that positive or significant results are more likely to get published than negative results, have many possible implications for researchers, patients and the general society. Over time, publication bias skews the scientific literature in favour of positive results which thus influence the design of new research projects and contributes to the reproducibility crisis that questions the validity of the scientific literature. In the field of oncology, this ultimately affects the treatment of cancer patients as clinical decision making rely on the scientific literature. The issue of publication bias seems to be even more evident for precision oncology and biomarker research, as aiming for perfection will leave less space for 'negative' results than in medical research in general. In addition, the complexity of precision oncology research that is based on a highly complex and heterogenous tumour biology will also be likely to generate more variations in the research outcomes which makes room for a more biased selection of results. Indeed, although biomarker and precision oncology research has received significant financial support recent years, still only a few biomarkers end up in the clinic, and even for the most successful biomarkers there are still challenges with defining biomarker cut-offs and deciding how different biomarker subgroups should be defined and treated.

Publication bias could be a consequence of either researchers deciding not to submit 'negative' results or the journals rejecting manuscripts where the results are 'negative', and it is likely that publication bias occurs at both these levels. Multiple actions have been suggested to reduce publication bias including clinical trial registration, forcing journals to report a balance between positive and negative results or make research results available elsewhere than in peer-reviewed journals. However, to address publication bias, we need to go to the root of the problem and convince researchers, journals/editors and the general community that negative results could be at least as important as positive results. Increased awareness about publication bias and methods developed to understand and address the problem, and to study the extent of publication bias is important in this context.

References

Angelis, C.D., J.M. Drazen, F.A. Frizelle, C. Haug, J. Hoey, R. Horton, S. Kotzin, et al. 2004. Clinical trial registration: A statement from the International Committee of Medical Journal Editors. *The New England Journal of Medicine* 351 (12): 1250–1251.

Baker, M. 2016. 1,500 scientists lift the lid on reproducibility. *Nature* 533 (7604): 452–454.

Bardy, A.H. 1998. Bias in reporting clinical trials. *British Journal of Clinical Pharmacology* 46 (2): 147–150.

Beautification and fraud. 2006. *Nature Cell Biology* 8 (2): 101–102.

Begg, C.B. 1985. A measure to aid in the interpretation of published clinical trials. *Statistics in Medicine* 4 (1): 1–9.

Blanchard, A. 2016. Mapping ethical and social aspects of cancer biomarkers. *New Biotechnology* 33 (6): 763–772.

Blanchard, A., and E. Wik. 2017. Chapter 1: What is a good (enough) biomarker? In *Cancer biomarkers: Ethics, Economics and Society*, ed. A. Blanchard and R. Strand, 7–24. Kokstad: Megaloceros Press.

Boisen, E. 1979. Testicular size and shape of 47,XYY and 47,XXY men in a double-blind, double-matched population survey. *American Journal of Human Genetics* 31 (6): 697–703.

Dickersin, K., and Y.I. Min. 1993. NIH clinical trials and publication bias. *The Online journal of current clinical trials* Doc No 50: [4967 words; 53 paragraphs].

Driessen, E., S.D. Hollon, C.L.H. Bockting, P. Cuijpers, and E.H. Turner. 2015. Does publication bias inflate the apparent efficacy of psychological treatment for major depressive disorder? A systematic review and meta-analysis of US national institutes of health-funded trials. *PLoS One* 10 (9): e0137864.

Easterbrook, P. 1987. Reducing publication bias. *BMJ* 295 (6609): 1347–1347.

Easterbrook, P.J., J.A. Berlin, R. Gopalan, and D.R. Matthews. 1991. Publication bias in clinical research. *The Lancet* 337 (8746): 867–872.

Editors. 1909. The reporting of unsuccessful cases (editorial). *The Boston Medical and Surgical Journal* 161: 263–264.

Else, H. 2019. What universities can learn from one of science's biggest frauds. *Nature* 570 (7761): 287–288.

Fanelli, D. 2009. How many scientists fabricate and falsify research? A systematic review and meta-analysis of survey data. *PLoS One* 4 (5): e5738.

FDA. n.d. "*H.R.3580 – 110th Congress (2007–2008): Food and Drug Administration Amendments Act of 2007 | Congress.gov | Library of Congress*." https://www.congress.gov/bill/110th-congress/house-bill/3580. Accessed 30 Oct 2020.

FDA-NIH Biomarker Working Group. 2016. *BEST (Biomarkers, EndpointS, and other Tools) Resource*. Silver Spring (MD): Food and Drug Administration (US).

Fleck, L.M. 2017. Chapter 5: Just caring: Precision medicine, cancer biomarkers and ethical ambiguity. In *Cancer biomarkers: Ethics, economics and society*, ed. A. Blanchard and R. Strand, 73–94. Kokstad: Megaloceros Press.

Fong, E.A., and A.W. Wilhite. 2017. Authorship and citation manipulation in academic research. *PLoS One* 12 (12): e0187394.

Franco, A., N. Malhotra, and G. Simonovits. 2014. Social science. Publication bias in the social sciences: Unlocking the file drawer. *Science* 345 (6203): 1502–1505.

Ghannad, M., M. Olsen, I. Boutron, and P.M. Bossuyt. 2019. A systematic review finds that spin or interpretation bias is abundant in evaluations of ovarian cancer biomarkers. *Journal of Clinical Epidemiology* 116: 9–17.

Helmer, S., D.B. Blumenthal, and K. Paschen. 2020. What is meaningful research and how should we measure it? *Scientometrics* 125 (1): 153–169.

Ioannidis, J.P. 1998. Effect of the statistical significance of results on the time to completion and publication of randomized efficacy trials. *The Journal of the American Medical Association* 279 (4): 281–286.

Kern, S.E. 2012. Why your new cancer biomarker may never work: Recurrent patterns and remarkable diversity in biomarker failures. *Cancer Research* 72 (23): 6097–6101.

Lakhtakia, R., and I. Burney. 2015. A Brief History of Breast Cancer: Part III – Tumour biology lays the foundation for medical oncology. *Sultan Qaboos University Medical Journal* 15 (1): e34–e38.

Levecque, K., F. Anseel, A. De Beuckelaer, J. Van der Heyden, and L. Gisle. 2017. Work organization and mental health problems in PhD students. *Research Policy* 46 (4): 868–879.

Lexchin, J., L.A. Bero, B. Djulbegovic, and O. Clark. 2003. Pharmaceutical industry sponsorship and research outcome and quality: Systematic review. *BMJ (Clinical Research Ed.)* 326 (7400): 1167–1170.

Lobo, F.S., S. Wagner, C.R. Gross, and J.C. Schommer. 2006. Addressing the issue of channeling bias in observational studies with propensity scores analysis. *Research in Social & Administrative Pharmacy: RSAP* 2 (1): 143–151.
Lowe, D. 2019. Graduate abuse I In the Pipeline. *Science Translational Medicine*, October 28. https://blogs.sciencemag.org/pipeline/archives/2019/10/28/graduate-abuse.
Marks-Anglin, A., and Y. Chen. 2020. A historical review of publication bias. *Research Synthesis Methods* 11 (6): 725–742.
Marquart, J., E.Y. Chen, and V. Prasad. 2018. Estimation of the percentage of US patients with cancer who benefit from genome-driven oncology. *JAMA Oncology* 4 (8): 1093–1098.
Medical Research Council. n.d. *Understanding Health Research · Common sources of bias*. https://www.understandinghealthresearch.org/useful-information/common-sources-of-bias-2. Accessed 18 Mar 2021.
Miyakawa, T. 2020. No raw data, no science: Another possible source of the reproducibility crisis. *Molecular Brain* 13 (1): 24.
Müller, M.J., B. Landsberg, and J. Ried. 2014. Fraud in science: A plea for a new culture in research. *European Journal of Clinical Nutrition* 68 (4): 411–415.
Nieminen, P., G. Rucker, J. Miettunen, J. Carpenter, and M. Schumacher. 2007. Statistically significant papers in psychiatry were cited more often than others. *Journal of Clinical Epidemiology* 60 (9): 939–946.
Pain, E. 2017. Ph.D. students face significant mental health challenges. *Science*, April 4.
Pannucci, C.J., and E.G. Wilkins. 2010. Identifying and avoiding bias in research. *Plastic and Reconstructive Surgery* 126 (2): 619–625.
Polley, M.-Y.C., and J.J. Dignam. 2021. Statistical considerations in the evaluation of continuous biomarkers. *Journal of Nuclear Medicine* 62 (4): 605–611.
Prinz, F., T. Schlange, and K. Asadullah. 2011. Believe it or not: How much can we rely on published data on potential drug targets? *Nature Reviews. Drug Discovery* 10 (9): 712.
Quirke, V.M. 2017. Tamoxifen from failed contraceptive pill to best-selling breast cancer medicine: A case-study in pharmaceutical innovation. *Frontiers in Pharmacology* 8: 620.
Scherer, R.W., J.J. Meerpohl, N. Pfeifer, C. Schmucker, G. Schwarzer, and E. von Elm. 2018. Full publication of results initially presented in abstracts. *Cochrane Database of Systematic Reviews* 11: MR000005.
Shields, P.G. 2000. Publication bias is a scientific problem with adverse ethical outcomes: The case for a section for null results. *Cancer Epidemiology, Biomarkers & Prevention* 9 (8): 771–772.
Simes, R.J. 1986. Publication bias: The case for an international registry of clinical trials. *Journal of Clinical Oncology* 4 (10): 1529–1541.
Slamon, D.J., B. Leyland-Jones, S. Shak, H. Fuchs, V. Paton, A. Bajamonde, T. Fleming, et al. 2001. Use of chemotherapy plus a monoclonal antibody against HER2 for metastatic breast cancer that overexpresses HER2. *The New England Journal of Medicine* 344 (11): 783–792.
Song, F., S. Parekh, L. Hooper, Y.K. Loke, J. Ryder, A.J. Sutton, C. Hing, C.S. Kwok, C. Pang, and I. Harvey. 2010. Dissemination and publication of research findings: An updated review of related biases. *Health Technology Assessment* 14 (8): iii–193.
Stebbing, J., and D.A. Sanders. 2018. The importance of being earnest in post-publication review: Scientific fraud and the scourges of anonymity and excuses. *Oncogene* 37 (6): 695–696.
Sterling, T.D., W.L. Rosenbaum, and J.J. Weinkam. 1995. Publication decisions revisited: The effect of the outcome of statistical tests on the decision to publish and vice versa. *The American Statistician* 49 (1): 108–112.
Sterling, T.D. 1959. Publication decisions and their possible effects on inferences drawn from tests of significance—Or vice versa. *Journal of the American Statistical Association* 54 (285): 30–34.
Twa, M.D. 2019. Scientific integrity and the reproducibility crisis. *Optometry and Vision Science* 96 (1): 1–2.
Ulaner, G.A., D.M. Hyman, D.S. Ross, A. Corben, S. Chandarlapaty, S. Goldfarb, H. McArthur, et al. 2016. Detection of HER2-positive metastases in patients with HER2-negative primary

breast cancer using 89Zr-Trastuzumab PET/CT. *Journal of Nuclear Medicine* 57 (10): 1523–1528.

Vera-Badillo, F.E., M. Napoleone, M.K. Krzyzanowska, S.M.H. Alibhai, A.-W. Chan, A. Ocana, B. Seruga, A.J. Templeton, E. Amir, and I.F. Tannock. 2016. Bias in reporting of randomised clinical trials in oncology. *European Journal of Cancer* 61: 29–35.

Woo, S.Y., and S. Kim. 2020. Determination of cutoff values for biomarkers in clinical studies. *Precision and Future Medicine* 4 (1): 2–8.

Wu, L., and X. Qu. 2015. Cancer biomarker detection: Recent achievements and challenges. *Chemical Society Reviews* 44 (10): 2963–2997.

Xiao, C., Y. Gong, E.Y. Han, A.M. Gonzalez-Angulo, and N. Sneige. 2011. Stability of HER2-positive status in breast carcinoma: A comparison between primary and paired metastatic tumors with regard to the possible impact of intervening trastuzumab treatment. *Annals of Oncology* 22 (7): 1547–1553.

Yi, M., D. Jiao, H. Xu, Q. Liu, W. Zhao, X. Han, and K. Wu. 2018. Biomarkers for predicting efficacy of PD-1/PD-L1 inhibitors. *Molecular Cancer* 17 (1): 129.

Zarin, D.A., T. Tse, R.J. Williams, and S. Carr. 2016. Trial reporting in ClinicalTrials.gov – The final rule. *The New England Journal of Medicine* 375 (20): 1998–2004.

Assessing the Cost-Effectiveness of Molecular Targeted Therapies and Immune Checkpoint Inhibitors

John Alexander Cairns

Introduction

Two major groups of therapies have in recent years been transforming the care of many cancer patients, namely molecular targeted therapies and immune checkpoint inhibitors. Molecular targeted therapies interfere with the molecules required for tumour growth and progression (Abramson 2018). Whereas immune checkpoint inhibitors mobilise the body's immune system to destroy cancer cells. The success of the latter is such that they have rapidly become some of the most widely prescribed anticancer therapies (Robert 2020).

This chapter aims to discuss the cost-effectiveness of such therapies. Rather than undertaking a systematic review of economic evaluations of these therapies, I will restrict my focus to those appraised by the National Institute for Health and Care Excellence (NICE), a public body whose positive recommendations are mandatory for the National Health Service (NHS) in England. This approach has a number of advantages. Firstly, it facilitates comparison of checkpoint inhibitors with molecular targeted therapies because the evidence submission from manufacturers and its subsequent critique by an independent Evidence Review Group focuses on a Reference Case (imposing a degree of uniformity in approach) and is conducted in the context of well-established appraisal methods (NICE 2013). Secondly, all of these assessments of the cost-effectiveness of these health technologies have been undertaken for a common purpose, namely, to determine which treatments will be made available by the NHS in England. Thirdly, the level of detail regarding the clinical effectiveness evidence and the economic modelling choices (and the impact of alternative choices) is much greater than is the case for journal articles.

J. A. Cairns (✉)
London School of Hygiene & Tropical Medicine, London, UK
e-mail: John.Cairns@lshtm.ac.uk

A. Bremer, R. Strand (eds.), *Precision Oncology and Cancer Biomarkers*, Human Perspectives in Health Sciences and Technology 5,
https://doi.org/10.1007/978-3-030-92612-0_11

This chapter is based on experience with respect to the appraisal of therapies for non-small-cell lung cancer because this area offers better opportunities than any other for the comparison of molecular targeted therapies with immune checkpoint inhibitors in terms of the nature of the available data and the methods of economic evaluation employed. The twenty-one appraisals of molecular targeted therapies in non-small-cell lung cancer that have been completed to date comprise six EGFR-TK inhibitors (gefitinib, erlotinib, afatinib, necitumumab, osimertinib and dacomitinib), five ALK inhibitors (crizotinib, ceritinib, alectinib, brigatinib and lorlatinib) and two ROS1 inhibitors (crizotinib and entrectinib). The fifteen appraisals of checkpoint inhibitors used in the treatment of non-small-cell lung cancer comprise the anti-PD-1 agents nivolumab and pembrolizumab, and the anti-PD-L1 agents atezolizumab and durvalumab. As can be seen from Table 1, in the treatment of non-small-cell lung cancer, the first guidance with respect to molecular targeted therapies was issued in 2010, whereas the first guidance with respect to immune checkpoint inhibitors appeared in 2017.

The chapter compares the economic evaluation of molecular targeted therapies and of immune checkpoint inhibitors by reviewing the challenges involved in assessing the cost effectiveness of these new agents. With respect to economic evaluation, the main differences arise from the nature and extent of the available clinical evidence used to model their cost-effectiveness and the number of clinical indications for use. The chapter then goes on to consider the specific arrangements for determining access to oncology medicines, highlighting how these differences, with respect to evidence and proposed use, lead to different experience for the two groups of therapies. Although all oncology medicines in England are subject to the same decision-making arrangements, these arrangements have different implications for checkpoint inhibitors owing primarily to their having many more clinical indications than targeted therapies.

Table 1 Targeted therapies and checkpoint inhibitors for treating non-small cell lung cancers by year of first NICE guidance

Year	Targeted therapies	Checkpoint inhibitors
2010	gefitinib	
2011		
2012	erlotinib	
2013	crizotinib	
2014	afatinib	
2015		
2016	ceritinib, necitumumab, osimertinib	
2017		nivolumab, pembrolizumab
2018	alectinib	atezolizumab
2019	brigatinib, dacomitinib	durvalumab
2020	entrectinib, lorlatinib	

Economic Evaluation of Targeted Therapies and Checkpoint Inhibitors

In this section I review the economic evaluation of checkpoint inhibitors and targeted therapies, with an emphasis on the differences between the two technologies.

The clinical pathways for non-small-lung cancer patients are fairly numerous even when restricting attention to systemic anti-cancer therapy. This is, in part, a result of the advent of molecular targeted therapies making it important to distinguish several distinct patient groups, and because the increasing number of treatments increases the number of lines of treatment available to patients. Starting with the comparators used when estimating cost-effectiveness, there is fairly clear evolution of comparators over time with respect to molecular targeted therapies. For example, with respect to EGFR-TK inhibitors when gefitinib was originally considered as a first line treatment the comparator was platinum doublet therapy, then when erlotinib was considered, gefitinib became the comparator, and for afatinib the comparators were erlotinib and gefitinib, then when dacomitinib and osimertinib were appraised the comparators were afatinib, erlotinib and gefitinib. Similarly, in the case of first line treatment of ALK+ patients, crizotinib was compared to pemetrexed and cisplatin, then ceritinib and alectinib were compared with crizotinib, pemetrexed plus carboplatin/cisplatin, then brigatinib was compared to alectinib, ceritinib and crizotinib.

A similar evolution with respect to the comparators used is not observed in appraisals of immune checkpoint inhibitors. This is partly because sequences of checkpoint inhibitors are not an option and partly because of a peculiarity of NICE appraisal methods, namely when drugs are reviewed following inclusion in the Cancer Drugs Fund "no changes to the scope of the appraisal will be considered" (para. 6.25) (NICE 2018). Thus, when nivolumab was re-appraised for advanced squamous non-small-cell lung cancer after chemotherapy and guidance was published in October 2020 (TA655), docetaxel remained the comparator, despite immunotherapies (atezolizumab and pembrolizumab) now being available for these patients (NICE 2020a).

Most of the appraisals have followed the same general approach to modelling costs and health outcomes. Twenty out of twenty-one appraisals of targeted therapies and fourteen out of fifteen appraisals of checkpoint inhibitors have featured partitioned survival models, the exceptions being TA192 (gefitinib) and TA578 (durvalumab). Partitioned survival models, unlike Markov models which are based on transitions between health states, use survival data directly to determine the time spent in different health states. In the case of cancer appraisals, the area under the overall survival curve is partitioned into time spent progression-free and time spent with progressed disease, using the progression-free survival curve. The two exceptions featured Markov models, gefitinib by virtue of its pre-dating the widespread enthusiasm for partitioned survival modelling (NICE 2010), and durvalumab because the manufacturer's attempts to implement partitioned survival modelling were thwarted by the extrapolated overall survival and progression-free curves crossing (NICE 2019a, b). Such crossing of curves, which is clearly impossible in practice since you need to be alive to enjoy progression-free survival, can arise when survival curves are fitted independently.

However, there were some differences in how survival was modelled. Cure rate models were never employed in the appraisal of the targeted therapies whereas, they were with checkpoint inhibitors. However, possibly surprisingly, only two of the fifteen appraisals featured formal modelling of cure rate models (atezolizumab in TA520 and durvalumab in TA578). On neither occasion was the committee persuaded by the manufacturer's analysis. The appraisal committee noted, in the case of atezolizumab, that the model had not been sufficiently justified by the company and the cure rate was not sufficiently supported by the evidence (NICE 2018), and in the case of durvalumab that the PACIFIC trial data "were too immature for a cure model to be robust" (NICE 2019a).

More substantial differences are apparent in the clinical data upon which cost-effectiveness was assessed and, in the methods, used to estimate the relative effectiveness of treatments. The key source of data, upon which the cost-effectiveness modelling for the intervention was based came from randomised controlled trials for all of the checkpoint inhibitors, whereas six of the twenty-one appraisals of targeted therapies were based on non-randomised data (single arm trials plus one randomised dosing study). There is a tendency for the checkpoint inhibitor trials to have larger sample sizes for the intervention, with on average, 303 patients receiving the checkpoint inhibitor as compared to 208 patients receiving the targeted therapies. There being fewer head-to-head trials in the case of targeted therapies and the trials enrolling fewer subjects possibly reflects the targeted nature of the therapies.

It might be expected that these two features would lead to increased reliance on indirect comparisons to establish treatment effects in the case of the targeted therapies. This is not supported by a comparison of the proportion of appraisals featuring such analyses. The proportion of appraisals where indirect comparisons were made by the manufacturers is broadly similar for the checkpoint inhibitors (67%) and targeted therapies (62%). However, the targeted therapies give the appearance of greater use of indirect treatment comparisons than is the case with the checkpoint inhibitors. There are six appraisals involving Matching Adjusted Indirect Comparisons in the case of targeted therapies and none for checkpoint inhibitors.

Population-adjusted indirect comparisons refer to the use of individual patient data to adjust for between-trial differences in the distribution of variables which influence outcome. The merits of direct comparisons in randomised head-to-head trials are well-established. Indirect treatment comparisons generate a range of response. In the *Cochrane Handbook for Systematic Reviews of Interventions* we are reminded that "Indirect comparisons are not randomized comparisons and cannot be interpreted as such. They are essentially observational findings across trials, and may suffer the biases of observational studies, for example due to confounding." (Higgins and Green 2011) Whereas Lu and Ades (2004) argue "… to ignore indirect evidence either makes the unwarranted claim that it is irrelevant or breaks the established precept of systematic review that synthesis should embrace all available evidence", making indirect treatment comparisons required rather than optional. Differing views notwithstanding, few if any would argue for indirect over direct comparisons. Population-adjusted indirect comparisons are an attempt to mitigate some of the limitations of indirect comparisons.

There are two main methods that use individual patient data to adjust for between-trial differences in the distribution of variables that influence outcome: Matching-Adjusted Indirect Comparison (MAIC) and Simulated Treatment Comparison (STC). For no obvious reason, use of the latter approach has been restricted to appraisals of treatments for urothelial carcinoma. Whereas, Matching-Adjusted Indirect Comparisons have been used more widely and have featured in six non-small cell lung cancer appraisals (brigatinib twice, ceritinib, entrectinib, lorlatinib and osimertinib). This can be contrasted with the practice observed in the conduct of the fifteen appraisals of checkpoint inhibitors where MAICs have not featured. The proximate explanation for this marked difference is that all of the checkpoint inhibitor appraisals drew their data primarily from randomised controlled trials, in contrast to the greater reliance placed on non-randomised studies in the appraisals of molecular targeted therapies. In addition, the population for targeted therapies is smaller and trials are also smaller which may make differences in baseline characteristics more substantial than in the case of checkpoint inhibitors.

Matching Adjusted Indirect Comparisons are a relatively new area of activity. They are particularly relevant when a manufacturer has individual patient data from a trial of its own drug but only has access to aggregate data for a drug to which it wishes to compare. They involve reweighting the population in a study where individual patient-level data are available to match more closely the aggregate data from another study. The central idea is that MAIC can allow comparison of trials where there are differences in the baseline characteristics of patients. We can distinguish "anchored" and "unanchored" indirect comparisons. Unanchored indirect comparisons are used where there is a disconnected treatment network or single arm studies. An unanchored MAIC assumes that all effect multipliers and prognostic factors are accounted for. Suppose you have individual data for a trial of A and aggregate data for another trial of B. The MAIC tries to generate the AB effect that would be observed in an A vs. B trial.

For example, in the appraisal of brigatinib for treating ALK-positive advanced non-small cell lung cancer after crizotinib, there was no clinical trial directly comparing brigatinib with the relevant comparator, ceritinib. The manufacturer estimated progression-free survival and overall survival using a hazard ratio from a meta-analysis of two MAICs. In the first, ceritinib data from the single arm ASCEND-2 were matched to the brigatinib arm of ALTA. In the second, the ceritinib arm of ASCEND-5 were also matched to the brigatinib arm of ALTA. The factors adjusted for in the MAICs comprised ECOG performance status, presence of brain metastases, age, crizotinib as last treatment before next TK inhibitor, gender, receipt of any prior chemotherapy, number of prior anti-cancer regimens and smoking history status.

All of the appraisals of checkpoint inhibitors have based their health state utility values on EQ-5D (Rabin and de Charro 2001) data collected in clinical trials. Whereas targeted therapies, while mainly relying on EQ-5D data collected in clinical trials, have also made use of mapping algorithms from QLQ-C30 (Aaronson et al. 1993) to EQ-5D on five occasions, and cancer-specific health state values on three occasions. This probably, in part, reflects changes in the conduct of economic evaluations over time and that several of the targeted appraisals predate consideration of the checkpoint inhibitors.

Arrangements Specific to Decision Making Involving Oncology Medicines

The appraisal by NICE of oncological treatments has differed significantly from the appraisal of non-oncological treatments. Since 2016, NICE has routinely appraised all oncology medicines, whereas this practice has only recently been extended to all medicines. *The 2019 Voluntary Scheme for Branded Medicines Pricing and Access* agreed that NICE would appraise all new active substances in their first indication (and usually any significant new therapeutic indication). This agreement, between the UK government, NHS England, the Association of the British Pharmaceutical Industry and individual manufacturers, also confirmed that appraisal timelines for non-oncology treatments would in future match those for oncology treatments, reflecting the longer lag in the past between the date of the marketing authorisation and any NICE guidance to the NHS.

There are two other important ways in which decision-making involving oncology therapies is clearly different from that for other medicines. Since January 2009, appraisal committees have been required to value the health benefits accruing to life-extending end-of life treatments more highly than those produced by treatments not qualifying for this status. The two criteria which currently must be met to be a life-extending end-of-life treatment are that, in the absence of the new treatment the patient group would have a short life expectancy, normally less than twenty-four months, and there is sufficient evidence to indicate that the treatment offers an extension to life, normally of at least three months. Note that there is no requirement that the treatment in question be a cancer therapy. However, to date a treatment has been determined to be a life-extending end-of-life treatment on 110 occasions, only one of which was not for advanced cancer. The single exception being nusinersen, a treatment for spinal muscular atrophy. The higher valuation of the health benefits in these cases has been implemented by using a cost-effectiveness threshold of £50,000 per QALY gained, instead of the cost-effectiveness threshold range of £20,000 to £30,000 used explicitly by NICE since 2004.

England has had a Cancer Drugs Fund since 2010 but it has had two distinct incarnations. The original Cancer Drugs Fund was introduced in 2010 as fulfilment of an election promise, that any cancer patient should be allowed any drug licensed in the previous five years, if sought by their doctor, even if NICE had determined that it did not represent good value for money for the NHS. The Cancer Drugs Fund was established by the new coalition government (with interim funding of £60 million, and from 1 April 2011, £200 million per annum) as an additional funding source for cancer drugs not routinely available through routine commissioning. Its operational management moved to NHS England in 2013. Concern over its financial sustainability led first to more rigorous assessment of candidate medicines, and then to proposals for an entirely new fund. This further instance of special treatment being extended to cancer therapies, offered poor value for the increasing proportion of NHS budget that it absorbed (Aggarwal et al. 2017).

The new Cancer Drugs Fund, launched in England on 1 July 2016, provides a mechanism by which access to some new therapies can be increased by offering appraisal committees an alternative to rejection, when there is an insufficiently compelling case for recommending that they are provided as part of routine clinical practice. If a committee, following its review of the evidence on clinical effectiveness and cost-effectiveness, does not recommend routine commissioning by NHS England, it can consider whether to recommend inclusion in the Cancer Drugs Fund. In making its decision the committee considers the following questions:

- Is the model structurally robust for decision making?
- Is there plausible potential to be cost-effective at the offered price?
- Could further data collection reduce uncertainty?
- Will ongoing studies provide useful data?
- Is Cancer Drugs Fund data collection via the Systemic Anti-Cancer Therapy relevant and feasible?

As well as tying inclusion in the Cancer Drugs Fund to the NICE appraisal of the medicine, the 2016 Cancer Drugs Fund addressed the budget sustainability issues of its precursor. First, as highlighted in the questions listed above, a selective approach to entry to the Cancer Drugs Fund is now practiced. Second, while in the fund the drug must be provided at a cost-effective price (a price informed by the committee's appraisal). Third, there is a budget cap supported by a provision that the Cancer Drugs Fund spend is shared across entrants when the cap is reached.

The first medicine to be made available through the new fund, in October 2016, was osimertinib for treating metastatic EGFR and T790M mutation-positive non-small-cell lung cancer (NICE 2016). The appraisal committee believed the ICER to be between £60,663 and £70,776 per QALY gained. Although, it met the criteria to be regarded as a life-extending end-of-life treatment, these ICERs were too high to recommend that it be routinely commissioned. The greatest area of uncertainty concerned the extrapolation of the overall survival data for patients treated with osimertinib. It was anticipated that this would be reduced over the next two years with follow-up of the AURA-3 trial. The other main uncertainty concerned the generalisability of the AURA-3 trial to English clinical practice. It was anticipated that data collected over the next two years would document treatment patterns and baseline patient characteristics, and in particular, provide more accurate estimates of the duration of treatment. The appraisal committee subsequently met in February 2020, and following review of the additional data, recommended routine commissioning of osimertinib for this indication (NICE 2020b).

There are some clear differences between the non-small-cell lung cancer appraisals of immune checkpoint inhibitors and of molecular targeted therapies with respect to life-extending end-of-life status and entry to the 2016 Cancer Drugs Fund.

If we compare the proportion of the two groups considered to meet the criteria to be assessed as life-extending end-of-life treatments, we find that in thirteen out of fifteen non-small-cell lung cancer appraisals (87 per cent), the checkpoint inhibitor was deemed to meet the criteria to be considered an end-of-life life-extending treatment with respect to at least one comparator. Whereas molecular targeted therapies were similarly considered to be life-extending end-of life treatments in ten out of twenty-one appraisals (48 per cent). There are two potential explanations for this marked difference, one is that checkpoint inhibitors are typically extending life more substantially than the molecular targeted therapies (making it more likely that the extension to life criterion is met). Another potential explanation is that the targeted therapies are being used earlier in the treatment pathway than the checkpoint inhibitors when life expectancy under current treatment is greater (making it less likely that they meet the life expectancy criterion).

Fifty-seven per cent of non-small cell lung cancer appraisals of checkpoint inhibitors, where inclusion in the 2016 Cancer Drugs Fund was an option, were subsequently included in the fund. Whereas only fourteen per cent of appraisals of molecular targeted therapies for non-small cell lung cancer have led to provision within the fund. There are a number of potential explanations for this differential recourse to the 2016 Cancer Drugs Fund. It may be that there is a higher likelihood that uncertainties regarding clinical effectiveness, and ultimately cost-effectiveness, can be satisfactorily reduced by inclusion in the fund in the case of checkpoint inhibitors, as compared to molecular targeted therapies. This might arise because the impact of checkpoint inhibitors, on some patients' health, is much more long-lived and the uncertainties are consequently greater but amenable to longer follow-up of trial participants.

There may also be an explanation in terms of drug pricing. The health benefits that the average patient can expect from a particular treatment are likely to vary by indication. With uniform pricing across indications there can be some indications where use of the medicine does not offer a favourable balance between the expected costs of treatment and the expected benefits, and others where use of the medicine provides satisfactory value for money. Indication-specific pricing offers the opportunity to provide patients with particular indications access to medicines that they might be denied under uniform pricing. Described thus indication-specific pricing appears something of a boon. It can also be described as price discrimination which transfers surplus from those paying for health care to the firms selling the medicines (Chandra and Garthwaite 2017). NHS England has been reluctant to engage in indication-specific pricing as part of routine commissioning, but the door is not firmly closed. The *2019 Voluntary Scheme on Branded Medicines* recognises uniform pricing as the norm but concedes "In cases where uniform pricing would lead to a reduction in total revenue for a medicine overall from the introduction of additional indications, other forms of commercial flexibility may be considered for medicines with a strong value proposition. In these cases, commercial flexibility would only be considered where the level of clinical effectiveness is highly differentiated, but substantial in all indications under consideration" (DHSC 2018, para. 3.36).

The potential relevance of uniform pricing is that the non-small-cell lung cancer molecular targeted therapies are characterised by a lower number of NICE-recommended indications (usually one or two, occasionally three) as compared to checkpoint inhibitors. NICE has issued guidance, for example, on sixteen occasions in the case of pembrolizumab and fifteen times for nivolumab, and many further appraisals are in development. Moreover, pembrolizumab has been appraised in seven clinical areas: colorectal cancer, head and neck cancer, Hodgkin lymphoma, melanoma, non-small-cell lung cancer, renal cell carcinoma and urothelial carcinoma, and nivolumab has been appraised in these seven clinical areas plus oesophageal cancer. This is in marked contrast to the twelve targeted therapies which, with a single exception, only have non-small-cell lung cancer indications. The sole exception, entrectinib, also has an indication for NTRK fusion-positive solid tumours. The potential foregone revenue to the manufacturer from setting a single uniform price would thus be expected to be less in the case of the targeted therapies. While a drug included in the Cancer Drugs Fund will need to be offered at a potentially cost-effective price, since it is not part of routine commissioning the scope for price to vary across indications is much greater. Consequently, manufacturers might have a stronger incentive to seek inclusion of their checkpoint inhibitors in the 2016 Cancer Drugs Fund.

The multiple therapeutic indications for some new drugs were noted in NICE's 2021 consultation on changes to its appraisal processes, in which they specifically highlighted pembrolizumab, nivolumab, and atezolizumab. They posed the question whether "the current approach to conduct a full technology appraisal for every new indication is proportionate or if an alternative simpler approach to evaluation ... may be more appropriate for technologies that are evaluated multiple times" and sought views concerning different approaches (NICE 2021). Thus, NICE's interest in considering alternative approaches stems primarily from constraints on the organisation's evaluative capacity, however, as we have seen, multiple indications also raise at least as important issues with respect to the pricing of these therapies.

Conclusions

While this review is based on experience in non-small-cell lung cancer it offers insights regarding the evaluation of targeted therapies and checkpoint inhibitors across oncology. The differences with respect to economic evaluation, such as, in terms of the clinical evidence underlying the modelling of cost-effectiveness, and the differences observed with respect to life-extending end-of-life status and with respect to inclusion in the Cancer Drugs Fund, are not anticipated to be specific to non-small-cell lung cancer but arise from differences between immune checkpoint inhibitors and molecular targeted therapies.

Although it appears that some of the differences between oncology treatments and non-oncology treatments are beginning to reduce, this looks like being achieved by moving appraisal and decision-making with respect to non-oncology drugs closer to that of oncology drugs. An increasing proportion of non-oncology treatments will be subject to appraisal by NICE. Also, the general election at the end of 2019 brought forward a proposal to extend the Cancer Drugs Fund to create a new Innovative Medicines Fund. Additionally, NICE have recently consulted on changes to their methods of appraisal, expressing a clear desire to replace the life-extending end-of-life criteria with a severity criterion (NICE 2020c). While many oncology medicines are still likely to qualify for a more favourable treatment of their health benefits, it is anticipated that they will be joined by a number of non-oncology medicines. Just how many non-oncology medicines and the new severity criteria are yet to be revealed. Thus, the study of the recent past of decision making with respect to oncology drugs is likely to be increasingly relevant to a much wider range of medicines, as differential weighting of QALYs and the use of managed access schemes are extended beyond oncology drugs.

This chapter describes where we are rather than how we got here. There is clearly a suggestion that the differences observed between the appraisal of molecular targeted therapies and that of immune checkpoint inhibitors derives from the more limited clinical data and the more restricted application of the targeted medicines. However, clinical studies are not designed, and indications are neither sought nor granted without human agency.

References

Aaronson, N.K., S. Ahmedzai, B. Bergman, et al. 1993. The European Organisation for Research and Treatment of Cancer QLQ-C30: A quality-of-life instrument for use in international clinical trials in oncology. *Journal of the National Cancer Institute* 85: 365–376.

Abramson R. (2018). Overview of targeted therapies for cancer. *My Cancer Genome*. https://www.mycancergenome.org/content/molecular-medicine/overview-of-targeted-therapies-for-cancer/.

Aggarwal, A., T. Fojo, C. Chamberlain, C. Davis, and R. Sullivan. 2017. Do patient access schemes for high-cost cancer drugs deliver value to society? Lessons from the NHS Cancer Drugs Fund. *Annals of Oncology* 28 (8): 1738–1750.

Chandra, A., and C. Garthwaite. 2017. The economics of indication-based drug pricing. *New England Journal of Medicine* 377 (2): 103–106.

Department of Health & Social Care and Association of the British Pharmaceutical Industry. 2018. *The 2019 voluntary scheme for branded medicines pricing and access*. https://www.gov.uk/government/publications/voluntary-scheme-for-branded-medicines-pricing-and-access.

Higgins J.P.T., and S. Green (eds.). *Cochrane Handbook for Systematic Reviews of Interventions Version 5.1.0* [updated March 2011]. The Cochrane Collaboration, 2011. Available from www.handbook.cochrane.org.

Lu, G., and A.E. Ades. 2004. Combination of direct and indirect evidence in mixed treatment comparisons. *Statistics in Medicine* 23 (20): 3105–3124.

NICE. 2010. *Gefitinib for the first-line treatment of locally advanced or metastatic non-small-cell lung cancer (TA192)*. https://www.nice.org.uk/guidance/ta192.

———. 2013. *Guide to the methods of technology appraisal 2013*. https://www.nice.org.uk/process/pmg9.

———. 2016. *Osimertinib for treating locally advanced or metastatic EGFR T790M mutation-positive non-small-cell lung cancer (TA416)*.

———. 2018. *Guide to the processes of technology appraisal*. https://www.nice.org.uk/process/pmg19.

———. 2019a. *Durvalumab for treating locally advanced unresectable non-small-cell lung cancer after platinum-based chemoradiation (TA578)*. https://www.nice.org.uk/guidance/ta578.

———. 2019b. *Position statement: consideration of products recommended for use in the Cancer Drugs Fund as comparators, or in a treatment sequence, in the appraisal of a new cancer product*. January 2019.

———. 2020a. *Nivolumab for advanced squamous non-small-cell lung cancer after chemotherapy (TA655)*. https://www.nice.org.uk/guidance/ta655.

———. 2020b. *Osimertinib for treating EGFR T790M mutation-positive advanced non-small-cell lung cancer (TA653)*. https://www.nice.org.uk/guidance/ta653.

———. 2020c. *The NICE methods of health technology evaluation: the case for change*. Published 6 November 2020. https://www.nice.org.uk/about/what-we-do/our-programmes/nice-guidance/nice-technology-appraisal-guidance/changes-to-health-technology-evaluation.

———. 2021. *Review of the health technology evaluation processes*. Published 4 February 2021.

Rabin, R., and F. de Charro. 2001. EQ-5D: A measure of health status from the EuroQol Group. *Annals of Medicine* 33 (5): 337–343.

Robert, C. 2020. A decade of immune-checkpoint inhibitors in cancer therapy. *Nature Communications* 11: 3801. https://doi.org/10.1038/s41467-020-17670-y.

Real-World Data in Health Technology Assessment: Do We Know It Well Enough?

Jiyeon Kang

Introduction: The Growing Importance of Data in Health Care

Oncology research has been accompanied by important health care innovations in cutting-edge technology over the last decades. In the era of information technology, a remarkable amount of biological clinical data has been generated. As the data accumulated, the demand for bioinformatics flared up. Advanced data science has rapidly emerged in oncology research, where it is used for the analysis and interpretation of biological data for cancer diagnosis and clinical treatment planning (Bayat 2002). The paradigm of cancer treatment has shifted from chemotherapy to immunotherapy, evolving into 'precision oncology' guided by highly sophisticated biomarkers. The growing precision medicine market is predicted to exceed USD 119 billion by 2026 (Ugalmugle and Swain 2020). Advances in technologies such as data science are leading the fourth industrial revolution. Data are not only pushing the technology forward but are also seen as a key factor to achieve success in the fourth industrial revolution. Data are expected to advance the development of new technology and industry like artificial intelligence (AI). Health care is one of the most commonly addressed applications of the technological data revolution. Not only contributing to developing new technologies, but data such as big data or real-world data (RWD) are also expected to help provide scientific and systematic evidence to policymakers by combining all available evidence.

As a way of example, in the United States, the 21st Century Cures Act enacted in 2016 was designed to help accelerate medical product development, bringing new innovations and advances into patients who need them faster and more efficiently (FDA 2020a). This Act also placed additional focus on the use of RWD to support the regulatory decision making, including approval of new indications for approved

J. Kang (✉)
London School of Hygiene and Tropical Medicine, London, UK
e-mail: Jiyeon.Kang@lshtm.ac.uk

A. Bremer, R. Strand (eds.), *Precision Oncology and Cancer Biomarkers*, Human Perspectives in Health Sciences and Technology 5,
https://doi.org/10.1007/978-3-030-92612-0_12

drugs (FDA 2020b). Three years after, the US Food & Drug Administration (FDA) approved the palboclib for treating male breast cancer based on RWD from electronic health records (Wedam et al. 2020). Rare diseases such as male breast cancer have the challenge of obtaining evidence from RCT, which is a big hindrance to developing drugs. However, in this case, RWD played an important role as primary evidence instead of RCTs in getting the FDA's approval. Using RWD, FDA enabled the early approval of new health technology, which fulfils the unmet medical needs. This example shows that the data utilisation impacts early stages of drug development and the whole process of development, such as drug approval and reimbursement decision making.

Data have therefore become a central topic in health care. The objective of this chapter is to look at the opportunities and challenges of using RWD in health technology assessments, in particular by using examples of the NICE technology appraisals. In order to do so, we will first look at the definition of RWD and big data in section "Real-world data and big data: some definitions". The following each sections "Health Technology Assessment in the era of information technology" and "Opportunities related to using RWD in HTA" will describe the opportunities and challenges of using RWD in health technology assessment (HTA) with detailed example, how manufacturers, evidence review groups (ERGs) and committee have used RWD in appraisals. In the last section, it will briefly emphasise the deliberation of using RWD based on understanding of its potentialities and limitations.

Real-World Data and Big Data: Some Definitions

When looking at the current trends in terms of data in health care, there are two key concepts – big data and RWD. Commonly, the terms RWD and big data are used interchangeably. However, the relationship between RWD and big data is not as straightforward. Although there is no consensus on their definitions and the boundary between big data and RWD, the two terms are not identical. The Head of Medicines Agencies and European Medicines Agency set up a joint task force for the best use of big data, including RWD such as electronic health records and data from patient registries in 2019 (HMA, EMA 2019). Broadly speaking, big data usually refers to "the explosion in quantity (and sometimes, quality) of available and potentially relevant data, largely the result of recent and unprecedented advancements in data recording and storage technology"(Gibbs and McKendrick 2015, 235). Characteristics of big data are summarised into volume (massive amounts), velocity (high-speed processing) and variety (heterogeneous data), the so-called 3Vs of big data (2020). On the other hand, RWD is data relating to patient health status and/or the delivery of health care routinely collected from a variety of sources, including electronic health records, medical claims and billing, product and disease registries, as well as health-related data from mobile devices (FDA 2018). Data such as NHS electronic hospital data, cancer registry data, claims data, and even patient-reported information collected from wearable devices are all RWD.

Big data and RWD are inherently complex, encompassing a wide range of information specific to the health and everyday life of individuals. This is an extensive definition that leaves room for interpretation. For example, NHS electronic hospital data and cancer registry data are routinely collected, large and unprocessed data can be used in potentially different ways. Therefore, they can be RWD as well as big data. Nonetheless, it is incorrect to say that all RWD is big data, even if the attributes of big data seem substantially transferable to RWD. Since big data collectively indicate all data, it is not necessarily RWD. For example, the Big Data Institute at the University of Oxford is working on clinical AI for the patient-centred management and treatment of chronic disease (Oxford Big Data Institute 2019). In order to understand the complexity of the disease, it includes clinical trials, NHS hospital data and all available forms of data from all over the world. Whereas these collective data are considered big data, it is wrong to classify these data as RWD. Similarly, RWD is not always big data. Data from Compassionate Use Programme (CUP) are an example. The CUP is the scheme, which allows the patient who cannot enter a clinical trial to use unauthorised medicine under strict conditions (EMA n.d.-a). The drug in development can be available to patients who are not eligible for clinical trials or use unapproved therapies (EMA n.d.-b). According to the definition, it is RWD, a retrospective observational cohort study routinely collected from the real-world. The CUP was used in appraising the clinical effectiveness and cost-effectiveness of idelalisib for treating refractory follicular lymphoma by the National Institute for Health and Care Excellence (NICE) in the UK. In the NICE appraisal of idelalisib (TA604), CUP data collected in the UK and Ireland (n = 65) was submitted in order to complement the evidence. It generates valuable information which belongs to a category of RWD in patient populations with unmet needs, and it informs future RWD use (Balasubramanian et al. 2016, 251). While the study furnished the information on the real-world evidence, the study's characteristics deviated from the attributes of big data in terms of volume, velocity, and veracity. In this chapter, the scope of data only focuses on RWD in order to understand the issues around RWD more comprehensively.

Health Technology Assessment in the Era of Information Technology

In this era of information technology, "evidence-based practice" has been a keyword. Evidence-based practice is the "integration of best research evidence with clinical expertise and patient values (De Brún 2013, 3)." When integrating the evidence, it is necessary to consider all available data in an unbiased, transparent and scientific manner. HTA is an example of evidence-based practice aiming to provide the best evidence to health care decision-makers. As briefly mentioned above, HTA is a systematic evaluation of short- and long term safety, clinical effects, and cost-effectiveness of health technology and technology-related social, economic, and ethical issue in terms of health care resource use (Henshall et al. 1997; Potter et al. 2008; WHO 2015). Over the last decades, HTA has become more critical as evidence-based decision making has become more prominent in the health system. Specialised HTA bodies such as NICE have worked hard to enhance the methods of synthesising the evidence. Based on the evidence, the NICE produces guidance, including technology appraisal guidance (TA guidance) and advice for health, public health and social care practitioners (National Institute for Health and Care Excellence (NICE) n.d.). In evidence-based medicine, randomised controlled trials (RCTs) are regarded as the highest level of evidence because they are designed to be unbiased and have less risk of systematic bias (Burns et al. 2011). In the drug approval process, RCTs have been mainly required to show efficacy and safety compared to a control group. Whilst RCTs are the gold standard of evidence for establishing efficacy, it is sometimes difficult to conduct RCTs. For example, the medicine treating rare disease has difficulty in conducting RCTs due to lack of appropriate trial design, proper measurements to complement the trial design, selection of the correct sample and ethical recruitment to participation (Augustine et al. 2013). Moreover, trial-based economic evaluation raises questions regarding generalisation, how representative is the trial of the patient population (Sculpher et al. 2004, 2). Health economic models require a range of data, not all of which are available from RCTs. It is being explored how RWD are able to supplement and enrich the evidence in the arena of HTA (Makady et al. 2017a). Consequently, HTA is reshaping to now be incorporating RWD as evidence, and critical questions are posed regarding the most appropriate ways to incorporate RWD as evidence in HTA. While the NICE has been already committed to embracing all available evidence to appraise innovative health technologies, they set out their ambitions to increase and extend the use of data, including RWD in the development and evaluation of NICE guidance. In February 2020, the NICE announced a statement of intent that a broader range of data will be utilised to address evidence gaps, including electronic health record data and RWD "looking at health and social care practice outside of trials, such as registries and clinical audits" (NICE 2020a, b). In November 2020, the NICE launched the consultation on reviewing their methods for health technology evaluation (NICE 2020b). In their proposal, the NICE explicitly addressed their preference for RCTs but also emphasised the role of the comprehensive evidence base, including non-RCTs and real-world evidence. NICE mentioned that "This type of evidence (real-world evidence) is an important topic, and NICE health technology evaluations are ambitious in ensuring that we make the most of this valuable resource" (Aggarwal et al. 2017; NICE 2020a).

Opportunities Related to Using RWD in HTA

RWD can be used in HTA in several ways. First, RWD can help to extrapolate the long-term survival curve after the trial period for economic evaluation. The NICE makes appraisal recommendations based on the cost-effectiveness or the estimated costs of interventions in relation to expected health benefits over the lifetime of patients (NICE n.d.). The health benefit, usually survival rate in oncology, is then extrapolated from clinical trials, as they only show the health outcome over a limited time period. As the clinical evidence from trials is often limited, the extrapolation is likely to be biased. In that case, RWD can positively supplement the extrapolation of the survival rate by documenting some patients' characteristics and clinical practice over a longer observation period. For example, in the NICE technology appraisal of pembrolizumab with carboplatin and paclitaxel or nab-paclitaxel for untreated metastatic squamous non-small-cell lung cancer (NSCLC) (NICE TA600 iussed in 2019c), the manufacturer used RWD to extrapolate overall survival (OS) in its submission. In the appraisal, the OS at cancer stage 4 was not available due to the small number of surviving patients. Therefore, the company mentioned that:

> It was considered necessary to assess longer-term OS for the trial chemotherapy arm using available population data for squamous NSCLC patients and compare to results from parametric fitting. (p: 141, company submission)

In order to assess the survival, the company analysed the real-world registry data, the US Surveillance, Epidemiology and End Results (SEER) database. The SEER database is an authoritative source for cancer statistics in the United States. It is considered to be the gold standard for data quality amongst cancer registries in the US and globally (Duggan et al. 2016, 4). As real-world and big registry data, the SEER registry is a large, population-based sample, which represents over one-quarter of the US population as well as has long follow-up periods. The company compared OS beyond 12 months, from the projection by the parametric fitting approach of trial data and the SEER population data in order to examine the potential bias of using the SEER registry. As SEER data provided long-term data, it was able to observe that mortality risks within SEER gradually declined until around year 10 and then appeared to stabilise in the range of roughly a 10% risk per year. The company addressed the potential over-estimation of long-term mortality when using available clinical trial data for the best fitting parametric extrapolation model. Therefore, the company's model used SEER data in both intervention and comparator arms. Although the committee preferred the model, which did not use the SEER database in NICE TA600, because of the absence of second-line treatment in the database, and too optimistic assumptions in the model, it shows how registry data can be used to estimate the long-term survival in HTA.

The second way RWD is used in HTA is when it provides information regarding the comparators such as choice of relevant comparators and treatment effects. As new health technologies become more sophisticated, they can have several comparators, and as the number of potential comparators increases, it becomes less likely that there are head-to-head RCTs comparing their clinical effectiveness. It is unavoidable to synthesise the evidence using other RCTs or other types of research for data for the comparators. In the technology appraisal of cabozatinib for treating previously treated advanced renal cell carcinoma (NICE TA463 issued in 2017b), four comparators, namely axitinib, everolimus, nivolumab and best supportive care,

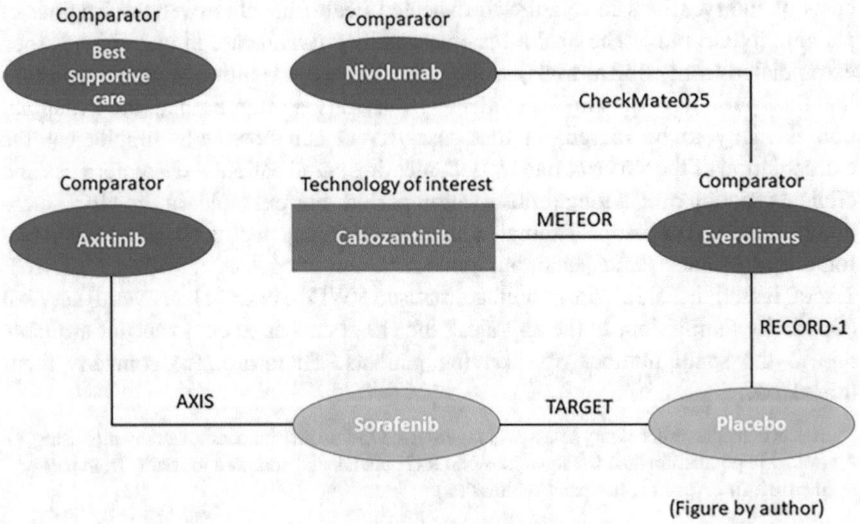

Fig. 1 The evidence network for clinical outcome of TA463

were selected. However, the METEOR phase 3 trial of cabozatinib only included everolimus in the comparator arm since nivolumab was approved during its trial. Due to the lack of comparator data, an indirect comparison was conducted based on available RCTs (Fig. 1). Combining RWD and RCTs in evidence synthesis has the potential to support the findings from RCTs, increase precision and enhance the decision-making process (Efthimiou et al. 2017).

Furthermore, in order to provide fast access to novel treatments, drug regulatory agencies like FDA and EMA accelerate approval based on the results of single-arm trials (Miller and Joffe 2011). The single-arm trial is the simplest trial design to obtain evidence of the efficacy of treatment among individuals with the targeted medical condition without randomisation and a control arm (Evans 2010, 1). Single-arm trials provide the outcome based on the hypothesis that they are also clinical trials designed to test the efficacy or safety of the intervention when there is no comparator. When the information of the comparator is not available in the same trial, the efficacy of comparators mostly comes from other data sources such as RWD or historical RCTs, which exploit the data of previously conducted RCTs (Zhang et al. 2010). NICE published an appraisal of axicabtagene ciloleucel (NICE TA559 issued in 2019a), in which an observational cohort study was used to provide data for the comparators. Axicabtagene ciloleucel is recommended for use within the Cancer Drugs Fund as an option for treating diffuse large B-cell lymphoma and primary mediastinal large B-cell lymphoma after two or more systemic therapies. It is an autologous anti-CD19 chimeric antigen receptor (CAR) T-cell therapy, which is an innovative technology that modified genetics. As this technology has been approved based on ZUMA-1, a single-arm study, comparator data needed to be taken from an alternative source, SCHOLAR-1. This database is a retrospective patient-level study with pooled data from two observational cohorts and follow-up of two large phase 3 RCTs. As the different patient characteristics in two data were likely to impact the clinical outcome, the company adjusted for patient performance status in order to exclude patient in SCHOLAR-1 who would not have been eligible for ZUMA-1. Even though it was noted that comparative-effectiveness results from single-arm studies were prone to bias, the committee concluded that using two single-arm studies was suitable and that it would consider the results of these studies in its decision making given the population characteristics (poor prognosis and vulnerability) and potential difficulties with randomisation. Another example is the guidance for venetoclax for treating chronic lymphocytic leukaemia (NICE TA487 issued in 2017d). The company set two comparators, including best supportive care and palliative care. Since, the clinical evidence for venetoclax came from one phase I (M12-175 trial) and two phase II (M13-982, M14-032) single-arm trials, the data for the comparator came from a different source. The company chose the UK CLL (chronic lymphocytic leukaemia) Forum registry data for survival outcome of palliative care in its submission. In the appraisal, the ERG comments that palliative care was not valid comparator as patients suitable for palliative care have the more advanced disease than those for whom ventoclax is an option. The committee concluded that best supportive care was a more appropriate comparator, and the evidence of palliative care was excluded in final decision-making. These examples show that RWD can help to fill the evidence gaps in some instances.

The third way RWD is used in HTA is when RWD supplements the information on a generalisation of evidence. Eichler et al. (2011) pointed out the limitation of RCTs with respect to the efficacy-effectiveness gap of results on the therapeutic efficacy of medicines from tightly controlled RCT settings and the effectiveness of medicine in the real-world. In the appraisal of afatinib for treating epidermal growth factor receptor mutation-positive locally advanced or metastatic non-small-cell lung cancer (NICE TA310 issued in 2014), the ERG highlighted that:

> in view of the important uncertainties around the roles played by specific mutations (singly or in combination) in determining clinical benefit from TKIs (tyrosine kinase inhibitors), it would be most valuable to have data from a long-term clinical registry of all UK patients treated with TKIs. Such a data source could provide a basis for research and audit to inform future assessments of TKIs in a UK specific population.

This implies that RWD such as clinical registry can help generalise the result of RCTs, by including some of the uncertainties, complexity and non-linearity that characterise the efficacy of a treatment in a real-world setting. Also, RWD could give additional information, which is able to reflect the current clinical practice. In appraisal of oncology medicine, the choice of comparators and subsequent treatments is important to populate the cost-effectiveness model as it impacts not only survival outcome but also the cost. Usually, the clinical guideline indicates the treatment line, which clearly informs which drugs are available in each treatment line. However, the treatments are not equally used in clinical practice. Some treatments can be more frequently used than others due to better compliance or clinical prognosis. Also, there is a lack of an established standard of care in the latter line of treatment. In these cases, RWD can provide a snapshot of drug usage. In the technology guidance for Ibrutinib for treating Waldenstrom's macroglobulinaemia (NICE TA491 issued in 2017e), the company demonstrated that the physician's choice is the most relevant comparator due to the lack of a standard of care in the clinical guideline. In order to try to delineate the composition of a physician's choice, a pan-European chart review was used. A medical chart review of Waldenstrom's macroglobulinaemia (WM) patients was used to generate data on epidemiologic/treatment patterns and efficacy outcomes for WM over a prolonged period of time, specifically on subsequent lines of treatment (i.e. 3L and 4L) because WM patients tend to receive multiple lines of treatment during their lifetimes. RWD would help to maintain the validity and generalisability of the evidence by capturing the current clinical practice.

Lastly RWD is used in HTA to appraise the treatments for rare diseases or conditions, the so-called orphan medicines (EMA n.d.-c). Orphan medicines have difficulties gathering the information to populate the economic evaluation model due to the small patient population. It is challenging to conduct good quality of RCTs. In most cases, the assumption of the model is based on the clinical experts' opinions. RWD can give a wide range of information required in the cost-effectiveness analysis. It could be a treatment effect of the comparator or the resource use data such as the frequency of hospitalisation. For example, among 1930 people diagnosed with follicular lymphoma annually in the UK, only 52 double refractory patients are eligible for the idelalisib (NICE TA604 issued in 2019d). The manufacturer of idelalisib submitted DELTA, a single-arm trial, as primary clinical evidence along with a comparator cohort created from the registry data (HMRN; haematological malignancy research network). The committee acknowledged that it was likely that the HMRN was the only source of comparative data available for the UK population, and agreed to accept the estimate of progression-free survival from HMRN even though HMRN data had limitations. RWD also can supplement the information on choosing a survival model of rare cancer. The choice of survival distribution model has a huge impact on the estimate of survival. It is critical to know how hazard is changed over time. However, clinical trials of treatment for a rare cancer are less likely to provide the full picture of changing hazard due to the small size of the trial population. RWD such as registry which the long-term observed outcome is available, can help validate the model assumptions, including the choice of survival model. Likewise, in the appraisal of precision medicines, RWD might be able to fill the evidence gap created by difficulties in showing the statistical significance due to small populations.

To summarise, RWD is used in HTA for four main reasons:

- To supplement the information when extrapolating the long-term survival curve after the trial period for economic evaluation.
- To help provide information about the comparators such as the choice of relevant comparators reflecting clinical practice and treatment effects.
- To help supplement the information on a generalisation of evidence which is hardly captured in RCTs.
- To help appraise treatments for rare diseases or conditions.

Challenges of Using RWD in HTA

Despite the opportunities of using RWD in HTA mentioned in the previous sections, and the growing hype regarding big data, RWD is not a panacea for the evidence paucity in HTA. Indeed, we lack an understanding of the potential benefits, risks and limitations of using RWD. The first important challenge to the use of RWD in HTA, and despite an increasing interest in RWD worldwide, is that there is no consensus on the precise contours of what constitutes RWD, and many different definitions can be found (Makady et al. 2017b). This is one of the significant obstacles to using RWD in HTA. Indeed, while the flexible definition of RWD allows representing different concepts or types of information, it also limits the potential role of RWD in HTA (Berger et al. 2017, 3). With the objective of strengthening the use of RWD in HTA, Makady and colleagues (2017a, b) proposed four broad categories to define RWD: (1) data collected in a non-RCT setting, (2) data collected in a non-interventional/non-controlled setting, (3) data collected in a non-experimental setting, and/or (4) remainders. Among these four categories, 'data collected in a non-RCT setting' was the most commonly used definition of RWD. These definitions focus on the setting of collecting data. However, definition of RWD by FDA highlights the way to collect the data. As an umbrella term, FDA defines that RWD is the data relating to patient health status and/or the delivery of health care, which is routinely collected from a variety of sources (FDA 2018). In most cases, two definitions get along well; however, some studies can be interpretated in different way by the choice of definition. For example, A. Lloyd and colleagues on health state utility value is frequently used in the NICE appraisals. In this study, they interviewed the general public to get access to some of the societal preferences about treatment of metastatic breast cancer. The study was designed to include 100 people in order to try and represent the preference of the general public once in the study period (Lloyd et al. 2006). Whilst the health utility values were collected outside clinical data, data about health status was not routinely collected. Depending on which definition is chosen, this study can be defined as either RWD or not. According to the definition of FDA, the data from this study is not RWD as the data is collected once outside clinical trial. On the other hand, it can be RWD as the study is non-RCTs. Without any consistency in the definition of RWD, the potential benefits of using RWD in HTA are weakened.

One of the main concerns of using RWD in the appraisal is the issue of confounding. In statistics, a confounding variable is a variable, other than the independent variables of interested that may affect the dependent variable. It can lead to erroneous conclusions about the relationship between the independent and dependent variables (McDonald 2009). RWD is prone to be manipulated and biased by residual confounding since it is hard to control all the confounding factors, including explicit factors as well as underlying factors, without randomisation (Grieve et al. 2016). It is inadequate to distinguish between the effect of the treatment, a placebo effect, and the effect of natural history (Evans 2010, 2). For example, patient's health status such as cancer stage and underlying health condition are highly likely to influence clinical outcomes. As the response rate to second-line treatment differs from first-line treatment, it is critical to understand the patient characteristics for precise assessment. The appraisal of Ibrutinib for treating relapsed or refractory mantle cell lymphoma (NICE TA502 issued in 2018) included the HMRN audit data for comparator data since the main clinical evidence was a single-arm trial. The HMRN data consisted of evidence from a unified clinical network operating across 14 hospitals in Northern England (Yorkshire). The company used data on the benefit of the comparator (R-chemo; rituximab + chemotherapy) from the HMRN audit of 118 patients with mantle cell lymphoma that had been treated with first-line treatment. However, the ERG had a concern about the evidence that the HMRN audit did not specifically relate to patients with relapsed or refractory mantle cell lymphoma. The ERG also highlighted that:

> It is also noteworthy that since this is not a trial, differences in outcomes between patients receiving R-chemo and those receiving chemotherapy alone may be subject to confounding. The HR (hazard ratio) reported in the audit includes adjustments only for age and sex. (ERG document, 67)

Another set of challenges to the use of RWD in HTA are the unanchored comparisons. Unanchored treatment comparison result from the network of studies being disconnected or single-arm studies (Phillippo et al. 2016). When treatment outcomes come from single-arm studies such as phase 1/2 clinical studies or observational studies, the comparison is unanchored. Unanchored comparison is highly likely to misguide the result as it is confounded by the differences between the two populations. Since the number of technologies in which single-arm trials are the primary clinical evidence has increased for drug approval and reimbursement assessment, the population adjustment methods such as matching adjusted indirect comparison (MAIC) and simulated treatment comparison (STC) were highlighted (Phillippo et al. 2016). The methods assume to take account of all effect modifiers and prognostic factors and control them. If the assumption fails, it will lead to a biased conclusion. In the appraisal of cemiplimab for treating metastatic or locally advanced cutaneous squamous cell carcinoma (NICE TA592 issued in 2019b), only two single-arm trials of cemiplimab were available. The comparator data were very limited. Therefore, a non-UK retrospective chart review study was included in company's base case. The study evaluated the outcome of patient who took systemic therapy reviewing patient hospital records (Jarkowski et al. 2016). The company tried to use STC and MAIC for indirect treatment comparison. However, it concluded to choose the naïve comparison due to the uncertainty around missing unmeasured prognostic factors and the validity issue of survival curve, which comes from significantly reduced effective sample size (65% of the original sample size). The committee noted that it was not methodologically recommended because outcomes were likely to be confounded by differences between the populations of the studies (Fig. 2).

Besides, when incorporating RWD into HTA, different approaches should be applied due to the variation of the contents in RWD. As RWD include the diverse type of data, each dataset has different attributes. It means that all RWD does not

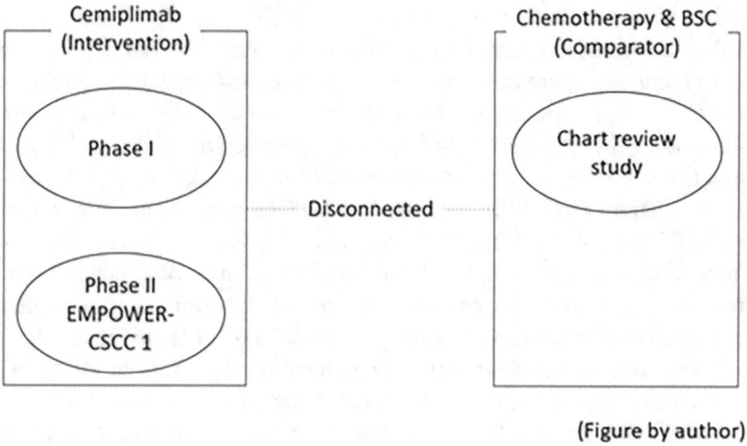

Fig. 2 Indirect treatment comparison of TA592

necessarily provide the same information. For example, Korean healthcare claims data collected by Health Insurance Review and Assessment Service (HIRA) are national-wide data of over 50 million people (Kim et al. 2020). The database includes the information of age, gender, diagnosis, and utility volume of medical intervention (diKhi n.d.). Although the data have useful information in terms of resource use, the clinical data of patients is not available. Inevitably, clinical research such as survival analysis using the claims data sets the operational definition by an individual researcher. It is a strong assumption that the effect modifiers are adjusted by variables defined operationally. Eventually, it could bring uncertainty into the appraisal. Therefore, several approaches of incorporating RWD by characteristics should be discussed in order to use RWD in the appraisals without distortion.

Moreover, the quality of RWD questions the reliability of the outcome as evidence. To evaluate the quality of RWD, we need to know precisely how the data has been collected and how it has been used in HTA. Due to the characteristics of observational studies, RWD has limitations in the quantity and quality of information. In the aforementioned venetoclax technology appraisal guidance (NICE TA487), the quality of data is the issue to include as the key evidence for decision making. The target population for the decision problem was stratified by 17p deletion/TP53 mutation group and failure of B-cell receptor pathway inhibitor (BCRi). Therefore, information on chromosomal abnormality and disease staging is essential. While the registry data have information on time from BCRi treatment failure to death, staging information is not complete. The lack of staging information introduced the significant mismatch between comparators group and intervention group. In company submission, it reported that:

> As patients without the deletion have a better prognosis than patients with the deletion, and given the fact that UK CLL forum data were not stratified by del(17p)/TP53 mutation, this may contribute to overestimating the survival of palliative care which appears better than BSC on the long term. (CS, 145)

Another challenge is that of generalisation. RCTs provide efficacy and safety data with relatively high internal validity, but their results may not be readily generalisable to a broader, more heterogeneous population (Makady et al. 2017b). RWD is expected to provide more information to reflect the clinical practice. Notwithstanding the expectation for improving the external validity of RCTs, RWD has limitations in terms of representativeness of the population. It is highly questionable whether all RWD fully capture a holistic picture of reality. For example, the GIDEON (Global Investigation of therapeutic DEcisions in hepatocellular carcinoma and Of its treatment with sorafeNib) study predominantly includes the Asian population to represent the general UK population in the evaluation of sorafenib for treating advanced hepatocellular carcinoma (NICE TA474 issued in 2017c). Since the treatment effect of sorafenib differed by global regions, it is questionable to use GIDEON data to predict the treatment effect in the UK population. Another example is ceritinib for previously treated anaplastic lymphoma kinase positive non-small-cell lung cancer (NICE TA395 issued in 2016). The manufacturer of certinib submitted additional real-world evidence (Gainor et al. 2015), which were medical records reviewed to determine OS and PFS (progression free survival) in patients who were treated with sequential crizotinib and ceritinib between 2008 and 2014. The ERG criticised that this retrospective non-randomised study did not clearly show how similar these participants are to those in the ceritinib studies. In the appraisal of nivolumab for treating relapsed or refractory classical Hodgkin lymphoma (NICE TA462 issued in 2017a), the generalisability of RWD into UK practice was questioned. The company used the Cheah et al. study for evidence on the clinical outcome estimates of comparator, OS and PFS. The data used in the study came from the American hospital database (Cheah et al. 2016). The committee considered whether the population and composition of treatments in the Cheah et al. study reflected clinical practice in the UK. It considered that the study population partially matched the population of interest. Furthermore, it deemed that the study may not reflect UK practice, notably regarding subsequent treatment rates of allogeneic stem cell transplant. Even if RWD is collected from routine practice, the context of collecting data could be different by country or region. The difference is likely to introduce a bias in representativeness. As the study type such as observational study does not guarantee the generalisability of the evidence, the clinical and social context should be carefully considered when using RWD.

To summarise, these are some of the key challenges related to the use of RWD in HTA:

- There is no consensus on the precise contours and definition of what constitutes RWD.
- RWD inherits the risk of confounder to see the causality.
- RWD is challenging to see the relative treatment effect due to the disconnection with other clinical studies.
- It is required to understand each dataset separately with a caveat that individual data categorised as RWD have different characteristics.
- Quality of RWD such as incompleteness is often questioned.

- RWD is not necessarily generalisable as it does not always reflect whole patients or up-to-date practices.

What Is Next: Do We Want RDW in HTA Be the Cynosure of All Eyes?

The hope for advances in health care through the use of big data, and more specifically RWD, is getting stronger. The pharmaceutical industry has been using RWD for decades to conduct post-market research, inform its decision-making, respond to requests from external stakeholders, and improve market positioning (Mckinsey&Company 2020). The advances in digital and advanced analytics allow RWD to be more employed in the health care ecosystems. The quality of RWD itself is also steadily enhanced as the research using RWD has increased drastically in the last decade (Booth et al. 2019; Evans et al. 2021). Growing interests in RWD clearly create more and more opportunities to generate new evidence from drug development to post-approval studies (Rudrapatna and Butte 2020). In the era of digitalised RWD, the progression in the use of RWD makes the public hold great promise in the ability of it to transform the entire health care system (Berger et al. 2015). The FDA approval of palbociclib shows that RWD can take a central role in the regulatory process. While the FDA accepted RWD in limited use, such as informing the prognosis or natural history of the disease, it was the first approval based on RWD in oncology. The FDA shows confidence that leveraging data such as RWD using modern techniques will unlock new insight and provide state-of-the-art tools to enhance public health. Such potentialities have inundated public discourses with optimistic narratives of cutting-edge scientific innovation and hopes for a complete cure for devastated cancer patients, praising modern science progress and putting forward opportunities for accessing innovative technology at the earliest possible time. However, it is still questioned that RWD can replace the state of RCTs (Ramagopalan et al. 2020). Without appropriate consideration of using RWD, the regulatory body has to take much greater level of uncertainty than the benefit of accelerating patient access to treatment.

In HTA process, diverse types of data such as RWD are already incorporated into evidence. As the interest in RWD is growing, the use of RWD in treatment effects receive more attention. But as discussed above, using RWD in HTA is intricate. Indeed, we have seen that there are many uncertainties related to what RWD can actually bring to HTA in particular, and health care systems in general. RWD are complex, difficult to grasp and to define, and it is also difficult to evaluate their quality. In addition, RWD has limitations to assess the relative treatment effects due to confounders and dysconnectivity. Therefore, it is crucial to think to what extent RWD can be incorporated in HTA, in which part of evidence synthesis it can actually contribute, and what are its limitations. Diverse definitions and data formations allow RWD to be used widely, but also present challenges of consistency, quality, generalisation, and purpose. It is important to critically scrutinise RWD and the hope it might convey, and carefully examine the quality of RWD as a source of evidence. RWD is not, and should not be considered as, an easy fix to the complex question of how to assess health technologies. Arguably, a more systematic approach to RWD could help enhance its robustness when used in HTA. In this chapter, I questioned the idea of RWD as a corrective measure, by critiquing the effectiveness of RWD utilisation in HTA. By critically questioning the drawbacks, limitations and challenges of using RWD in HTA, we can expect to have a more balanced and responsible use of RWD in the future, that does not overpromise results that are unachievable in a context of high uncertainties and complexity. It would also help form more realistic expectations of what RWD can and cannot bring to HTA and health care in general. And fundamentally, before expanding the use of RWD in technology assessment, we should think about exactly what are RWD, to what purpose we want to use them, and how we can meaningfully evaluate their quality. From that line of critical questions, we would be able to think more realistically about the practical benefits and challenges to incorporate these data into evidence synthesis, and to what extent RWD actually contributes to the evaluation of new technology.

References

Aggarwal, A., T. Fojo, C. Chamberlain, C. Davis, and R. Sullivan. 2017. Do patient access schemes for high-cost cancer drugs deliver value to society? – Lessons from the NHS Cancer Drugs Fund. *Annals of Oncology* 28 (8): 1738–1750. https://doi.org/10.1093/annonc/mdx110.

Augustine, E.F., H.R. Adams, and J.W. Mink. 2013. Clinical trials in rare disease: Challenges and opportunities. *Journal of Child Neurology* 28 (9): 1142–1150. https://doi.org/10.1177/0883073813495959.

Balasubramanian, G., S. Morampudi, P. Chhabra, A. Gowda, and B. Zomorodi. 2016. An overview of compassionate use programs in the European union member states. *Intractable and Rare Diseases Research* 5 (4): 244–254. https://doi.org/10.5582/irdr.2016.01054.

Bayat, A. 2002. Science, medicine, and the future: Bioinformatics. *British Medical Journal* 324 (7344): 1018–1022. https://doi.org/10.1136/bmj.324.7344.1018.

Berger, M.L., C. Lipset, A. Gutteridge, K. Axelsen, P. Subedi, and D. Madigan. 2015. Optimizing the leveraging of real-world data to improve the development and use of medicines. *Value in Health* 18 (1): 127–130. https://doi.org/10.1016/j.jval.2014.10.009.

Berger, M., G. Daniel, K. Frank, A. Hernandez, M. McClellan, S. Okun, M.Overhage, et al. 2017. *A framework for regulatory use of real-world evidence*. Available at https://healthpolicy.duke.edu/sites/default/files/atoms/files/rwe_white_paper_2017.09.06.pdf.

Booth, C.M., S. Karim, and W.J. Mackillop. 2019. Real-world data: Towards achieving the achievable in cancer care. *Nature Reviews. Clinical Oncology* 16 (5): 312–325. https://doi.org/10.1038/s41571-019-0167-7.

Burns, P.B., R.J. Rohrich, and K.C. Chung. 2011. The levels of evidence and their role in evidence-based medicine. *Plastic and Reconstructive Surgery* 128 (1): 305–310. https://doi.org/10.1097/PRS.0b013e318219c171.

Cheah, C.Y., D. Chihara, S. Horowitz, A. Sevin, Y. Oki, S. Zhou, and N.H. Fowler. 2016. Patients with classical Hodgkin lymphoma experiencing disease progression after treatment with brentuximab vedotin have poor outcomes. *Annals of Oncology* 27 (7): 1317–1323. https://doi.org/10.1093/annonc/mdw169.

De Brún, C. 2013. *The information standard guide finding the evidence-7th*. NHS.

diKhi. n.d. *Healthcare Big Data platform*. Available at https://hcdl.mohw.go.kr/BD/Portal/Enterprise/DefaultPage.bzr. Accessed 3 Aug 2020.

Duggan, M.A., W.F. Anderson, S. Altekruse, L. Penberthy, and M.E. Sherman. 2016. The surveillance, epidemiology, and end results (SEER) program and pathology: Toward strengthening the critical relationship. *American Journal of Surgical Pathology* 40 (12): e94–e102. https://doi.org/10.1097/PAS.0000000000000749.

Efthimiou, O., D. Mavridis, T.P.A. Debray, M. Samara, M. Belger, G.C.M. Siontis, S. Leucht, and G. Salanti. 2017. Combining randomized and non-randomized evidence in network meta-analysis. *Statistics in Medicine* 36 (8): 1210–1226. https://doi.org/10.1002/sim.7223.

Eichler, H.G., E. Abadie, A. Breckenridge, B. Flamion, L.L. Gustafsson, H. Leufkens, M. Rowland, C.K. Schneider, and B. Bloechl-Daum. 2011. Bridging the efficacy g-effectiveness gap: A regulator's perspective on addressing variability of drug response. *Nature Reviews Drug Discovery* 10 (7): 495–506. https://doi.org/10.1038/nrd3501.

EMA. n.d.-a. *Compassionate use*. https://www.ema.europa.eu/en/human-regulatory/research-development/compassionate-use. Accessed 23 July 2020.

———. n.d.-b. *Compassionate use*. https://www.ema.europa.eu/en/human-regulatory/research-development/compassionate-use. Accessed 1 May 2020.

———. n.d.-c. *Orphan medicine*. https://www.ema.europa.eu/en/glossary/orphan-medicine. Accessed 11 Aug 2020.

Evans, S.R. 2010. Clinical trial structures. *Journal of Experimental Stroke and Translational Medicine* 3 (1): 8–18. https://doi.org/10.6030/1939-067X-3.1.8.

Evans, S.R., D. Paraoan, J. Perlmutter, S.R. Raman, J.J. Sheehan, and Z.P. Hallinan. 2021. Real-world data for planning eligibility criteria and enhancing recruitment: Recommendations from the clinical trials transformation initiative. *Therapeutic Innovation & Regulatory Science* 55 (3): 545–552.

Favaretto, M., E. De Clercq, C.O. Schneble, and B.S. Elger. 2020. What is your definition of Big Data? Researchers' understanding of the phenomenon of the decade. *PLoS One* 15 (2): e0228987.

FDA. 2018. *Framework for FDA's real-world evidence program*. https://www.fda.gov. Accessed 1 May 2020.

———. 2020a. *21st Century Cures Act*. https://www.fda.gov/regulatory-information/selected-amendments-fdc-act/21st-century-cures-act. Accessed 14 Mar 2021.

———. 2020b. *Real-world evidence*. https://www.fda.gov/science-research/science-and-research-special-topics/real-world-evidence. Accessed 6 May 2020.

Gainor, J.F., D.S.W. Tan, T. De Pas, B.J. Solomon, A. Ahmad, C. Lazzari, F. de Marinis, et al. 2015. Progression-free and overall survival in ALK-Positive NSCLC patients treated with sequential Crizotinib and Ceritinib. *Clinical Cancer Research* 21 (12): 2745–2752. https://doi.org/10.1158/1078-0432.CCR-14-3009.

Gibbs, W.J., and J.E. McKendrick. 2015. *Contemporary research methods and data analytics in the news industry*. Hershey: IGI Global. https://doi.org/10.4018/978-1-4666-8580-2.

Grieve, R., K. Abrams, K. Claxton, B. Goldacre, N. James, J. Nicholl, M. Parmar, et al. 2016. Cancer drugs fund requires further reform. *BMJ* 354: i5090. https://doi.org/10.1136/bmj.i5090.

Henshall, C., W. Oortwijn, A. Stevens, A. Granados, and D. Banta. 1997. Priority setting for health technology assessment. Theoretical considerations and practical approaches. Priority setting Subgroup of the EUR-ASSESS Project. *International Journal of Technology Assessment in Health Care* 13 (2): 144–185.

HMA, EMA. 2019. *HMA-EMA Joint Big Data Taskforce – Summary report*. https://www.ema.europa.eu/en/about-us/how-we-work/big-data. Accessed 30 Apr 2020.

Jarkowski, A., R. Hare, P. Loud, J.J. Skitzki, J.M. Kane 3rd, K.S. May, N.C. Zeitouni, et al. 2016. Systemic therapy in advanced cutaneous squamous cell carcinoma (CSCC): The Roswell Park experience and a review of the literature. *American Journal of Clinical Oncology: Cancer Clinical Trials* 39 (6): 545–548. https://doi.org/10.1097/COC.0000000000000088.

Kim, S., M.-S. Kim, S.-H. You, and S.-Y. Jung. 2020. Conducting and reporting a clinical research using Korean healthcare claims database. *Korean Journal of Family Medicine* 41 (3): 146–152. https://doi.org/10.4082/kjfm.20.0062.

Lloyd, A., B. Nafees, J. Narewska, S. Dewilde, and J. Watkins. 2006. Health state utilities for metastatic breast cancer. *British Journal of Cancer* 95 (6): 683–690. https://doi.org/10.1038/sj.bjc.6603326.

Makady, A., R.T. Ham, A. de Boer, H. Hillege, O. Klungel, W. Goettsch, and GetReal Workpackage 1. 2017a. Policies for use of real-world data in health technology assessment (HTA): A comparative study of six HTA agencies. *Value in Health* 20 (4): 520–532. https://doi.org/10.1016/j.jval.2016.12.003.

Makady, A., A. de Boer, H. Hillege, O. Klungel, and W. Goettsch. 2017b. What is real-world data? A review of definitions based on literature and stakeholder interviews. *Value in Health* 20 (7): 858–865.

McDonald, J.H. 2009. *Handbook of biological statistics*. Baltimore: Sparky House Publishing.

Mckinsey&Company. 2020. *Creating value from next-generation real-world evidence*. https://www.mckinsey.com/industries/pharmaceuticals-and-medical-products/our-insights/creating-value-from-next-generation-real-world-evidence. Accessed 19 Mar 2021.

Miller, F.G., and S. Joffe. 2011. Balancing access and evaluation in the approval of new cancer drugs. *JAMA : The Journal of the American Medical Association* 305 (22): 2345–2346. https://doi.org/10.1001/jama.2011.784.

National Institute for Health and Care Excellence (NICE). n.d. *About | NICE*. https://www.nice.org.uk/about. Accessed 14 May 2019.

NICE. 2014. *TA310 Afatinib for treating epidermal growth factor receptor mutation-positive locally advanced or metastatic non-small-cell lung cancer*. NICE.

———. 2016. *TA395 Ceritinib for previously treated anaplastic lymphoma kinase positive non-small-cell lung cancer*. NICE.

———. 2017a. *TA462 Nivolumab for treating relapsed or refractory classical Hodgkin lymphoma*. NICE.

———. 2017b. *TA463 Cabozantinib for previously treated advanced renal cell carcinoma*. NICE.

———. 2017c. *TA474 Sorafenib for treating advanced hepatocellular carcinoma*. NICE.

———. 2017d. *TA487 Venetoclax for treating chronic lymphocytic leukaemia*. NICE.

———. 2017e. *TA491 Ibrutinib for treating Waldenstrom's macroglobulinaemia*. NICE.

———. 2018. *TA502 Ibrutinib for treating relapsed or refractory mantle cell lymphoma*. NICE.

———. 2019a. *TA559 Axicabtagene ciloleucel for treating diffuse large B-cell lymphoma and primary mediastinal large B-cell lymphoma after 2 or more systemic therapies*. NICE.

———. 2019b. *TA592 Cemiplimab for treating metastatic or locally advanced cutaneous squamous cell carcinoma*. NICE.

———. 2019c. *TA600 Pembrolizumab with carboplatin and paclitaxel for untreated metastatic squamous non-small-cell lung cancer*. NICE.

———. 2019d. *TA604 Idelalisib for treating refractory follicular lymphoma*. NICE.

———. 2020a. *NICE's methods of technology evaluation – Presenting a case for change*. https://www.nice.org.uk/news/article/nice-s-methods-of-technology-evaluation-presenting-a-case-for-change.

———. 2020b. *Reviewing our methods for health technology evaluation: Consultation*. https://www.nice.org.uk/about/what-we-do/our-programmes/nice-guidance/chte-methods-consultation. Accessed 14 Mar 2021.

———. n.d. *7 assessing cost effectiveness*. https://www.nice.org.uk/process/pmg6/chapter/assessing-cost-effectiveness.

Oxford Big Data Institute. 2019. *BDI and Sensyne Health create new research alliance*. https://www.bdi.ox.ac.uk/news/bdi-and-sensyne-health-create-new-research-alliance. Accessed 19 Mar 2021.

Phillippo, D.M., A.E. Ades, S. Dias, S. Palmer, K.R. Abrams, and N.J. Welton. 2016. *NICE DSU technical support document 18: Methods for population-adjusted indirect comparisons in submissions to NICE report by the decision support unit*. www.nicedsu.org.uk. Accessed 3 Aug 2020.

Potter, B.K., D. Avard, V. Entwistle, C. Kennedy, P. Chakraborty, M. McGuire, and B.J. Wilson. 2008. Ethical, legal, and social issues in health technology assessment for prenatal/preconceptional and newborn screening: A workshop report. *Public Health Genomics* 12 (1): 4–10. https://doi.org/10.1159/000153430.

Ramagopalan, S.V., A. Simpson, and C. Sammon. 2020. Can real-world data really replace randomised clinical trials? *BMC Medicine* 18: 13.

Rudrapatna, V.A., and A.J. Butte. 2020. Opportunities and challenges in using real-world data for health care. *Journal of Clinical Investigation* 130 (2): 565–574.

Sculpher, M.J., F.S. Pang, A. Manca, M.F. Drummond, S. Golder, H. Urdahl, L.M. Davies, and A. Eastwood. 2004. Generalisability in economic evaluation studies in healthcare: A review and case studies. *Health Technology Assessment* 8 (49): 1–192.

Ugalmugle, S., and R. Swain. 2020. Precision medicine market size to exceed $119 Bn by 2026. *Global Market Insight*. https://www.gminsights.com/pressrelease/precision-medicine-market. Accessed 16 July 2020.

Wedam, S., L. Fashoyin-Aje, E. Bloomquist, S. Tang, R. Sridhara, K.B. Goldberg, M.R. Theoret, L. Amiri-Kordestani, R. Pazdur, and J.A. Beaver. 2020. FDA approval summary: Palbociclib for Male patients with metastatic breast cancer. *Clinical Cancer Research* 26 (6): 1208–1212. https://doi.org/10.1158/1078-0432.CCR-19-2580.

WHO. 2015. *WHO | HTA definitions*. https://www.who.int/health-technology-assessment/about/Defining/en/. Accessed 13 May 2019.

Zhang, S., J. Cao, and C. Ahn. 2010. Calculating sample size in trials using historical controls. *Clinical Trials* 7 (4): 343–353. https://doi.org/10.1177/1740774510373629.

Just Caring: Precision Health vs. Ethical Ambiguity: Can we Afford the Ethical and Economic Costs?

Leonard M. Fleck

It is easy to see "precision medicine" and "precision health" complementing one another. We want precision medicine available to us when we are unfortunate enough to be faced with a life-threatening cancer that could be cured or managed with a targeted cancer therapy. At the same time, we would rationally prefer to take advantage of whatever medicine might offer us that would prevent the emergence of that cancer in the first place or treat it in its earliest stages, which is the goal of precision health. However, precision medicine and precision health can just as easily and realistically be seen as competing for resources with one another, as we will explain below.

Just Caring: Cancer, Targeted Therapies, and Cost Control

The fundamental ethical and economic problem with health care today is that we have limited resources (money) to meet virtually unlimited health care needs. From an ethical perspective, this is what I refer to as the "Just Caring" problem (Fleck 2009). What we would identify as health care needs have multiplied exponentially as a result of very costly, life-prolonging medical technologies that have been developed over the past fifty years, targeted cancer therapies being one pre-eminent example.[1] This has strained social budgets in both the United States and the

[1] More than 90 targeted cancer therapies or immunotherapies have received approval from the Food and Drug Administration (FDA). The median list price for a course of treatment or a year of treat-

L. M. Fleck (✉)
Center for Ethics, College of Human Medicine, Michigan State University, East Lansing, MI, USA
e-mail: fleck@msu.edu

A. Bremer, R. Strand (eds.), *Precision Oncology and Cancer Biomarkers*, Human Perspectives in Health Sciences and Technology 5,
https://doi.org/10.1007/978-3-030-92612-0_13

European Union, as much recent research has demonstrated (Vokinger et al. 2020; Wilking et al. 2017; Hofmarcher et al. 2020; Peppercorn 2017; Yabroff et al. 2019; Bender 2018; Leopold et al. 2018).[2] Relieving that strain means controlling health care costs by controlling access to these very expensive life-prolonging technologies, especially in clinical circumstances where the clinical benefit is very marginal relative to costs. This is health care rationing that has ethically substantial consequences. Hence, the ethical challenge is to determine how such limitations can be justly decided. Having said that, we might not see very easily how this challenge has anything to do with pursuing precision health at the expense of precision medicine and unlimited access to these targeted cancer therapies.

Prevention is supposed to be quite inexpensive: do not smoke, consume alcohol in moderation, use sunscreen, get a reasonable amount of exercise, eat a healthy diet. However, in spite of adhering with saintly devotion to these health directives, some individuals will still find themselves faced with a cancer diagnosis. This could be for genetic reasons, or environmental reasons, or just the random breakdown of cellular machinery. Roughly 40% of Americans will develop a cancer over the course of their life. The projections in the United States for 2020 are that 1.8 million individuals will be diagnosed with cancer; cancer deaths will be a little over 600,000 (National Cancer Institute 2020). Comparable figures for Europe are 3.9 million cancer diagnoses in 2018 and 1.9 million deaths (Ferlay et al. 2018). These cancer deaths are the product of metastatic disease. Metastatic cancer is a terminal condition. The targeted cancer therapies and immunotherapies that are at the leading edge of cancer research are all used to treat metastatic disease. However, none of them can justifiably claim to yield a cure. In the vast majority of circumstances these interventions yield marginal gains in life expectancy measurable in months, though for a small fraction of patients there will be extra years of life. From a social point of view, it is unclear that this represents either a wise or just use of social resources.

ment is now $150,000 (Manz et al. 2019). At the upper end of these therapies is CAR T-cell therapy, a form of immunotherapy that has a front-end cost of $475,000 with $200,000 in addition (or more) to manage the predictable complex side effects of this therapy (cytokine release syndrome or various neurotoxicities).

[2]The primary messages from these articles would be the following: (1) The very high prices of these drugs do not reflect the very marginal clinical value produced in most cases; (2) The projected aggregated social costs of these targeted therapies is not sustainable; (3) If nothing is done to control these escalating prices and costs, both social equity and solidarity will be unjustly compromised. The following passage (Bach 2019) summarizes nicely the problem: "Introductory prices of cancer drugs have risen more than 100-fold since 1965. The trend is unabating. Prices were up last year and again this year. Prices are increasing not only for new drugs as they first enter the market, but also for drugs already in use, often in cases where there is no suggestion that the treatment is any better than originally thought." To illustrate this last point, imatinib [Gleevec®] was introduced in 2002 at a price of $36,000 per year. In 2017 the price had risen to $146,000. To be sure, unlike the vast majority of targeted cancer therapies, imatinib has been very beneficial for the vast majority of patients needing this drug. Still, that does not justify that price increase over the years.

Cancer: Finding the First Cell/Preventing Future Cells

If cancer cannot be completely prevented by good health behavior, and if metastatic disease is essentially incurable, then the next best preventive strategy would be to identify and attack cancer at its earliest possible stages. This is the perspective embraced by Dr. Azra Raza in her recent book, *The First Cell* (Raza 2019). Dr. Raza is an oncologist who has been in practice for more than thirty years. Her cancer research has been focused on myelodysplastic syndromes [MDS], a pre-leukemic condition. Her husband's research was in the same area, though, ironically, he died of acute myelogenous leukemia [AML], which happens in about 33% of patients with MDS. The basic thesis of her book is that we are wasting tens of billions of dollars every year on cancer therapies that are extraordinarily costly and that yield only marginal gains in life expectancy and maximal increases in suffering (physiological, psychological, financial and social). She believes that these same resources should be redirected to destroying cancer in its earliest stages, those "first cells," through multiple preventive strategies. On one level, this is an eminently reasonable position for which she advocates. On another level, this is an ethically radical proposal, given that she wishes to redirect tens of billions of dollars from aggressive life-prolonging care to a preventive strategy that would reduce the need for such aggressive life-prolonging care. This generates the key questions we will address in this chapter: Is this strategy ultimately ethically defensible? Is this a strategy that ought to be embraced by a "just" and "caring" society, given that the consequence of embracing such a strategy would be the "premature" death of hundreds of thousands of metastatic cancer patients each year, most of whom would be denied extra months of life?[3]

A first response to this ethics problem might be that it represents a false choice. We should be doing whatever is possible to prevent cancer, or to attack it in its earliest stages, but if those efforts fail, then we certainly ought to pursue therapeutic life-prolonging options for those unfortunate patients who are faced with metastatic disease, even if those efforts might never be curative. Why would Raza not endorse that view? Here is one scenario Raza has in mind that would suggest a response to this question. She suggests that "everyone from birth to death is regularly screened for the first appearance of cancer cells in the body." Once those cancer cells had been identified "protein markers would be identified, providing a zip code for the cancer cells. A tube of blood from the individuals would be obtained, and T cells would be isolated, activated, and armed with the address for the cancer based upon the unique protein bar code and the RNA signature it expressed" (Raza 2019, 238).

[3]To be clear, I am using the term "premature" to characterize a death when we have the medical technology that would provide additional length of life for a patient facing an imminent death but that technology is denied to the patient for whatever reason. Relative to that future point in time when a patient's death would occur as a result of the life-prolonging technology, their death now would be "premature." I am using this term in a purely descriptive way with no implied prescriptive judgment. That is, I am not saying in any specific circumstances whether that "premature" death is right or wrong.

This scenario has something of a futuristic quality to it.[4] However, what we do have in reality is a new "liquid biopsy" test introduced by a company named GRAIL (Oxnard 2019). This test can detect twenty different cancers in their earliest stages by examining cell-free DNA in the blood. More recent news reports suggest that the test might be capable of detecting fifty different cancers as well as the source of that cancer in 89% of cases. To be clear, this is not a perfect test. Dana-Farber reports that the test has a sensitivity of 32% for a Stage I cancer, 76% for Stage II, 85% for Stage III, and 93% for Stage IV (Oxnard 2019). The test is far from ready for clinical deployment. Consequently, no price has been announced for the test. However, another company offers a liquid biopsy test for eight cancers, which is priced at $500. This figure is likely too low for the GRAIL test. Recall in the quotation above Dr. Raza imagining that individuals would be "regularly screened" with such a test. What might that mean?

Though cancer is much more of a threat to older individuals, a significant number of young adults are diagnosed with a cancer, and some of them will die from that cancer. Given that everyone has a 40% lifetime risk of a cancer diagnosis, it would not be unreasonable to have this screening test on an annual basis. If the test is only offered to adults over twenty-one, that would be 198 million individuals in the United States. At a cost of $500 per test, the potential cost to "society" would be $99 billion per year.[5] That figure brings into sharp focus Dr. Raza's imperative that funding this preventive effort ought to come from what she regards as wasteful and marginally effective spending on these targeted cancer therapies for metastatic disease. This, in turn, raises a number of ethics issues which must be addressed.[6]

We may start by noting that Dr. Raza deserves to be ethically commended for requiring for this preventive effort come from *within* current cancer spending, as

[4]The idea of "screening from birth to death" has a very therapeutic aura around it, but there are some ethical dark spots in that aura. Most cancers arise spontaneously as a result of environmental factors, such as melanoma. Some cancers are hereditary. If we did Whole Genome Sequencing [WGS] of infants at birth, we would identify cancers to which individuals were genetically susceptible (likely later in life). These hereditary cancers would have some probability factor attached to them. If someone has the BRCA1 gene associated with breast cancer, their lifetime risk could be anywhere from 40% to 85%. Children would not be told that as children. However, they would be told that as young adults, though that overrides their putative right "not to know." However, once they know, if they have a certain psychological disposition, they will want frequent testing for any sign of that early cancer. That has societal cost consequences as well as harmful psychological consequences for these patients. In addition, identifying a hereditary cancer through WGS has implications for the vulnerability of relatives, who or may not want that information and related anxieties.

[5]The population of the European Union (still counting the UK) is about 520 million, of which 334 million would be adults over age 21. For that population the hypothetical cost of annual testing with the Grail liquid biopsy would be $167 billion per year.

[6]To avoid confusion, Dr. Raza herself says nothing about the Grail test. She is, however, an advocate for a radical re-distribution of cancer treatment resources toward prevention. It is only for the sake of discussion and analysis in this chapter that I am attributing to her an endorsement of what I will call the Grail testing protocol as one possible incarnation of that for which she is a clear advocate.

opposed to taking that money from heart disease, or resources dedicated to treating addictions or mental illness, or some other disease category. Her basic argument is that most of the targeted cancer therapies represent low-value care, too little good for too much money. We can readily imagine a reasonable argument for saying that it would be unjust to take resources from some other area of health care need where the care provided there was of high-value in order to purchase in its place low-value cancer care. For now, we will simply pass over this issue and focus on ethics issues that arise within the context established by Dr. Raza's proposal.

Trading off Identified Lives and Statistical Lives: Ethical Issues

Perhaps the most salient ethical concern would pertain to the implied tradeoff between the *identified* lives of patients with metastatic cancer and the *statistical* lives that we would hope to save through annual preventive liquid biopsies (Daniels, 2012). The defining feature of statistical lives is that they are nameless and faceless. This is typically true both before the fact and after the fact when considering preventive measures. If we install guardrails at dangerous curves on state highways, we might save fifty lives per year. One year later we might see a decline in fatal slides off the highway of forty-five lives. We would have no idea who those forty-five individuals might have been whose lives were saved, or if the guardrails were necessarily what made that difference. By way of contrast, we know with perfect clarity the identities of individuals with metastatic cancer whose lives were extended (if only for months) as a consequence of their having received one of these targeted cancer therapies. Grail's liquid biopsy may correctly identify individuals with a very early stage cancer (no clinically evident symptoms). That cancer may eventually manifest itself symptomatically, at which point it will most likely be effectively treated (minimal likelihood of recurrence). This is why we currently have 17 million cancer survivors in the United States today (Simon 2019). Projections put that survival figure at 21.7 million by 2029. This projection would not include any assumption about the successful clinical deployment of Grail's liquid biopsy.

We need to emphasize that the vast majority of these cancer survivors will ultimately die of something other than their cancer. Why does this matter? It matters because Grail cannot claim that its liquid biopsy would have saved all these lives. Most of these lives would have been saved by current cancer therapies provided at the appearance of clinical symptoms. This point is important for judging whether Grail's liquid biopsy represents high-value care. If Grail's liquid biopsy reduced the number of metastatic cancer deaths by 70%, that would constitute significant evidence for thinking of the test as representing high-value care. However, that will still leave us with both ethical and economic concerns.

Keep in mind that we are assessing Dr. Raza's proposal (as I have constructed it). The hypothetical under discussion says that we would reduce the number of metastatic cancer deaths in the United States by 400,000 annually. That would still leave 200,000 individuals with a terminal metastatic condition. Dr. Raza would provide those individuals with comfort care, but they would be denied these extraordinarily expensive targeted cancer therapies that for most of them would only yield extra months of life. These are clearly *identified* individuals, most of whom might desperately wish to gain whatever additional life might be possible with access to these targeted therapies. We are trading off for their sacrifice of that additional life indefinite gains in life expectancy (something close to or better than a normal life expectancy) for these 400,000 other individuals who also would otherwise have died prematurely from their metastatic disease. From a purely rational, utilitarian maximization perspective, this looks like an eminently reasonable tradeoff. However, those 400,000 individuals are *statistical* lives.

It may sound odd to say that these are statistical lives, but the fact is that we have no ability to identify who those 400,000 individuals might be. Consequently, they have something of an abstract, ghostly status. This is certainly true from a psychological perspective, compared to those patients with metastatic cancer who want to live longer. To clarify, 1.8 million individuals in the United States are diagnosed with cancer each year. Some of them are diagnosed with metastatic disease. However, the vast majority of those cancer diagnoses will be treated with current therapies. Some portion of those individuals, despite treatment with curative intent, will go on to have metastatic disease. Most of those metastatic cases will not be predictable at the time of initial treatment, even though that treatment might have been early in the disease process. Others might have gone on to have metastatic disease, except that they were treated early and effectively. At this point in time these are all statistical lives.

Eventually, however, in this hypothetical example we end up with 200,000 identifiable individuals with metastatic disease. Is it just and justified that we would deny these individuals our very expensive targeted therapies, as proposed by Dr. Raza, because that $99 billion was allocated to the preventive effort which (by hypothesis) saved 400,000 lives that otherwise would have died prematurely? That brings us back to the question regarding the ethical weight that ought to be attached to statistical lives versus identifiable lives when we are allocating resources for purposes of saving/prolonging lives.

Should identifiable lives (patients with metastatic cancer) have more moral weight (greater just claims to health care resources) than statistical lives (patients at risk of a premature death from cancer)? Brock (2015) would answer this question negatively.[7] He considers and rejects a number of arguments in support of the

[7] See also Paul Menzel (2012). He will defend a position similar to Brock's view. Menzel writes, "If any of us loses our life from lack of adequate prevention, surely, it seems, we lose something *just as valuable* as we do if we lose our life from lack of effective treatment" (2012, 194). This too seems to support Dr. Raza's view. More broadly, these same issues will arise in the context of environmental debates. How much ethical weight should be given to the rights and interests of future generations when we (in the present) need to assess a range of more or less costly options for addressing any number of environmental challenges?

opposite conclusion. I will consider his arguments in the specific context of this chapter. He first calls attention to the Rule of Rescue. The Rule of Rescue might seem to require prolonging the lives of the identifiable metastatic cancer patients, in part because we have at hand the capacity to do so readily with these targeted cancer therapies. By way of contrast, 99% of the liquid biopsies performed annually would have saved no one since the results would be negative for any cancer cells. However, the typical application of the Rule of Rescue involves huge societal expense to rescue trapped miners deep underground (or others in various dire circumstances, such as the cave rescue in Thailand).[8] It would be tragic if we (society) were aware of these situations but had no ability at all to prevent that loss of life. However, it would be unconscionable if we had the ability to intervene but simply ignored the plight of those desperate individuals and allowed them to die. This is the moral logic that says we must do what we can to prolong the lives of those metastatic cancer patients.

Notice that I did correctly describe this latter situation by saying that our goal would be to *prolong, not save,* the lives of these patients. In the typical successful rescue situation individuals have their lives saved; they have an indefinite life expectancy, as in the Thai cave rescue story. That is precisely what is not true with regard to our metastatic cancer patients. Consequently, the moral logic embedded in the Rule of Rescue does not apply in this situation. Perhaps a minor correction is in order when I assert that it is "not true" that these metastatic cancer patients can be saved. In fact, a very small percentage of these patients, sometimes referred to as "super responders", will gain extra years of life from one of these targeted cancer therapies. They may well die of something other than their cancer ten or more years from now.[9] What should we conclude from this? Does this mean that the Rule of Rescue does apply, even though the rescue effort in the vast majority of cases will be unsuccessful? Does this mean that these rescued identifiable lives outweigh the *merely* statistical lives we would hope to save with our liquid biopsy intervention?

This last question brings us to a second consideration by Brock that would favor allocating life-prolonging resources toward identified lives over statistical lives, namely, the uncertainty associated with preventive efforts.[10] We might invest $100 million in an anti-smoking campaign, hoping to save 100,000 lives from lung cancer. The campaign might fail abysmally because smoking is addictive and woven in complex ways into the lives of individuals. Lots of uncertainty seems integral to many preventive efforts. By way of contrast, we are certain to get some benefit in

[8] See Cheung and Wong (2018), The full story of Thailand's extraordinary cave rescue. BBC News.

[9] There is ongoing research aimed at determining why these individuals are so fortunate (Cavallo 2018). One possible explanation is that the evolution of their cancer is genetically stable. In other words, as the cancer proliferates, and millions of daughter cells are generated, the genetic driver of that cancer remains the same in all those daughter cells. This is what would allow a targeted cancer drug designed to attack the genetic driver of the cancer to kill or control that cancer even in its metastatic form. This is an unusual circumstance. The enormously more common scenario involves genetic heterogeneity among the metastatic cancer cells, which is the primary reason why no single targeted cancer therapy can defeat metastasized cancer (Turajlic et al. 2019).

[10] To be clear, Brock's ultimate conclusion is that identified lives are not intrinsically more ethically valuable (and worthy of costly life-prolonging resources) than statistical lives.

the form of additional life when we provide various targeted therapies to individuals with metastatic lung cancer.[11] However, making an analogy with an anti-smoking campaign would be misleading in this case. By hypothesis, we are certain that 400,000 lives will be saved from a cancer death through reliance on annual screening with this liquid biopsy. We are uncertain which particular lives will have been saved, but those statistical lives are not *merely* statistical lives. Those are real lives (Menzel 2012) that should have substantial moral weight in allocating life-saving resources, as Dr. Raza would insist. In addition, I must emphasize again that in almost all cases no lives are *saved* among the metastatic cancer patients. Only a marginal gain in life expectancy is achieved compared to the 400,000 lives correctly described as being saved.

A third ethically relevant consideration that comes into play is *urgency of need*. Metastatic cancer patients who have failed several therapeutic regimens surely have an urgent need for one of these targeted cancer therapies since they have exhausted all other therapeutic possibilities. In contrast, those individuals who would be discovered to have a relatively early stage cancer through the liquid biopsy would have a substantially less urgent need because they would have many other options for treating their cancer. The conclusion we are supposed to accept, contrary to Dr. Raza, is that these metastatic cancer patients have a stronger just claim to costly targeted therapies than the statistically possible cancer patients have a claim to an annual liquid biopsy.

Daniels (2015) offers a supporting argument to this last conclusion. He refers to this as the "concentration of risk" argument. He asks us to imagine Alice, who has a life-threatening infectious disease certain to kill her unless she receives five life-saving tablets of some drug. Five other women have been exposed to Alice. There are only these five tablets. If each of these women receive one preventive tablet, their lives are certain to be saved. If all the tablets are given to Alice, then one of these five women will end up dying as a result of the exposure. It appears we have one death either way. Should we just flip a coin? Daniels opposes that idea and asserts that 100% of the risk of death is concentrated in Alice, whereas only a 20% risk of death is associated with each of the five women. Therefore, Alice has the strongest just claim to the five tablets. Again, there is a dis-analogy between the Alice example and our 200,000 metastatic cancer patients.

[11] A critical assumption in this sentence is that the targeted therapy provided to these lung cancer patients is correctly matched to the mutation that is the driver of that cancer. However, that requires reliance upon an appropriate biomarker test to make that connection. However, as Seo (2017) has pointed out the validity of most biomarkers test is very uncertain. She writes, "The lack of evidentiary standards is evaluating the clinical effectiveness of a biomarker test is one of the main limiting factors in relation to the integration of a biomarker test into clinical use. Clear evidence requirements should be formulated. No consensus currently exists on methodological approaches in measuring the clinical effectiveness of cancer biomarkers for targeted therapies" (32). We need to add that, as things are now, a valid biomarker may still result in the use of a targeted therapy with uncertain effectiveness at the level of the individual patient. This uncertainty complicates both ethical and economic judgment.

Alice has her life saved, but our 200,000 metastatic cancer patients are still doomed to die (with few exceptions) in a relatively brief period of time. We might be tempted to call attention to that small cadre of super-responders among those 200,000 metastatic patients. Those individuals can be correctly regarded as being Alice-like; their lives will be saved. However, returning to Dr. Raza's main point, if saving the Alice-like super responders required re-allocating that $99 billion to these targeted cancer therapies for metastatic cancer patients, then we would be sacrificing the 400,000 Alice-like lives that (by hypothesis) would be saved by investing in population-wide screening with our liquid biopsy for the sake of that small cadre of super responders. Those may well be statistical lives, but they are not *merely* statistical lives. They have as much moral worth as the lives of any of those metastatic cancer patients.

Brock next calls our attention to the "aggregation" problem as the basis for (in our case) preferring to allocate resources to the metastatic cancer patients as opposed to screening with the liquid biopsy for preventive purposes. In brief, the aggregation problem involves giving very high priority to a small number of patients who will derive very substantial benefits over a very large number of other patients who will only receive a small benefit with some limited budget. For example, we can spend $5 million to save ten patients by implanting in each of them an artificial heart (without which they would all have died in six months), or we can use that same money to stabilize 10,000 sprained ankles. We will stipulate that treating the sprained ankles will yield many times the health benefits (by whatever measure) than saving the lives of those ten patients needing the artificial heart. Still, our fundamental moral intuition would be that such a trade-off would be unconscionable. Once again, however, it can be argued that there is a dis-analogy here. It is a relatively small benefit that accrues to our metastatic cancer patients and an enormously greater benefit that would be denied to those 400,000 individuals whose cancer will go undetected without the liquid biopsy and result in their premature deaths. We may not know the identity of those individuals, either before the fact or after the fact. Nevertheless, that fact does not alter the moral equation.

Finally, there is the argument that the "medically least well off" ought to have priority for limited life-prolonging resources over those who are relatively healthy but at risk for serious illness. When stated in this very abstract form, this argument would strike many as being eminently reasonable. However, specification regarding patients who are among the medically least well off will yield results that are far from being either fair or reasonable. The assumption has to be that the medically least well off have some capacity for significant benefit if society makes available the necessary resources. Patients in a persistent vegetative state or in the late stages of dementias are clearly among the medically least well off. Just as clearly, they are incapable of significant benefit beyond bare life maintenance at costs in excess of $100,000 per year. Daniels (1985) has argued that what health care justice requires is protecting fair access for all to the normal opportunity range of a society. These are patients who are entirely outside that opportunity range. They have no capacity to participate in life.

Our metastatic cancer patients are also rightly thought to be among the medically least well off. Most of them would still have some access to the opportunity range of a society, though that access will be limited. They are not like patients with very advanced dementia. In most cases they will have had access to multiple cancer therapies that had already provided them with extra years of life that otherwise would have been denied them. In that respect they have not been unjustly ignored. Their just claims to needed health care have been met. If these targeted cancer therapies yielded, say, five extra years of life on average at a cost of $150,000 per life-year gained, we would have a much more difficult problem in determining whether that $99 billion should not be spent on them or should be spent instead on cancer screening with the liquid biopsy. However, given the factual scenario that is current cancer care, as Dr. Raza would attest, the benefits for the vast majority of these patients are marginal. If we were to forbid allocating that $99 billion to liquid biopsy screening, we would be sustaining an annual population of 600,000 or more individuals with metastatic cancer who would be for that last year of life among the medically least well off, instead of reducing that population to 200,000 per year by using the liquid biopsy screening technology. On the face of it, that outcome would not seem to be either reasonable or just, whether we were egalitarians or utilitarians or prioritarians in our understanding of health care justice.

Can We Just Abandon Metastatic Cancer Patients to Save Money?

Up to this point I have presented what I would regard as the most compelling arguments in support of Dr. Raza's position. However, there are ethically problematic features of her position that we now need to consider. In spite of all the arguments above, it will still feel ethically awkward to deny social funding for targeted cancer therapies or immunotherapies for those 200,000 (hypothetical) metastatic cancer patients who would not have benefitted from having annual access to our liquid biopsy screening protocol (Verweij 2015).[12] Though I have emphasized marginal gains in life expectancy along with a small cadre of super responders who would gain multiple extra years of life, that is ultimately an inaccurate characterization. There will be a continuum of gains in life expectancy. The most unfortunate patients

[12] Verweij appeals to the value of solidarity as the legitimate ethical basis for supporting the rule of rescue: "Diverting resources from rescue to prevention might be rational if the sole aim is to save as many lives as possible, but it would in fact negate the importance of the fact that people are standing together, sharing hope and fear, and supporting each other in the face of – and the fight against – disaster" (2015, 145). Having said this, Verweij emphasizes that this comment applies to our mining rescue and cave rescue cases. In health care, when we must live with limited budgets, Verweij emphasizes considerations of fairness and justice must dictate the allocation of resources between treatment and prevention.

may suffer harm; others will gain nothing; others will gain a few months; still others will almost gain a year; and more fortunate patients will gain an extra year or two. This makes it a lot harder to endorse with ethical equanimity redistributing all our life-prolonging cancer resources to the preventive liquid biopsy strategy. A simple solution would be to just pour more money into cancer treatment. However, that means we would fail to take seriously the "Just Caring" problem. There are limits to what any society can afford to spend on meeting health care needs, given multiple other legitimate and compelling non-health care social needs. In addition, I am unaware of any arguments that would justify spending unlimited sums of money to meet the needs of cancer patients, as opposed to patients with any of a number of other life-threatening medical conditions.

Cancer is not ethically special. Dr. Raza is clearly correct in calling attention to the vast sums of money we currently spend on low-value care for metastatic cancer patients. The problem, of course, as things are now, is that we only know after the fact that what we provided was low-value care because we spent $50,000 for a patient who gained only three extra months of life. I will put aside for the moment issues related to efforts at identifying who such patients might be before the fact through the use of predictive biomarkers. Instead, I want to consider the challenge that the liquid biopsy strategy I have described also represents low-value care. This might sound odd since the scenario I sketched suggested that 400,000 lives annually would be spared by such a strategy from progressing to a terminal metastatic disease state. I will remind the reader that this was an arbitrary number, not based upon any empirical evidence at all. The objective was simply to establish an initial framework for ethical analysis.

My critic will point out that if we are doing 198 million liquid biopsies each year, more than 99% of them will be negative at a cost of $99 billion. That will strike many as an obvious instance of low-value care. In addition, it is far from clear that simply calling attention to the 400,000 lives annually spared from a premature cancer death because of this preventive effort would sufficiently justify this massive expenditure of resources.[13] The obvious solution would be to be far more selective in the population screened with our liquid biopsy strategy. Here are some possible options: (1) Screen only those individuals and first-degree family members where there has been a family history of cancer. (2) Add to (1) individuals with established behaviors likely to result in increased risk for various cancers, such as smokers or individuals with significant sun exposure. Roughly 40% of all cancers are attributed to behavioral choices by individuals. (3) Add to (1) and (2) individuals who have been diagnosed with a cancer and successfully treated (the assumption being that they are likely at increased risk of cancer recurrence). (4) Add to (1), (2), and (3) individuals who have compromised immune systems that might increase their risk of cancer. (5) Add to all the previous categories

[13] Louise Russell is one researcher who has critically assessed the widespread belief that preventive care saves money. She has argued that, in many circumstances, prevention is just not cost-effective. She would likely conclude that with regard to our annual liquid biopsy screening proposal. See her book *Is Prevention Better Than Cure?* (2010).

individuals above the age of 50, the assumption being that cancer is most often a disease for which older individuals are at risk. Note: In all these cases we are assuming that the cost of this testing would be a social expense. Individuals outside these categories would be free to obtain this testing at their own personal expense.

Trying to identify categories of individuals who would have the most reasonable and strongest just claims to this annual liquid biopsy screening would likely have a strong arbitrary component, as we discuss below. We might imagine avoiding the need to draw these lines if we could show that our initial proposal made economic sense when viewed in total. If we are saving 400,000 lives from a premature death from metastatic cancer each year, then we are also saving the cost of treating those metastatic cancers. We can assume that cost might be $100,000 per metastatic cancer patient saved. If so, that savings would amount to $40 billion. That would still leave $60 billion in screening costs each year for which there were no offsetting economic gains.

Though ethicists would likely not endorse this next point, economists would note that those 400,000 individuals are still going to die, and it would be extremely unlikely that they would die cheaply. Those deaths would most likely be of other chronic degenerative conditions with multiple years of high health costs. A substantial number will develop dementia and require at least a couple years of long-term care at $100,000 per year. Consequently, if our concern with controlling health care costs is about *all* health care costs, then the bottom-line savings would be substantially less than $40 billion. For the sake of argument, let us say that the net savings would be $20 billion.

Someone might then argue that $20 billion in savings ought to be used to provide $100,000 worth of targeted cancer treatments for each of the 200,000 individuals each year who would still end up with metastatic cancer in spite of the screening program. However, our erstwhile economist will again call our attention to an awkward economic fact, namely, that the savings we would expect to achieve by not having to pay for targeted cancer therapy for 400,000 individuals will only be realized more than two decades into the future. The costs of doing the screening will all be incurred in the present and represent additions to current health care costs. The assumption behind this conclusion is that it would be unethical for reasons of both justice and compassion to deny the current 600,000 patients who will die this year from their metastatic cancer the targeted cancer therapies and immunotherapies that can give them additional months (sometimes years) of life in order to offset the current costs of the proposed liquid biopsy screening program. In other words, transitioning from our current metastatic cancer therapeutic protocols to the biopsy screening protocol proposed by Dr. Raza would be ethically, economically, and politically problematic. Is there a just and reasonable way of addressing the "transition" challenge?

The Transition Challenge: Efficiency versus Compassion

Menzel (2012) has addressed the transition challenge. He addressed it very broadly, maybe too broadly. This is the question that Menzel started with: "*Should we provide relatively inefficient treatment to identifiable individuals at relatively high risk for relatively immediate harms rather than more efficient preventive care to equally identifiable individuals at lower risk for more distant harms*" (214; his italics). He defends what he calls the equivalence thesis, namely, that the lives on either side of the equation are equally worthy of being saved through the allocation of social resources. However, the identifiable individuals at high risk can only have their needs met *inefficiently.* In other words, this is an unwise use of social resources, which is what Dr. Raza would argue. Society, he contends, in the form of future generations, would be much better off if the resources now being used inefficiently to purchase marginal gains in life expectancy were to be used instead in the preventive mode to prevent the need for the future inefficient uses of those resources. However, he argues, we cannot afford to do both things during some sort of transition period. This has the ethically problematic consequence of sacrificing the lives of a whole generation (in our case) of metastatic cancer patients who would all be denied access to these expensive targeted therapies in order to eliminate the need for such therapies for future generations. This does not appear to be either fair or compassionate.

Menzel asks us to consider the alternative. If we agree that it would be wrong to sacrifice this generation for the sake of future generations, then how would we ever make the transition that he would argue is rationally, ethically, and economically required? We would be stuck in the present circumstances. We would be compassionate to the current generation of metastatic cancer patients but we would be sustaining the pain and suffering and costs associated with metastatic cancer for numerous future generations who would not otherwise need that compassion because they would not have to endure this suffering if we now made a different decision aimed at preventing the need for such care in the future. Menzel writes: "By spending at a lower health productivity rate for that prevention than we do for treatment, we would fail to decrease the incidence of these very diseases for future generations, condemning more people than necessary to never being situated where they can find in their rational self-interest to vote for a more limited priority for treatment, with all the benefits of such a policy" (2012, 214). Menzel sees the "long run" perspective associated with prevention as justifying the "unfairness" objection raised in connection with denying life-prolonging care to a current generation of seriously ill patients. This is essentially a utilitarian perspective.[14]

[14] An alternative response to the unfairness objection has been offered by Callahan (1990). This is the "ragged edge" problem, which very often occurs when we must make a rationing or allocation decision fairly. "We can accept the ragged edge, not because we lack sympathy for those on it, but because we know that, once a ragged edge is defeated, we will then simply move on to still another ragged edge, with new victims---and there will always be new victims. It is a struggle we cannot win....We can ask [instead], not how to continually push back all frontiers, smooth out all ragged

One way of reading Menzel is to imagine a very costly preventive intervention that is completely successful in eliminating that disease from future generations. This reading makes sense since Menzel thinks of this situation as requiring tragic sacrifice from just one generation of patients. However, that is not the scenario that we have sketched with regard to our annual liquid biopsy protocol. In our scenario we do save 400,00 individuals every year from dying as a result of metastatic cancer. That would still leave those 200,000 individuals each year who will die as a result of their metastatic cancer. Consequently, what we would have to be willing to sacrifice are those 200,000 individuals every year far into the indefinite future, as opposed to "just" one generation of metastatic cancer patients. That seems to require much more in the way of ethical justification than the scenario Menzel has in mind. We can readily imagine the situation Menzel envisions as being tragic, ethically necessary to effect a greater life-saving result, but still regrettable. It is much more difficult to justify as a "regrettable tragedy" a situation that occurs repeatedly and persists far into the indefinite future without any obvious effort to end it.

If we consider the situation from a very "raw" utilitarian perspective, then the argument can be made that we *save* 400,000 lives with our liquid biopsy protocol while "only" giving up 200,000. In addition, the 400,000 lives are restored to having an indefinite life expectancy while the vast majority of the 200,000 will have lost less than a year of potential life if provided with some targeted cancer therapy. Still, perhaps 20% of those 200,000 individuals, would end up losing prematurely anywhere from one to ten extra years of life (the higher numbers pertaining to the super responders). Of course, as things are now, we would not know who those individuals were who had so much more to lose without treatment. However, to tolerate this outcome as a society, we would have to harden our hearts, close our eyes, and numb the compassion-generating portion of our brains. That, by itself, suggests the need for more critical and creative thinking. How, then, can we effect the transition Dr. Raza recommends in a way that is congruent with what a just and caring society ought to be?

Whole Genome Sequencing: Another Precision Health Ethical Challenge

That brings us back to the need to slim down dramatically the liquid biopsy screening protocol. Can that be done in a way that is at least "roughly just"? We noted above that one group of patients who would seem to have top priority for access to these liquid biopsies would be those identified as being at risk for hereditary cancers. That represents only about 10% of all cancers (National Cancer Institute 2017). How would we imagine identifying them? The short answer would be that we do

---

edges, but how to make life tolerable on the ragged edges" (1990, 65). That is where the importance of palliative care comes in for a just and caring society.

Whole Genome Sequencing [WGS] of almost all Americans alive today and every child born each year. Such sequencing with professional analysis and interpretation and counseling would cost about $5000 per person, or $1.5 trillion for 300 million Americans today. In addition, it would cost $20 billion per year to do WGS of each birth cohort. In the real world we do not have the technological capacity or personnel that would be needed to accomplish this task, never mind managing such huge economic costs. As with our liquid biopsy screening protocol, we would have to slim down and prioritize who would have access to such WGS at social expense. How can that be done fairly?[15]

I would certainly not endorse as "just enough" the libertarian view that it should be up to individuals with their own resources to determine whether such WGS was "worth it" to them. This would clearly disadvantage the financially less well off who would be at risk of a premature death from metastatic cancer that could have been avoided with access to WGS and/or our proposed liquid biopsy. What about an egalitarian perspective? There are several varieties of egalitarianism. Broadly speaking, egalitarians are committed to equal concern and respect for all. Would that mean everyone has an equal claim to WGS to establish their vulnerability to cancer? That would be a practical impossibility and ethically indefensible, given limited resources for meeting unlimited health care needs, not just cancer-related needs. Egalitarians will generally accept the view that greater health needs justly command more social resources (appendicitis commands more resources than a sprained ankle). How serious or urgent is the need for WGS to establish lifetime cancer risk relative to managing heart disease or Parkinson's or multiple sclerosis or dozens of other chronic degenerative conditions? This would not be an easy or obvious answer for an egalitarian.

Keep in mind that if the whole US population underwent WGS, 60% of those individuals would never develop cancer over their entire lifetime.[16] Given that statistic, utilitarians would not endorse WGS for the population as a matter of social justice. Prioritarians want to ensure that a society meets the just claims for needed health care for those who are "medically least well off." We might imagine that the 10% of the population at risk for some hereditary cancer would fit that criterion. That would certainly be true if we restrict ourselves to considering only patients at risk for cancer. However, this is where prioritarians might find themselves internally conflicted. There are the metastatic cancer patients who desperately need the targeted cancer therapies for some additional gain in life expectancy (sometimes a

[15] As I suggested in footnote #4, we should not assume that WGS is an ethically unalloyed good. There are numerous ethical pitfalls associated with its use for various purposes, whether preventive, predictive, therapeutic, or reproductive. An excellent summary of those issues is provided by Dondorp and deWert (2013).

[16] That 60% figure might be misleading. That is roughly the number of individuals who will develop cancer sometime during their life. However, the large majority of those cancers will be the result of environmental exposure, i.e., choosing to smoke, or lots of sun exposure without protection. None of these cancers would be identified through WGS. A roughly correct figure would be 5% to 10% of cancers would be hereditary and identifiable through WGS.

significant gain in life expectancy, as discussed above), not to mention all the other patients at substantial risk for premature death because they are in the advanced stages of some chronic degenerative condition. These are all patients with *urgent, imminent, actual* health care needs, as opposed to the *potential* health care needs of individuals with a hereditary risk of cancer. In the debate over the moral weight that should be attached to identified lives as opposed to statistical lives for purposes of allocating life-prolonging resources, prioritarians would generally strongly endorse giving more weight to those identified lives. What we need to keep in mind is that having a hereditary risk for cancer does not necessarily mean that one will actually have cancer during one's lifetime. Women with a BRCA1 mutation will have a lifetime risk of breast cancer in the 40% to 85% range, depending upon which of several hundred mutations in the gene they might have.

For the moment, let us put aside the practical and ethical challenges of determining who would have a just claim to WGS at social expense. We would (presumably) want to identify the 10% of the population at risk for a hereditary cancer, even though it is a high lifetime risk, but not a certainty. However, we would also have the capacity to determine a polygenic cancer risk score for virtually everyone. Such a score would be the product of literally hundreds of genetic variants in an individual each of which might increase very slightly their lifetime risk of cancer. Their lifetime risk would likely not be as great as someone at risk for a hereditary cancer. They might be told that their lifetime risk was "just average," that is, at 40%. Would they feel that they had a just claim with regard to our annual liquid biopsy? An enormous number of people would get a result like that, which would defeat the need to reduce the use of annual liquid biopsies at social expense. A significant number of individuals would also be told that their lifetime risk of cancer was below average, say at 20% or 30%. That is not zero. Some of these individuals will develop a cancer. Most of them will have that cancer caught at an early and treatable stage. Others will be less fortunate. They will progress to metastatic cancer and a premature death. So, one of the things we need to keep in mind, whether for the average or below average cancer risk patients, is that the fewer annual liquid biopsies we do, the greater (statistically) will be the increase in patients with metastatic disease. Should we see this as being unjust, ethically problematic? This is what is referred to in the literature as the "ragged edge" problem (Callahan 1990, chap. 2) or the "cut-off" problem (Rosoff 2017). No perfectly rational or perfectly just rationale can be offered for drawing a line at one point rather than another point. A line has to be drawn (as we have argued above) because we have only limited resources for meeting unlimited health care needs. There will always be patients with health needs or health risks just below that line who could make a reasonable just claim for the resources being provided to those above the line. The most we can reasonably hope for in this regard is rough justice.[17]

[17] Space does not permit any lengthy discussion regarding acceptable forms of rough justice. However, a quick illustrative example would be helpful. If a line were drawn that systematically resulted in individuals who were already among society's least well off (health-wise and economically) being further discriminated against, then that would be an unacceptable form of rough jus-

Lots of things muck up our intuitions regarding what is ethically required of a just and caring society for purposes of allocating resources toward precision health. We noted earlier that roughly 40% of cancers are linked to individual behavioral decisions (Mendes, 2017). This is something that would get the attention of luck egalitarians, who are committed to a responsibility-sensitive conception of justice. We might imagine luck egalitarians would endorse WGS that provided a polygenic cancer risk score, primarily as a way of educating everyone regarding their cancer risk. Unfortunately, most humans do not seem to be open to educational efforts to alter pleasurable behavior that represents a threat to health. What we might imagine instead as being more likely is that individuals with average or above average or somewhat below average cancer risk scores would be the first to demand annual liquid biopsy screening with the hope that this would catch an early treatable cancer without having to change the behavior that might have generated that cancer. Given this scenario, luck egalitarians would likely oppose WGS for a population at social expense as both wasteful (unjust) and irresponsible.

Economists again contribute to mucking up our intuitions regarding preventive efforts to reduce the incidence of fatal cancers in our society. We want efficiency, justice and compassion reflected in our efforts to prevent (reduce) the incidence of fatal cancers. We noted above that the total annual cost of providing access to a liquid biopsy as a screening tool in the US would be $99 billion. We hypothesized (for the sake of argument) that this effort would save 400,000 individuals annually from a premature death from cancer. That yields a cost-per-life-saved of about $250,000. This is not an unreasonable figure.[18] What makes that even more reasonable is if the average gain in life expectancy for those 400,000 individuals is twenty years. That means the cost-per-life-year-saved is $12,500. Compare that to the 200,000 individuals per year who would die from metastatic disease but whose lives could be extended (mostly briefly) if they were provided with some of these targeted cancer therapies. If each of those individuals were given a $100,000 targeted cancer therapy, then for those individuals who gained only three extra months of life, the cost-per-life-year-saved would be $400,000. This is the sort of number that would support Dr. Raza's proposal for shifting resources away from treating these patients toward the preventive efforts represented by the liquid biopsy screening protocol. However, a significant number (at least 20%) of those 200,000 metastatic cancer patients would gain at least an extra year of life for that $100,000. This is very close to what we spend per year for end-stage renal patients needing dialysis. From the

tice. An acceptable form of rough justice would put all identifiable social groups at roughly equal risk of being on the "wrong side" of that resource allocation line. In other words, the result should approximate the result of a pure lottery.

[18] Individuals with end-stage renal failure requiring dialysis will cost the federal government today $90,000 per year with an average survival of seven years. That represents a cost-per-life-saved of about $630,000. Individuals with HIV requiring a four-drug combination will have annual drug costs of about $35,000 per year and can gain thirty extra years of life. Stribild® is a four-drug combination in a single pill with a cost of $3550 per month, or $42,600 per year. That would yield a cost-per-life-saved of over $1 million. (Silverman 2016).

perspective of health care justice and compassion, that makes it much more difficult to justify sacrificing these lives for the sake of putting in place the proposed preventive effort. What makes it even more difficult to justify such sacrifice would be the 5% to 10% of patients in this category who would gain multiple extra years of life from access to one or more targeted therapies (albeit at a cost of at least $100,000 for each of those extra life-years gained). What is the right thing to do, all things considered?

Answering this question is made more difficult because, for the most part, we do not know before the fact which individuals with metastatic cancer are likely to live an extra year or more with the help of a targeted therapy. This is mostly a true statement. However, much research regarding cancer biomarkers is resulting in our identifying before the fact patients who are more likely to benefit substantially from access to a targeted cancer therapy. One type of targeted cancer therapy are the drugs known as checkpoint inhibitors, such as nivolumab and pembrolizumab. These drugs target PD-1 and PD L-1 (programmed cell death protein 1), whose job it is to regulate the immune system from over-reacting and generating one of a number of immune disorders. However, cancer cells can use these proteins to hide themselves from the immune system. The checkpoint inhibitors are intended to suppress PD-1 and make the cancer cells more visible to the immune system. For some types of cancer, higher levels of PD-1 expression are a good biomarker of a more effective response to these checkpoint inhibitors. The actual literature in this regard is mixed (Dudley et al. 2016; Ugurel et al. 2020; Yi et al. 2018). What if, however, a slight majority of cancer patients with these higher levels of expression gain one or two extra years of life, not just seven months? Do all these patients then have a just claim to have access to these checkpoint inhibitors at social expense? High levels of tumor mutational burden are also a good (not perfect) biomarker for a more effective response to these drugs (Chan et al. 2018).

These are just illustrative examples I have offered. I could have picked a dozen others. The core ethics question is this: If we develop the capacity to identify with a high degree of likelihood, using various biomarkers, patients who are most likely to gain an additional year of life from access to various targeted cancer therapies or immunotherapies, should we use that capacity to separate moderate and strong responders from marginal responders for purposes of allocating these therapies at social expense? In other words, would considerations of justice and compassion justify this rationing practice as part of an effort to balance providing limited resources to both prevention (our liquid biopsy protocol) and treatment?

We need to consider one more possible scenario. The basic premise behind Dr. Raza's proposal is that the vast majority of metastatic cancer patients today achieve only marginal gains in life expectancy with access to these targeted cancer therapies, which is why she believes we ought to pursue aggressive cancer prevention rather than aggressive treatment. However, that might be changing, perhaps significantly. Many researchers (Cajal et al. 2020; Prasetyanti and Medema 2017; Dagogo-Jack and Shaw 2018) now believe that because of the genetic heterogeneity of cancer, and its evolving nature within an individual, multiple targeted therapies need to be used, either in combination with one another or sequentially, a strategy which has yielded considerable success in the treatment of HIV. This can increase substantially the cost per patient per year to achieve several extra years of life with "managed" metastatic disease (nothing curative). For example, Workman et al. (2017) note that the cost of combining nivolumab and ipilimumab would be priced at about $252, 000 per year for the treatment of advanced melanoma. More recently, Larkin et al. (2019) show a five-year overall survival rate of 52% for these same patients with this same combination of targeted therapies. Again, these examples are only intended to be illustrative of the direction of advances in cancer care at present.

What I ask you to imagine is that future research is successful in finding combinations of these targeted therapies that yield three to five additional years of life for 75% of that batch of 200,000 metastatic cancer patients. How should that alter the balance in the distribution of social resources between our liquid biopsy prevention strategy and the effective treatment needs of these 150,000 metastatic cancer patients? My hypothetical scenario is that we are using combination cancer therapy to achieve this result at a cost of $200,000 per patient per year for an average of a three-year gain. That would amount to $30 billion for that first cohort of patients, $60 billion for that second cohort, and $90 billion for that third cohort and every year thereafter. To be clear, that $90 billion is *only* for the care of these metastatic cancer patients, not any other cancer patients who are treated and cured in any given year. This is roughly the amount we would have to pay every year for our annual liquid biopsy screening proposal of every adult American. This also assumes that we would provide comfort care only for that other 50,000 metastatic cancer patients each year who were not candidates for any of these more successful therapies. That is, we would not provide them with low-value targeted cancer therapies (which would raise the same ethics issues as earlier discussed, albeit on a smaller scale).

As a matter of health care justice, would we be ethically obligated to increase by $90 billion per year what we spend on cancer, keeping in mind that whatever the moral logic was that justified an affirmative answer to this question would have to provide a similar answer in every other area of medicine where lives could be prolonged at a similarly high cost? That would represent a rejection of the basic premise behind the "Just Caring" problem. We have only limited resources to meet virtually unlimited health care needs. Alternatively, we could trim the costs of our liquid biopsy protocol. Let us say that we would screen annually only half the American population, thereby saving $50 billion per year.[19] Unless we somehow managed to choose the exactly correct 50% who were at the highest risk for cancer that became metastatic, we would in fact increase the number of patients who would eventually be faced with metastatic disease.

Simple math (under this scenario) would suggest that we would increase the number of annual metastatic cancer cases by 200,000. We will assume instead that

---

[19] Alternatively, we could screen everyone every two years. That would save the same amount of money. However, that would be neither just nor rational, given that we would know in many cases before the fact that certain identifiable population groups were at greater risk for cancer in the future.

we are cleverer than that, and consequently, the increase would be 125,000 (for the sake of argument). 25,000 of that annual cohort would be added to the 50,000 who would receive comfort care only; no targeted therapies or immunotherapies at social expense. The other 100,000 would receive the advanced targeted therapies described above at an annual cost of $20 billion for the first cohort, $40 billion for the second cohort and $60 billion for the third cohort and every year beyond that, which would really amount to $150 billion for each of the out years under this scenario for treatment and $50 billion for the ongoing preventive screening effort. Again, this would be in addition to whatever the current annual costs for cancer treatment are. What choices should a "just" and "caring" society make that has only limited resources for meeting virtually unlimited health care needs? This question is asked with regard to the use of social resources, not private resources. A companion question might be: What choices with regard to all of our scenarios above may a "just" and "caring" society make that would relegate the cost of those choices to individual willingness and ability to pay? I take it that a society that left access to treatment for a heart attack or appendicitis to individual ability to pay would be correctly judged to be both unjust and uncaring. Likewise, a society that refused to pay for purely cosmetic procedures (not needed as a result of disease or accident) would not be open to justified moral criticism. Where do our questions fit between these two poles?

Rational Democratic Deliberation: Not Precision Ethics But "Roughly Just"

We have raised a large number of ethically challenging questions above; we have not offered to defend any particular response to those questions. This is because, as I have suggested, there is no "most just" or "most reasonable" response to most of these questions. Many trade-offs are possible that would be "just enough" and "reasonable enough." To be clear, unjust and unreasonable trade-offs are possible as well. The trade-offs are among the core values that define competing conceptions of distributive justice, as well as with other fundamental social values. The trade-offs we have in mind pertain to policy choices, not choices made by individuals. Consequently, a sufficient level of social agreement is necessary for these policies to be both fair and legitimate. I have argued elsewhere (Fleck 2009, chap. 5) that this agreement should be achieved through fair and inclusive processes of rational democratic deliberation governed by what I refer to as "constitutional principles of health care justice." Before elaborating a bit on that, let us review our key questions for democratic deliberation.

- Should resources currently used to treat metastatic cancer patients be redirected to efforts at either preventing the emergence of cancer or identifying and treating it in its earliest stages?
- Should identified lives of patients with metastatic cancer be given equal moral weight for the distribution of life-prolonging resources with the statistical lives

(future possible patients) whose early cancer can be prevented from progressing to metastatic disease?
- Should resources be allocated in a more balanced way between treating metastatic cancer patients and minimizing through prevention the number of cancer patients who progress to metastatic disease? If so, what justice-relevant criteria should be used to limit our screening efforts and to limit our treatment efforts?
- What is the fairest and reasonably cost-effective way of meeting the "transition challenge" as we sought to shift resources from aggressive treatment of metastatic disease to prevention aimed at reducing the number of future patients with metastatic disease?
- Should we do Whole Genome Sequencing of every American early in life to establish a lifetime cancer risk score that would then be used to identify individuals most likely to benefit from more targeted preventive efforts?
- Should we as a society invest in more research aimed at identifying biomarkers that would allow us to identify and predict more reliably which individuals with metastatic cancer would gain the most in life expectancy if provided with the relevant targeted therapies at social expense?
- If we were successful in improving the survival of 75% of metastatic cancer patients for an average gain in life expectancy of three years at a cost per patient of $600,000 for those three years, should the resources needed to cover that expense come from the preventive screening efforts we have described?

Why do we need rational democratic deliberation to address these issues? The short answer is that all these questions are about public goods whose fair distribution or resolution cannot be fairly or adequately addressed through any private decisional mechanism. Does the Grail protocol that I described represent high-value care that would justify a very high level of social investment? This is not a question that can be answered fully by asking the relevant experts to work out the cost-effectiveness equations. Multiple other social values would need to be considered and trade-offs assessed that are not in the realm of any particular area of expertise. To be sure, lots of relevant expertise needs to be introduced into the social conversation, but the conversation itself should be a matter of inclusive rational democratic deliberation.

In the European context we would be asking the question whether a commitment to solidarity required public funding for the Grail protocol *and permitted reducing or eliminating funding for targeted cancer therapies for patients with metastatic cancer*. In the American context we would be asking whether a commitment to individual liberty meant that individuals should make the decisions for themselves whether it was worth it to them to pay for annual cancer screening with Grail's liquid biopsy. Alternatively, should annual liquid biopsy screening be seen as a public health measure, a dramatic measure to reduce premature deaths from cancer, for much the same reasons that we carefully assess the processing of our food supply or the introduction of pharmaceuticals to protect the health of all. This is a matter of equal concern and respect for all. However, this would be an additional $99 billion-dollar cost in the United States.

Where should that money come from? Should that money come from additional taxes or increased insurance costs? This is one option. However, it would raise the political and ethical question of whether giving cancer this "special" or "supreme" health status would be justified relative to many other possible health investments in other disease areas where we might be able to save more lives at a lower cost. The alternative proposed by Dr. Raza would have us simply take those funds from what we now spend to provide these targeted cancer therapies and immunotherapies to metastatic cancer patients. It is easy to imagine the reluctance (maybe horror) many would experience in contemplating that option. Of course, nothing requires us to do annual screening with our liquid biopsy for everyone. We could limit that screening to some range of high-risk cancer groups in order to protect funds for treating metastatic cancer patients. However, we could then imagine the anxiety that would provoke in many Americans, knowing that their lifetime risk of cancer was 40% (maybe higher) and that the cancer that might afflict them did not present symptoms until a very advanced stage, such as pancreatic cancer. Finally, we might ask, given the emotional overtones expressed in these last few sentences, how we could possibly have a *rational, civil, mutually respectful* conversation about such controversial and emotionally charged issues. How could self-interest not corrupt and disrupt the possibility of such a conversation?

John Rawls (1971) introduced into discussions of political philosophy the notion of a "veil of ignorance." Individuals behind the veil of ignorance would know nothing about their own social or economic or health status. We could imagine this as 330 million life slots in America today. Individuals would, however, know all the possible political, economic, and social structures and policy options that could constitute the basic structure of their society as well as the life prospects for various individuals living within that basic structure. Individuals behind the veil of ignorance would be charged with determining what the principles of justice should be that would determine the basic structure of society. They would know that, after they had achieved agreement, they would be randomly assigned to one or another personal identity in that society. Their life prospects in that identity might be reasonably good or somewhat less good. Their motivation behind the veil of ignorance would be to choose principles of justice that would maximize as much as possible in that framework life prospects for those who were socially and economically among the least well off.

Rawls has been criticized on the grounds that the veil of ignorance thought experiment is totally unrealistic because everyone knows what their interests are. Consequently, no such social conversation regarding justice is possible under that scenario since most people would argue for policies that would protect their personal interests. However, without going into any defense of Rawls' overall views, it is factually true that the vast majority of us at any point in our life are largely behind a health status veil of ignorance. Even at my advanced age, I have no idea what my most likely health risks are or the most likely cause of my future death. Alternatively, we can pretend that in my early twenties I underwent a genetic test that indicated I had a 70% chance that I would die of some specific cancer before age 60. Would I then want all sorts of social resources allocated for research and treatment of my specific cancer?

In purely private moments I might answer that last question affirmatively. But if I am part of a social policy conversation regarding the allocation of health dollars for social health insurance, as well as prevention and research, I would be reminded by my fellow citizens that, if I seemed too single-minded an advocate for my cancer, many members of my family, as well as friends and co-workers, were vulnerable to many other health problems that could result in their succumbing to a premature death. I would also be reminded that I likely had forty years ahead of me and that I was vulnerable to lots of other diseases or serious injury related to accidents for which social resources would need to be allocated. It would be irrational for me to focus exclusively on my risk of that cancer. In addition, it would be unkind for me to ignore the health risks to which all whom I cared about were vulnerable.

I would also be reminded that I was engaged in this very broad social conversation regarding the allocation of health care resources. We could allocate as much money as we wished to meeting health care needs, not just my health care needs, but everyone else's health care needs as well. However, if we wanted to increase the size of the health care budget to cover anything and everything in the way of treatment that might be offered by contemporary medicine, then I would have to be willing to pay unlimited sums as taxes or insurance premiums to make that possible. On the other hand, if I want limits on that budget and my wallet, then I would have to work with everyone else who is part of this social conversation to establish what sort of health needs will justify accessing that social budget to meet those needs.

I do understand what health needs are, and how unmet health needs can greatly disrupt or shorten a life. Consequently, I want allocation policies that are fair, compassionate, and that represent a wise use of limited resources. I want those policies to reflect the best medical and scientific knowledge that is available so that we are funding therapies that are most likely to be effective. I am certainly inclined to be mindful of the health care needs of my friends and family and acquaintances. It is more difficult to be very mindful of the health needs of the numerous faceless strangers who make up our society. However, I will be reminded that I am a complete stranger to all of them as well. I do endorse the view that every member of our society is entitled to equal concern and respect. I can imagine a situation in which at age sixty I am afflicted with the cancer that I have feared. I am not ready to die. I can imagine a targeted cancer therapy that would cost $200,000 and offer me only a 25% chance of six extra months of life. I would want to have that paid for from this insurance pool. Given a commitment to equal concern and respect, that would mean that I would have to be willing to absorb those same costs for all the other metastatic cancer patients who would want that costly treatment for a very uncertain marginal gain in life expectancy. If I thought that was a poor expenditure of my money for those others who are strangers to me, then they would have the right to make the same judgment regarding the cancer treatment I want for myself since I am just as much a stranger to them.

What this last paragraph illustrates is how, in reality, we can begin to achieve social agreement regarding health care priorities and corresponding limits.

Individuals must be willing to be reasonable. If I want to be treated justly, then I must be willing to treat others justly as well. What counts as being "just enough" will have to be articulated through this process of rational democratic deliberation.

What keeps this deliberative process from becoming biased, dominated by special interests that skew the results unfairly? First, the relevant medical and scientific facts (such as they are at a point in time) must be rationally respected. Some facts are mushy, such as survival curves with most of these targeted cancer therapies, which adds to the complexity of social decision making through democratic deliberation. Second, there are what I (Fleck 2009, chap. 5) refer to metaphorically as "constitutional principles of health care justice." These principles are intended to prevent majoritarian abuse and tyranny of those who are not capable of defending their basic interests. For example, in the United States today with employer-based insurance, many of these policies include very high deductibles and co-pays, especially with regard to these targeted cancer therapies. What that means in practice is that well-paid managers and executives can afford those co-pays and deductibles. Hence, they have effective access to these therapies. Ordinary workers would find it impossible to meet those requirements. Hence, they have no practical access to these therapies, though a portion of the cost of that insurance will be paid by them. In other words, they will be subsidizing access for the very well off. That would violate "equal concern and respect" as a constitutional principle of health care justice, which is to say that such a policy would not be an option for democratic deliberation, much less legitimation. It represents a form of exploitation.

Third, Rawls' notion of "wide reflective equilibrium" constrains democratic deliberation as well (1971, 1996).[20] Complex policy choices typically have wide-ranging dispersed consequences. What needs to be avoided are policy choices that generate a more unjust situation than the situation a policy change was intended to correct. If enormous social resources flow into cancer treatment, research, and prevention and yield mostly marginally beneficial results, other areas of medicine can justly inquire why comparable resources are not available for treatment, research, and prevention where there is a greater likelihood of more substantial health outcomes. This is the sort of "imbalance" that needs to be avoided or corrected in order to maintain overall a wide reflective equilibrium with respect to the just allocation of health care resources. In the earlier portions of this chapter I have tried to illustrate the sort of imbalances between our liquid biopsy protocol (precision prevention) and targeted cancer therapies (precision medicine) that must be addressed.

Let me conclude with an illustration of how a deliberative question might be talked through. I assume we cannot afford to do that annual liquid biopsy for all adult Americans at social expense. I also assume we would not endorse providing only comfort care for metastatic cancer patients in order to afford the liquid biopsy protocol (given that some patients might gain several extra years of life from one or

[20] Rawls himself offers only a very sketchy description of wide reflective equilibrium here and there in his writings. One of his students was Norman Daniels (1996) who has elaborated considerably on that notion and its application. It has been a focal point for philosophic discussion for the past thirty years.

another targeted therapy). I also assume we would not endorse providing unlimited access to all cancer therapies, no matter how high the cost, no matter how marginal the benefit. All these assumptions are ethical and economic; all (I believe) are reasonable (even if not self-evident). That means we need limits and compromise in all three regards.

Can we come to agreement regarding when it is ethically permissible to allow access to either advanced cancer therapies or our preventive liquid biopsy on the basis of an individual's willingness and ability to pay from their own resources? I think we could agree that if we can identify individuals at reasonable cost who are at significantly elevated risk for cancer, then those individuals ought to have access at social expense to our liquid biopsy annually. We will recall that any random American has a 40% lifetime risk of cancer. That is a significant number. However, if that is a number that triggers anxiety in a large portion of the population who demand annual liquid biopsy testing for a potential cancer at social expense, then we would be compelled to spend (wastefully) that $99 billion per year. What can be pointed out in the deliberative process is that 70% of individuals diagnosed with cancer will have it treated successfully and will not die of their cancer. We noted earlier that 40% of cancers are linked to behavioral choices by individuals. Increased efforts at public health education in this regard would be much less expensive than funding annual liquid biopsies.

Given this background, it would be neither unreasonable nor unjust to expect that individuals with very high anxiety levels regarding cancer could pay from their own resources the cost of annual liquid biopsy screening. Somewhat wealthier individuals would make this choice. Somewhat poorer individuals could not make that choice. Does that represent an injustice? I would argue that it does not represent an injustice because this is a very low-value intervention relative to all the other health care needs poorer members of our population might have to which they would have just claims, such as effective treatments for early stage cancers. In addition, the poor are not made worse off by the purchases of these liquid biopsy tests by the financially well off. Of course, we have to consider the fact that not providing this test at social expense will increase the number of individuals with metastatic disease relative to the 400,000 hypothetical individuals saved from metastatic disease in my scenario. What does a just and caring society owe those unfortunate individuals?

We owe these unfortunate individuals effective and cost-effective cancer treatments that yield significant benefit. As things are now, there are somewhat costly and somewhat effective treatments for many forms of metastatic disease. The very costly targeted therapies and immunotherapies are typically offered after these prior lines of treatment have been used, though many researchers and oncologists would like to see these targeted therapies become first-line treatment for metastatic disease. This might make medical, ethical and financial sense in some range of cases. This can then be seen as the trade-off for individuals who would have given up on endorsing social payment for the liquid biopsy option. Still, the implication of this view is that not everyone diagnosed with metastatic cancer will have access to these very expensive targeted cancer therapies at social expense. To preserve fairness and objectivity in this regard, we ought to fund research aimed at identifying reliable predictive biomarkers that would identify before the fact patients most likely to achieve substantial benefit from one or more targeted therapies, the precise definition of "substantial benefit" being left to the deliberative process. Again, the wealthy could buy access to targeted cancer therapies likely to be only very marginally beneficial. This does not represent an injustice to the non-wealthy who are no worse off as a result of permitting such purchases.[21]

The limits we would collectively place on accessing targeted cancer therapies at social expense should be congruent with comparable limits we place on accessing comparable life-prolonging therapies in many other areas of medicine. As noted above, this is what would be required for maintaining a just reflective equilibrium in the deliberative process. Finally, creative options are possible. We could endorse a policy of permitting individuals who were hyper-anxious about their cancer risks (without any objective basis for that anxiety) to have annual liquid biopsies at social expense with the understanding that they would give up their right to expensive targeted therapies should they still be unfortunate enough to end up with metastatic disease. Speaking personally, I would not see this as a wise trade-off. However, in a liberal, pluralistic society that places a high value on maximizing individual liberty, so long as that liberty is not used to violate the equal rights of others or public interests, this might be an option that should be permitted.[22]

In conclusion, it is reasonable for us, future possible cancer patients, to want both precision medicine and precision health. However, if we want both to the maximal degree that is technologically possible, we will create ethical, economic, and political challenges that would be ethically disruptive, economically unsustainable, and politically divisive. If we want a society that is just and caring, given limited resources and unlimited health care needs, we will need to define limits and legitimate trade-offs that are reasonable and "just enough." Competing theories of justice, as articulated by philosophers, will be too abstract for the inherent complexities associated with health care rationing and priority-setting in the real world. What we require instead are fair, well-structured and inclusive processes of rational democratic deliberation to address these issues. Such processes are most congruent with what a liberal, pluralistic, tolerant democratic society aspires to be. The role of philosophers in this process is to guide the construction of public reason, that is, the rational capacities and value commitments necessary for sustaining effective civil discourse regarding the most controversial social problems a democratic society

---

[21] If we allowed the wealthy to purchase access to transplantable organs, which are absolutely scarce, unlike cancer drugs, that would be unjust because it would make the non-wealthy less well off as a result. Access to transplantable organs must be based on criteria that are not wealth dependent.

[22] This can get complicated. What should a just and caring society do if an individual has chosen this option at age twenty-one but at age forty-five realizes this was not a wise choice? He no longer wants society to pay for these annual liquid biopsies. Does he then have a just claim to social payment for these targeted cancer therapies, should he end up with a metastatic cancer? What if he has this awakening at age sixty? These are complexities that would have to be considered as part of the deliberative process.

must address. Precision medicine, unguided by just public reason, will yield unhealthy public policy and noxious injustices in our health care system.

References

Bach, P.B. 2019. Insights into the increasing costs of cancer drugs. *Clinical Advances in Hematology & Oncology* 17: 287–288, 298.

Bender, E. 2018. Cost of cancer drugs: Something has to give. *Managed Care Magazine.* https://pubmed.ncbi.nlm.nih.gov/29763403/. Accessed 14 May 2020.

Brock, D. 2015. Identified versus statistical lives: Some introductory issues and arguments. In *Identified Versus Statistical Lives: An Interdisciplinary Perspective*, ed. G. Cohen, N. Daniels, and N. Eyal, 43–52. New York: Oxford University Press.

Cajal, S.R., M. Sesé, C. Capdevila, T. Aasen, L. Mattos-Arruda, S.J. Diaz-Cano, J. Hernández-Losa, and J. Castellví. 2020. Clinical implications of intratumor heterogeneity: Challenges and opportunities. *Journal of Molecular Medicine* 98: 161–177.

Callahan, D. 1990. *What Kind of Life: The Limits of Medical Progress*. New York: Simon and Schuster.

Cavallo, J. 2018. Unraveling the mystery of what gives exceptional responders their superpower: A conversation with Isaac S. Kohane, MD, PhD. *The ASCO Post.* https://www.ascopost.com/issues/august-25-2018/unraveling-the-mystery-of-what-gives-exceptional-responders-their-superpower/. Accessed 23 May 2020.

Chan, T.A., M. Yarchoan, E. Jaffee, C. Swanton, S.A. Quezada, A. Stenzinger, and S. Peters. 2018. Development of tumor mutation burden as an immunotherapy biomarker: utility for the oncology clinic. *Annals of Oncology* 30: 44–46.

Cheung, H., and T. Wong. 2018. The full story of Thailand's extraordinary cave rescue. *BBC News.* https://www.bbc.com/news/world-asia-44791998 Accessed 23 May 2020.

Dagogo-Jack, I., and A.T. Shaw. 2018. Tumor heterogeneity and resistance to cancer therapies. *Nature Reviews: Clinical Oncology* 15: 81–94.

Daniels, N. 1985. *Just Health Care*. Cambridge: Cambridge University Press.

———. 1996. *Justice and Justification: Reflective Equilibrium in Theory and Practice*. Cambridge: Cambridge University Press.

———. 2012. Treatment and prevention: what do we owe each other? In *Prevention vs. Treatment: What's the Right Balance?* ed. H. Faust and P. Menzel, 176–193. New York: Oxford University Press.

———. 2015. Can there be moral force in favoring an identified over a statistical life? In *Identified Versus Statistical Lives: An Interdisciplinary Perspective*, ed. G. Cohen, N. Daniels, and N. Eyal, 110–123. New York: Oxford University Press.

Dondorp, W.J., and G. deWert. 2013. The 'thousand-dollar genome': An ethical exploration. *European Journal of Human Genetics* 21: S6–S26.

Dudley, J.C., M.-T. Lin, D.T. Le, and J.R. Eshleman. 2016. Microsatellite instability as a biomarker for PD-1 blockade. *Clinical Cancer Research* 22: 813–820.

Ferlay, J., M. Colombet, I. Soerjomataram, T. Dyba, G. Randi, M. Bettio, A. Gavin, O. Visser, and F. Bray. 2018. Cancer incidence and mortality patterns in Europe: Estimates for 40 countries and 25 major cancers in 2018. *European Journal of Cancer* 103: 356–387.

Fleck, L.M. 2009. *Just Caring: Health Care Rationing and Democratic Deliberation*. New York: Oxford University Press.

Hofmarcher, T., P. Lingren, N. Wilking, and B. Jönsson. 2020. The cost of cancer care in Europe 2018. *European Journal of Cancer* 129: 41–49.

Larkin, J., V. Chiarion-Sileni, R. Gonzalez, J.-J. Grob, P. Rutkowski, C.D. Lao, C.L. Cowey et. al. 2019. Five year survival with combined nivolumab and ipilimumab in advanced melanoma. *New England Journal of Medicine* 381:1535-1546.

Leopold, C., J.M. Peppercorn, S.Y. Zafar, and A.K. Wagner. 2018. Defining value of cancer therapeutics – A health system perspective. *Journal of the National Cancer Institute* 110: 699–703.

Manz, C.R., D.L. Porter, and J.E. Bekelman. 2019. Innovation and access at the mercy of payment policy: The future of chimeric antigen receptor therapies. *Journal of Clinical Oncology* 38: 384–386.

Mendes, E. 2017. *More than 4 in 10 Cancers and Cancer Deaths Linked to Modifiable Risk Factors*. American Cancer Society. https://www.cancer.org/latest-news/more-than-4-in-10-cancers-and-cancer-deaths-linked-to-modifiable-risk-factors.html. Accessed 22 May 2020.

Menzel, P. 2012. The variable value of life and fairness to the already ill: Two promising but tenuous arguments for treatment's priority. In *Prevention vs. Treatment: What's the Right Balance?* ed. H. Faust and P. Menzel, 194–218. New York: Oxford University Press.

National Cancer Institute. 2017. *The Genetics of Cancer*. https://www.cancer.gov/about-cancer/causes-prevention/genetics. Accessed 22 May 2020.

National Cancer Institute: Surveillance, Epidemiology, and End Results Program. 2020. *Cancer State Facts: Cancer at Any Site*. https://seer.cancer.gov/statfacts/html/all.html.

Oxnard, G. 2019. New blood test capable of detecting multiple types of cancer. *Dana-Farber Cancer Institute News Release* (September 28). https://www.dana-farber.org/newsroom/news-releases/2019/new-blood-test-capable-of-detecting-multiple-types-of-cancer/.

Peppercorn, J. 2017. Financial toxicity and societal costs of cancer care: Distinct problems require distinct solutions. *The Oncologist* 22: 123–125.

Prasetyanti, P.R., and J.P. Medema. 2017. Intra-tumor heterogeneity from a cancer stem cell perspective. *Molecular Cancer* 16: 41–50.

Rawls, J. 1971. *A Theory of Justice*. Cambridge: Harvard University Press.

———. 1996. *Political Liberalism*. New York: Columbia University Press.

Raza, A. 2019. *The First Cell, and the Human Costs of Pursuing Cancer to the Last*. New York: Basic Books.

Rosoff, P. 2017. *Drawing the Line: Healthcare Rationing and the Cutoff Problem*. New York: Oxford University Press.

Russell, L. 2010. *Is Prevention Better Than Cure?* Washington, DC: Brookings Institution Press.

Seo, M.K. 2017. Economic evaluations of cancer biomarkers for targeted therapies: Practices, challenges and policy implications. In *Cancer Biomarkers: Ethics, Economics and Society*, ed. Anne Blanchard and Roger Strand, 25–38. Kokstad: Megaloceros Press.

Silverman, E. 2016. Gilead's new price hikes on HIV drugs anger AIDS activists. *Stat News*. https://www.statnews.com/pharmalot/2016/07/05/gilead-hiv-aids-drug-prices/. Accessed 28 May 2020.

Simon, S. 2019. Population of US cancer survivors grows to nearly 17 million. *American Cancer Society.* https://www.cancer.org/latest-news/population-of-us-cancer-survivors-grows-to-nearly-17-million.html. Accessed 19 May 2020.

Turajlic, S., A. Sottoriva, T. Graham, and C. Swanton. 2019. Resolving genetic heterogeneity in cancer. *Nature Reviews Genetics* 20: 404–416.

Ugurel, S., D. Schadendorf. K. Horny, A. Sucker, S. Schramm, J. Utikal, C. Pföhler et. al. 2020. Elevated baseline serum PD-1 or PD L-1 predicts poor outcome of PD-1 inhibition therapy in metastatic melanoma. *Annals of Oncology* 31:144-152.

Verweij, M. 2015. How (not) to argue for the rule of rescue: Claims of individuals versus group solidarity. In *Identified Versus Statistical Lives: An Interdisciplinary Perspective*, ed. G. Cohen, N. Daniels, and N. Eyal, 137–149. New York: Oxford University Press.

Vokinger, K.N., T.J. Hwang, T. Grischott, S. Reichert, A. Tibau, T. Rosemann, and A.S. Kesselheim. 2020. Prices and clinical benefit of cancer drugs in the USA and Europe: A cost-benefit analysis. *Lancet Oncology* 21: 664–670.

Wilking, N., G. Lopes, K. Meier, S. Simoens, W. Van Harten, and A. Vulto. 2017. Can we continue to afford access to cancer treatment? *European Oncology & Haematology* 13 (2): 114–119.
Workman, P., G.F. Draetta, J.H. Schellens, and R. Bernards. 2017. How much longer will we put up with $100,000 cancer drugs. *Cell* 168: 579–583.
Yabroff, K.R., T. Gansler, R.C. Wender, K.J. Cullen, and O.W. Brawley. 2019. Minimizing the burden of cancer care in the United States: Goals for a high-performing health care system. *CA: A Cancer Journal for Clinicians* 69: 166–183.
Yi, M., D. Jiao, H. Xu, Q. Liu, W. Zhao, X. Han, and K. Wu. 2018. Biomarkers for predicting efficacy of PD-1/PD L-1 inhibitors. *Molecular Cancer* 17: 129–143.

Rationing of Personalised Cancer Drugs: Rethinking the Co-production of Evidence and Priority Setting Practices

Eirik Joakim Tranvåg and Roger Strand

Introduction

In this chapter we will address the challenge of rising health care costs, how countries have developed systems and institutions for systematic priority setting and how these rationing decisions are taken with increasing uncertainty, fuelling public controversy. While personalised medicine is seen as a potential solution to this, we argue that due to some inherent traits it may also contribute to more uncertainty and controversy. The current system and strategies for priority setting might not be sufficient to address this. First we introduce concepts from science and technology studies and post-normal science in order to analyse the situation with a new perspective, and secondly we offer some new thoughts that might promote a fair and sustainable public priority setting practice in the future.

E. J. Tranvåg
Centre for Cancer Biomarkers CCBIO and Bergen Centre for Ethics and Priority Setting, Department of Global Public Health and Primary Care, University of Bergen, Bergen, Norway
e-mail: eirik.tranvag@uib.no

R. Strand (✉)
Centre for Cancer Biomarkers, Centre for the Study of the Sciences and the Humanities, University of Bergen, Bergen, Norway
e-mail: roger.strand@uib.no

A. Bremer, R. Strand (eds.), *Precision Oncology and Cancer Biomarkers*, Human Perspectives in Health Sciences and Technology 5,
https://doi.org/10.1007/978-3-030-92612-0_14

Personalised Cancer Care Increases the Health Gap

The sustainability of publicly financed health care systems are challenged by increasing costs. Well-known drivers of this health gap are an increasingly aging and sick population, higher expectations of what the health care system can do, and the development of new diagnostics and treatments. Advances in medical science and technology result in an even larger range of potentially beneficial treatments. Moreover, we live in a world in which medical innovation to a large extent is organised as a rent-seeking activity performed by private enterprise. As a consequence, medical progress also tends to lead to more expensive treatments. This general phenomenon holds very much true for new cancer drugs.

Health care systems across the world struggle to manage the escalating cost of new cancer drugs (Fojo and Grady 2009; Sullivan et al. 2011). Most new drugs for treatment of advanced cancers offer only a modest benefit to patients, while costs are far from modest (Saluja et al. 2018). Yearly treatment costs above 100,000 USD is a rule rather than an exception; some treatments cost far more. Kymriah, a CAR-T therapy for acute lymphoblastic leukaemia in children, was launched by Novartis with a list price of 475,000 USD (Prasad 2018). Anyone can sympathize with the child and the parents for whom this drug might be perceived as the last hope. However, there are opportunity costs, i.e., the costs of foregoing health benefits that could have been obtained if that money was spent elsewhere. The level of potential public spending associated with very costly cancer drugs is likely to cause poorer treatment and more suffering for other patients both within and outside of the sector of cancer care.

In Norway, priority setting in the specialized health care sector is guided by three principles: health benefit, resource use, and severity of disease (Meld. St. 34 (2015–2016) 2016). These criteria were unanimously endorsed by the Norwegian parliament in 2016, after a process that started 3 years earlier when an official committee on priority setting was established (NOU 2014:12 2014). This was the third such committee in Norway, illustrating a decade-long tradition of systematic priority setting discussions. In the white paper it is clearly stated that "equal cases shall be treated equally" (p. 11) and also that "…transparency and user participation will be central values" (p. 11). Another important feature is the distinction between individual and group level decisions, where the latter involve quantifying the criteria using quality adjusted life years and cost-effectiveness estimations.

Many publicly financed health care systems, like Norway, and also the UK, Sweden and in many member states of the European Union, have established governmental policies and institutions for health care priority setting. Within these institutions, procedures for evaluation and appraisal of new drugs have been developed to ensure that public money is spent in accordance with rules or criteria for priority setting. Typically, and in line with the principle of equal treatment, drugs included in the public health care scheme are held against an equal standard, independent of drug type and targeted patient groups. This systematic approach is based on theories and models from medical ethics, distributive justice and health economics, and is meant to enact basic ethical values by providing health care in a reasoned, reasonable and (tentatively) transparent manner. Impartiality and treating equal patients equally are key ethical considerations that are meant to be universal and uncontroversial (Kieslich et al. 2016).

Controversy is nevertheless common in many countries, and not the least with respect to cancer drug pricing and rationing (Gross and Gluck 2018; Wilson et al. 2008; Aggarwal et al. 2017). In the case of Norway, media studies indicate that public controversies are ubiquitous, to the extent that there have been years with new media stories about cancer patients who have been denied publicly paid access to a new treatment (Stenmarck et al. 2021).

Controversy as such is not a sign that anything is wrong. Health care rationing is an important political issue upon which there is legitimate disagreement. For the actors in the supply chain there is considerable economic interest; for individual patients the stakes may be a question of (prolonged) life or death. Indeed, a certain level of public contestation can be seen as healthy, as a sign of a vital democracy. In our opinion, the real cause of concern is rather the spiralling costs and the increasing unsustainability of public health care systems. The unsustainability seems in some cases to be aggravated by the nature of the surrounding public controversies, which strains the system, lead to ad hoc measures and exceptions from priority setting principles that drive costs up the spiral. Furthermore, the drivers of unsustainability on the public-political side seem to work in synergy with equally important drivers on the side of medical science and technology. We shall take some care to explain what we mean by that claim.

We noted above that medical progress tends to increase rather than decrease health care costs by at least two mechanisms: increase in range and volume of treatments, and the capitalist logic whereby a new product, sometimes medically superior to existing treatments, will be of higher worth and as a rule will be priced higher than its predecessors. A third mechanism, peculiar to the current trend towards personalised medicine, is that a larger share of new treatments are "tailored", aiming to prescribe "the right drug to the right patient at the right time". In other words, there are more new drugs that sell in relatively small volumes and fewer blockbuster drugs that, by economies of scale, may be sold at lower prices (Duffy and Crown 2008). A perverse effect is that very high list prices make negotiations for discounts widespread, which again implies less transparency in priority setting when governments agree to keep discounts confidential (Tranvåg 2019).

More to the core of personalized medicine, however, there is a proliferation of diagnostics schemes that each target smaller groups of patients defined by ever finer diagnostic criteria and biomarker characterisations. From a purely scientific point of view, this development promises higher precision in identifying patients and to better match drugs with their responders, and by avoiding ineffective and costly treatment of non-responders as well as toxicity and side effects. Indeed, the latter years the imaginary of "precision oncology" has gained traction. According to this imaginary, at least in its purest expressions, one may arrive at an exact scientific characterization of the molecular basis of disease in each individual and thereby devise the precise molecular cure or treatment.

It is outside the scope of this study to discuss the eventual realism of the reductionist imaginary of precision oncology. However, in the context of health care priority setting, personalisation has as a matter of fact so far often implied the opposite of precision. Personalised medicine leads to a higher number of treatments to test and finer stratification of patient groups, which both imply that clinical trials are done with fewer patients in each group, faster, and with more surrogate endpoints (Schork 2015; Chen et al. 2019). In this way, the development towards personalized medicine poses risks to methodological validity and a weakening of the evidence base (Moscow et al. 2018). For priority setting institutions the number of new drugs to assess have grown, whilst the evidence base for the assessment has gradually become increasingly thin and provisional (Davis et al. 2017; Naci et al. 2019; Tranvåg et al. Submitted).

At the same time nearly every new drug is met with a claim that the drug is highly beneficial to some small and narrowly defined subgroup of patients, and so science, industry and the public put high pressure on authorities to approve these drugs. And then, if the drug is approved for some small subgroup, there are always "ragged edges" around the definition of that group and always possible to make claims of scientific uncertainties in order to argue that the drug should also be made available to those who now find themselves excluded by the first limited approval (Fleck 2010). Such claims are well suited for news media because they typically concern a small number of individuals and allow for news coverage in terms of storylines about individuals at risk. This is an example of a new type of synergy between personalized medicine and personalized politics that focuses on the tragedy of the individual terminal patient, what Brekke and Sirnes (2011) called "the hypersomatic individual".

In sum, the development towards personalized cancer medicine poses new challenges and increases the pressure on institutions of health care rationing. At one level, more and better business-as-usual could appear to solve these challenges: Clearer and better specified criteria for priority setting; stronger demands on the pharmaceutical industry to present methodologically strong evidence; integrating real world evidence; international collaboration between governments to refuse secret price negotiations with the industry; better education of citizens so that they understand the realities of opportunity costs and the need for rationing. If all of this worked well for priority setting between groups of thousands of patients, it may also work for groups with dozens of patients by increasing the effort on all sides. Let us call this Plan A.

The authors of this chapter are not convinced that Plan A will work. If we were, there would be no need for the chapter; then we might as well leave our governments to continue as before. At least in the case of Norway, there is little sign of anything but Plan A on the side of governmental policy. Still, the level of controversy does not seem to decline, in an age where erosion of public trust in political and governmental institutions has been seen in many sectors. The rest of this chapter is devoted to our reasons for why Plan A might not work, and our suggestions for a possible Plan B.

Why Plan A Might Not Work and Why the Problem Is Connected to Biomarkers

Above, we delineated a Plan A for health care rationing in the age of personalised medicine, namely to strengthen its frameworks and institutions without much need to rethink its practices, or medical and scientific practices for that matter. At the same time, we opened up for the possibility that Plan A might not work. From a sociological perspective one might state, for instance, the quite obvious fact that governmental institutions in modern societies do not operate in isolation from the sectors that they govern and the public on behalf of whom they govern, but that they are in fact in need of some sort of legitimacy vis-à-vis both. Controversy and contestation can be a sign of vitality – but not without limits. There is a question of how much tension an institution can live with and how much power it will be able to gain.

Taking the immanent perspective, it is possible to give a more principled argument for why personalised medicine may create a need to rethink a priority setting strategy based on impartial and equal standards. The argument does not per se go against the rationality or desirability of such standards, but rather shows how the scientific development threatens to undermine the possibility of enacting them.

While the exact content of such standards may vary, some common features may be distinguished of the type of rationing principles that we are discussing here. First, they are not entirely casuistic and pragmatic. It would be entirely possible to organise health care rationing in terms of case-by-case deliberation and decision-making, say, performed by a sovereign committee whose composition secured some sort of legitimacy by its representativity. Such entities exist in health care systems; clinical ethics committees and internal review boards may resemble this extreme type of procedural legitimacy. However, this is not how health care rationing at the governmental level tends to be organized. Instead, it is designed to ensure some degree of distributive justice by making decisions with regard to groups rather than individuals, and by aiming to treat the groups fairly so that they receive whatever proportion of the health budget that is considered to be fair.

For a priority setting approach based on impartial assessments of different patient groups to work, a number of assumptions have to be made. One needs some form of generic accounting of resource use (e.g. monetary costs) and of health benefits (e.g. quality-adjusted life years (QALYs)) in order to make comparisons across patient groups. These measures of cost and benefit can be adjusted with some form of distributional aspects, typically by some estimate of need (in Norway, including severity of disease). The overall framework does not have to be utilitarian – it could be based on needs or capacities rather than utilities in the strict sense, and it could be adjusted with deontological principles about the duty to provide life-saving emergency treatments (as the end of life-criterium in the UK) – but it will have to be similar to utilitarianism in the sense that the right decision will be one that maximises some balance between overall health benefit and a fair distribution of health benefit.

Moreover, it will have to satisfy the requirement often alluded to by John Rawls' concept of "the veil of ignorance": fair principles for a just society can be agreed if no one know which status and interests they will have, thereby making decisions neutral and separated from self-interests (Rawls 1999). As a consequence, priority setting decisions should expressly *not* be based on nepotistic interest or undue discrimination. Examples of what is meant by the latter, are easy to give: For instance, it would be undue discrimination if the procedures or outcomes of the rationing process result in a systematic favouring of men rather than women; of Caucasians rather than Asians; of rich rather than poor people; of young rather than old persons, and so on. Along the lines of the sociological perspective we mentioned earlier we may note that such health care rationing systems *de facto* are at odds with social reality, in which the interests of, say, Caucasian rich men often are favoured over most other groups. In this sense the principles are ideals of a modernist, human rights- and Enlightenment-based type, trying to improve the social world by institutionalizing and enacting moral principles.

A crucial working assumption for such priority setting strategies to work is that it is possible to distinguish between legitimate and illegitimate discrimination of patient groups. Central in the priority setting frameworks in Norway and the UK is the principle of equal treatment. This states that persons that are equal in all ethically relevant characteristics must be treated equally, and that persons that are unequal in some ethically relevant characteristic can be treated unequally. Most people see gender, political views, religious convictions and sexual orientation as ethically irrelevant, and therefore as illegitimate grounds for unequal treatment. Need, severity of disease and benefit of treatment are by most seen as ethically relevant characteristics in priority setting decisions and may give reasons for a legitimate discrimination of patient groups. An example of such legitimate discrimination is to provide targeted treatment to patients with an EGFR (epidermal growth factor receptor) mutation and not to an otherwise similar group of patients who lacks the mutation.

But for this working assumption to hold, two conditions must be in place: First, the methods we use to estimate or predict benefit, need or some other ethical relevant characteristic must be of good enough quality, and second; the classification of groups should be independent from and uncorrelated with the classification of social groups.

In previous studies (Tranvåg et al. 2018, 2021) we have shown how this former condition can break down in clinical practice, in ways that are relevant to priority setting. Patient age is a well-suited example: On one hand, age discrimination is by most seen as *prima facie* morally unacceptable. On the other, patient age can be a highly informative and useful piece of information in clinical decision-making and may provide relevant information about risk and potential benefit. Therefore, age is used in multiple ways in which it is not easy to separate the descriptive, "objective" function from the normative function. For instance, clinical knowledge about how tough it typically is for an 85-year-old person to recover from surgery or live well with the side effects of a highly toxic cancer drug, may blend into the overall question of whether it is medically worthwhile to give the treatment – also in the absence of a scientific evidence based on the question.

In Tranvåg et al. (2021) clinicians were presented with hypothetical priority setting decisions. A high chronological age was found to be the single most important patient characteristic that influenced the doctors' priority setting decision for a new cancer drug. In the same survey the patients' smoking status was considered as an irrelevant characteristic for priority setting, despite it being a piece of information that may be relevant to patient-centred clinical decisions. It may very well be that whether a patient smokes or not would have had clear prognostic and predictive value of high relevance to many priority setting decisions. However, in real life this could often lead to allegations of undue discrimination.

A problem posed by personalised cancer medicine, and even more so by the imaginary of precision oncology, is that the information being used to stratify the patients is becoming massive and comprehensive. The working assumption that patient groups are uncorrelated with social groups is likely to fail more often. Moreover, as patient stratifications are being used in the arguably social system of health care priority setting, they take on social meaning and can become social groups. In a hyper-connected world with social media, one can easily envisage that patient subgroups can form their own communities, say, a community for those who score slightly below the threshold for being regarded as PD-L1 positive with respect to a certain treatment. Now, if their PD-L1 status is not only a negative predictive biomarker but also a negative prognostic biomarker for their condition, they could make the claim that they as a group are faring worse than the PD-L1 positive and accordingly are being unduly discriminated against if PD-L1 status is the unique criterion for denying them access to treatment. Adding the endless possibilities of combining biomarkers into batteries, there are equally endless possibilities of forming such imagined communities around claims of illegitimate discrimination.

A central challenge for the current priority setting strategy when faced with personalised medicine at a full scale, is that new ways of organizing clinical trials, with small groups of patients, surrogate endpoints and short follow-up time makes the evidence used for decisions uncertain. At some point it will no longer be meaningful nor ethically acceptable to classify patients into different groups and give them unequal treatment based on biomarkers for which the quality of prediction is very uncertain. If precision diagnostics are not precise enough to stratify patients into smaller groups in an ethically acceptable way, priority setting based on such stratification cannot be ethically acceptable either.

While the scenario laid out above is not full reality as of yet, it is the case that the public controversies witnessed especially since the entrance of costly immunotherapies against cancer indeed already do contain claims of undue discrimination. The typical proponents are not necessarily arguing against any form of rationing or cost control. Rather, they make a claim of being equivalent with those who got the drug or being different from those who should not get it. What we are arguing, is that the presence of such arguments is related to scientific progress and scientific literacy, and that there is reason to believe that the trend towards personalised medicine will make such arguments ever more frequent.

In order to sum up and characterize our argument, Plan A will work if the pressures against health care priority setting can be resolved by shifting the power balance, strengthening the priority setting institutions, reducing the power of big pharma and educating the public. This may very well be enough. But we, both as scientists and as a society in general, are obliged to think further. Plan B will be needed if the problem runs deeper and undermines the very assumptions upon which priority setting is built. What we believe is undermining these assumptions, is a blind spot of ethics and economics, namely the trajectory of the scientific development. To borrow a pair of concepts from the French sociologist Michel Callon (1998), the *frame* provided by priority setting principles is being *overflowed*, and this is the deep cause of contestation and controversy.

The Co-production Perspective as an Analytic Tool

> Increasingly, the realities of human experience emerge as the joint achievements of scientific, technical and social enterprise: science and society, in a word, are co-produced, each underwriting the other's existence. (Jasanoff 2004, 17)

During the final decades of the twentieth century, scholarship on science, technology and society advanced the understanding of how scientific, technological and societal development processes are "co-produced", how they are causally entangled into each other. This insight did not come easy; most philosophy of science used to emphasize the autonomy of science from society, and most modern institutions were built upon the assumption of that autonomy. Indeed, Bruno Latour (1993) argued that the efforts to conceptually demarcate between science and politics (and by implication, between nature and culture) are not only key to modern societies but constitute a type of work (of purification) that is necessary to enable and justify the massive production of linkages between science and politics (and nature and culture) that is characteristic of these societies. "Being modern" is to believe in the fiction that science and politics are independent; this belief is what allows us to create the reality that science and politics, and nature and culture, become ever more entangled, to the extent that cancer patients may be enrolled into the forefront of international research as well as becoming the subject of headline news and parliamentary debate. Part of that fiction is also to believe that facts and values are wholly independent and can be assessed independently from each other.

In reality however, value choices are embedded into scientific methodologies, such as when clinical endpoints are chosen, and conversely. Furthermore, factual matters influence value choices, for instance by changing the (actual or potential) option space (Hofmann et al. 2018). As long as these dependencies between science and technology and its interactions are not noted and pointed out, the assumption of their non-existence may be upheld and the modern institutions that are built upon this assumption may continue to appear functional. The moment they are noted and pointed out, however, disturbance arises: uncertainties, controversies, contestation and loss of legitimacy. These are expressions of the modern frame being overflowed.

While this may sound terrible to a Cartesian mind, the achievement of scholars such as Callon, Jasanoff and Latour has been to show how overflowing is the rule rather than the exception, and how all institutional arrangements will have to be seen as dynamic and situated, that is, contingent to their context in time and space.

To our knowledge, this co-production perspective has not been overly prominent in scholarly debates on health care priority-setting. A notable exception is Tiago Moreira (2011), who analysed cases of controversies surrounding the UK National Health Service and its advice authority NICE (National Institute of Health and Clinical Excellence). The main analytical concept in Moreira's study is that of *uncertainty,* defined as "the non-determinate or unsettled quality of a statement or a knowledge claim" (p. 1335). Along the lines with our description above, uncertainty is seen as a key expression of overflow of the frame: Controversy and contestation can be analysed in terms of claims of uncertainty, and such claims, if successful, can lead to a change in principles and practices of priority setting. Moreira presents two dimensions with regard to which such uncertainty claims were made in the case studies investigated: standards and disease-specific knowledge. For instance, both with respect to cancer drugs and dementia, the standards themselves were subject to problematisation. It was argued that the standardised metric of QALYs was inappropriate to deal fairly with the particular suffering and needs of cancer and dementia patients. What Moreira finds, is that in such cases an *exception* from priority setting rules is a likely outcome. Indeed, in the UK such exceptions have been introduced both for dementia and cancer. In other cases, the target of the claims of uncertainty are disease-specific knowledge, for instance whether a particular drug works well, and for whom. In such cases, an *impersonal rule* may be the likely outcome (for instance a threshold for allowable expenses per QALY gained). Finally, Moreira finds cases where uncertainty claims are successfully made in both the general and the disease-specific dimensions, and where a deliberative, pragmatic approach may be sought to provide justification in procedural fairness in the relative absence of authority based in scientific certainty.

We noted earlier that a possible blind spot of ethics is its tendency to take for granted the description of matters of fact as provided by science, or rather, to take for granted the possibility and desirability of science providing such descriptions. Within such working assumptions, the mechanics of priority-setting can work to calculate what it the just and fair solution given the matters of fact. The advantage of the co-production perspective as provided by science and technology studies (STS) is that it offers a more nuanced and complex analysis in which scientific descriptions are also seen as dynamic, as provided by actors in contestation with other actors, and as something that can be deconstructed as well as reconstructed in the course of action. Equipped with this analysis, nobody ought to be surprised by the presence of controversies.

Hofmann et al. summarised the Scylla of what they called "the traditional positivist account" with the Charybdis of "the social constructivist account" as the choice between simplistically "evaluating facts" and an equally simplistic approach of "facting values" (Hofmann et al. 2018). The Charybdis can be sensed in Moreira's quasi-normative conclusions in which the pragmatic, deliberative approach is seen as a solution that contributes to social robustness of the priority setting process. Effectively, the argument can be seen as going from "is" to "ought" as much as in a positivist account: Because various actors as a matter of fact challenged expert knowledge and hence created uncertainty about them, the decision process ought to take into account that uncertainty by taking a broad participatory, inclusive and deliberative approach in hybrid expert-lay fora whereby one may aim for a pragmatic balance between the various claims of matters of fact and value.

An important critique of this approach of "facting values" is that some claims may be truly unreasonable, uninformed or even not put forth in good faith. Sometimes a governmental body will dismiss such protests as unreasonable or unfair, and proceed notwithstanding the controversy. And sometimes they may be right in doing so, as when citizens want a new facility for waste management but Not In My Back Yard ("NIMBY") or when they want health care rationing but not for their own disease. And in that case, what we called Plan A above is warranted.

The positions of Scylla and Charybdis juxtaposed by Hofmann et al. (2018) are not entirely men of straw. Historically, there was some degree of political resonance between the science critique conveyed in the social constructivist heydays of STS (see e.g. Collins and Pinch 1993) and the agenda of citizen empowerment through public participation. Both movements grew out of the same political sources in the late 1960s (Sardar 2015). In this sense, a genealogical line can be drawn to Moreira's conclusion that pragmatic balance through hybrid fora is a way forward for health care priority setting and all the way back to Sherry Arnstein (1969) and the *ladder of participation*, where citizen control reigned highest in the hierarchy of public participation and where consultation and providing information were considered inferior and symbolic forms. It would not do Moreira's analysis justice, however, to ascribe to it the somewhat romantic views on citizen control of the 1960s. Rather, when discussing the possible Plans B for health care priority setting, we should enter into finer detail of the purpose of the participation.

Andy Stirling distinguished between instrumental, substantive and normative rationales for public engagement (Stirling 2008). The instrumental rationale is to use public engagement as a vehicle for apparent legitimacy, as when a lay person is included in an ethics committee more or less as a hostage, without much opportunity to influence the processes and outcomes. This rationale seems to be close to how lay persons are acting in the New methods system in Norway, as observers without any influence on the actual decision making. What Stirling calls the normative rationale, is the one of deliberative democracy: That certain processes and institutions may suffer from democratic deficit, and that public engagement may correct that deficit and the power imbalance that comes with it.

Somewhat in between, the substantive rationale is the idea that decision outcomes may be substantively *better* by opening up the processes to broader participation. At its core, the substantive rationale is consistent with Jürgen Habermas' ideas of discourse ethics and universal pragmatics: decisions get better if every argument is listened to and considered. In practice, however, the argument of the substantive rationale is often more specific and involves a critique of technocracy, how expert knowledge entails a risk of tunnel vision and that broader participation can improve decision-making by including experiential knowledge and a broader range of values and perspectives.

The Achillean heal of this substantive rationale is the notion of "better" – in what sense is the decision imagined to become "better" by changing the process? If "better" simply means more desirable from a certain actor's point of view, the substantive rationale is undistinguishable from the instrumental one, and the participation was actually non-participation in Sherry Arnstein's definition. Similarly, if "better" simply means more democratic, this would be equal to the normative rationale.

The discourse ethics tradition, with philosophers such as Habermas and Karl-Otto Apel, would translate "better" into some notion of validity or criterion of truthfulness, consistency or objectivity, connecting it to the ideal, "herrschaftsfreie Diskurs" that by respectful listening and talking moves towards consensus. In real life, however, instead of consensus, it appears that modern societies are moving into a phase in which ever more classes of decision problems are plagued by persistent controversies. Appraisals and priority setting of expensive cancer drugs seems to be such a class of problems (Strand 2017).

Silvio Funtowicz and Jerome Ravetz (1985) offered an analysis of such problems, an analysis that later was to be associated with the concept of "post-normal science". In their analysis, these types of controversies are typically not resolved by scientific and technical attempts to reduce uncertainty; rather these attempts are themselves politicised and may as well end up increasing the controversy. The normative suggestions by Funtowicz and Ravetz were similar to those of Moreira: Broadening the perspective with respect to who can bring relevant knowledge and values to the table – "extending the peer community" in their terms – and preparing for processes of sustained and inclusive deliberation. However, their rationale was not an idealist belief in the herrschaftsfreie Diskurs or a true democracy. Rather, it was based in the more pragmatic solution that otherwise the controversy will simply not go away by itself.

Central to the idea of post-normal science is to let go of unrealistic hopes of attaining certainty and truth about the issue at stake and rather aim for a set of knowledge and value claims of mutually acceptable quality for the involved parties. In this framework, "quality" is to be understood as fitness for purpose; and part of the deliberation process is to decide on the acceptable purposes. While being formulated within a type co-production perspective, the idea of quality as fitness for purpose actually gives more guidance than the usual "broaden the participation" and offers one possible middle route between the Scylla and Charybdis mentioned above – it guides us towards what "better" decisions could look like.

Let us recall the problem set out in the first two sections of this chapter: Not only the increasing health gap due to scientific and technological advances in a particular political economy, but also a concomitant deterioration of the evidence base for priority setting decisions, as clinical trials get smaller and faster. And even worse, as patient stratification becomes ever more fine-grained, it will be increasingly difficult to distinguish between due and undue discrimination.

In line with Moreira we can conclude that the problem is likely to imply persistent controversy. It will continue to be possible to raise uncertainty claims both with respect to standards and disease-specific knowledge; it is likely to become ever easier. The isolation between governmental priority-setting bodies and the political and public spheres that contest these bodies, is likely not to work much longer. Some new forms of broader deliberation are needed, and some kind of "pragmatic balance" is called for. But how?

The post-normal answer would be that the deliberation should aim at clarifying the purposes and revisit the knowledge base with respect to its fitness for purpose. It is not enough to make an uncertainty claim; in principle everything in this world can be questioned and called into uncertainty. What is needed, is to deliberate in good faith whether the uncertainty prohibits a decision, or whether the decision problem can be revisited and reframed. In this sense, fitness and purpose come together, in what we earlier called a *frame*. Part of these deliberations would deal with the question of standards, which means that the production of such standards – currently the work of health care ethicists, health economists and other experts –will have to be discussed. Also, how disease-specific knowledge is produced must be addressed. This means that trial design, how research and development of new drugs are organized and also how the whole political economy of drug development is set up, ought to be deliberated.

Sustainable Future Imaginaries for Cancer Drug Priority Setting

The post-normal question to rationing of cancer drugs is accordingly how the problem could, might and ought to be reframed (Stenmarck et al. 2021; Strand 2017). To ask such questions is to engage in socio-technical imagination (Jasanoff and Kim 2009), that is, to explore visions of future desirable scientific, technical and societal orders. In line with Jasanoff's co-production perspective, we insist that such orders, real or imagined, are scientific, technical and societal *at the same time;* they are co-produced in the sense of being produced together. Accordingly, if Plan A is merely to adjust and strengthen the institutions and practices of health care priority setting, the co-production perspective suggests that Plans B may reimagine the whole constellation of medical research, technology and practice together with the institutions and practices of priority setting. Plans B allow us to refuse to take for granted the current political economy of science and technology and the currently dominating reductionist imaginaries of personalised cancer medicine and precision oncology. The future *may be otherwise,* and we are entitled to imagine and strive for different futures.

We find it useful to distinguish between co-production (different things being produced together) and co-creation (different actors producing the same things). This distinction is in line with Jasanoff's perspective, although individual authors have defined these terms differently. At the same time, the co-production perspective lends itself naturally to the idea of co-creation and that the involvement of a wider range of actors may lead to more democratic, more socially robust and ultimately *better* sociotechnical imaginaries and realities by extending the peer community. There are, in other words, two fronts that would have to be opened up in parallel in order to reimagine future cancer care, research and priority setting. The post-normal co-creation perspective suggests a front in terms of procedure, while the co-production perspective would also suggest a front in terms of substantive matter.

The procedural dimension of more sustainable imaginaries is not too difficult to envisage. For example, there will be no co-creation of knowledge in the Norwegian system which Moreira labels as "the public education model" (p. 1334) where expert assessments are the only valid source of knowledge and public engagement is close to what Stirling describes as an instrumental rationale for public engagement. Further, as decisions are to be neutral and rational, this effectively excludes patients, relatives, many health care workers and others who have their own skin in the game or the skin of those they care for. Currently, their opinions and viewpoints are dismissed within the frame of priority-setting institutions as "emotional" and accordingly biased. No wonder, then, that the frame is regularly overflowed:

> From the perspective of the suffering patient, the arguments about bias effectively imply that those who are most affected by the decisions are excluded from taking part in them. Even worse: It is exactly the fact of being directly and severely affected by the decision that disqualifies them from taking part in it. […] Modern society empowers them to create their own careers, families, households and living conditions and democratically influence the development of their own communities and societies. However, when they arrive at the critical point in their lives – perceived as a life and death decision over a certain immunotherapy – they no longer have a say as citizens at the general level of priority setting because they are affected and therefore not impartial and rational. (Strand 2017, 136–7)

If one rethinks how the role of patients in priority setting decisions could be, we could imagine that very ill, perhaps terminally ill cancer patients were invited to deliberative processes that would decide if their treatment could be prioritized. If the deliberation was real and not just what Arnstein called manipulation or placation, one would have to meet their demands with real counterarguments. Those included in the deliberation would have to accept that valid arguments against giving priority to their potential life-saving drugs was articulated. Essentially, one truth that would have to be put on the table is that no man is an island and that someone's suffering and death is in fact no tragedy for society. Humans are mortal and death is part of life, an insight that modern society is trying to neglect, hide and forget.

Indeed, the current ecosystem of rationing decisions in publicly financed healthcare systems is remarkably poor. Those who are directly affected are represented by proxies – patient organizations and of course also the pharmaceutical industry, that presents itself as patient guardians while at the same time have their own legal obligations to maximize profit for their owners. Those who are indirectly affected – citizens who pay taxes and may have other and competing welfare needs – are also represented by proxies in the form of governmental actors and institutions. Indeed, the latter may be seen to try to represent everybody's interest and well-being. On the top of all this, there are confidential drug prices, censoring of published documents and unavailability of public justifications for decisions. This is the frame the overflowing of which still surprises some. From the co-creation perspective, bearing Stirling's normative rationale in mind, one would suggest to open up these processes to become inclusive and transparent.

When arguing as we do above for procedural reform and with a normative rationale, we should be careful not to be too directive in terms of the substantive issues. Indeed, what counts as *better,* is something for the deliberative processes in extended peer communities, and ultimately, society at large, to decide. Still, as we insisted above, the co-production perspective offers the view of scientific, technical and societal matters as entangled into each other and dependent on one another. Rather than "facting values" one can and, we believe, ought to open up the many value-laden assumptions underlying how personalised cancer medicine is imagined, practiced and governed.

A question that was already indicated above, is why at all society should devote so much public spending to cancer medicines or to health in the first place. Another question is why treatment innovations have to be so expensive. The answer may be found in the political economy of the research and innovation system of the health sector, which is characterized by the choice to allow private companies to seek high profits in return for the promise of fast innovation. Without going into a sweeping critique of capitalism, one could very well imagine that priority setting policies and research and innovation policies were coordinated to obtain policy coherence around the goal of reducing cost and finding a sustainable path for personalised medicine. Currently, public money subsidizes research that produces modestly beneficial drugs that the public health care system later has to buy for what many would say are perversely high prices.

A different trajectory could even be included under the increasingly important research policy concept of "openness", as in open access and open science. One could imagine policies by which public research funding was provided to develop biomarkers that predict toxicity and poor effect, and that could discover new indications for old drugs that no longer was under patent protection. By ceasing to accept the discourse of urgency and the imperatives argued for by industry, sometimes in coalition with patient organizations, other innovation trajectories could be sought – for effective prevention, for repurposing, for biomarkers to reduce ineffective or harmful treatment, for adaptive treatment regimes and other innovations that resist commodification and accordingly could be more affordable.

We present such imaginations while stating that we do not wish to be overly directive. This may seem strange to the reader: How can one deliver a proposal that, if implemented, would be likely to cause multinational pharmaceutical companies to go bankrupt, and still pretend to be careful and cautious? Our reply is to be found in the post-normal perspective: We are not experts who are speaking truth to power. First of all, we do not know better than everybody else. As combined citizens and researchers, we express our position for others to engage in and discuss. Secondly, even if we were under the illusion of being omniscient, we would not consider such changes, concomitantly in the political economy of research and the governance of the health sector, as nowhere viable if they were to be enforced by a technocratic process. Co-production is de facto political; this is why there is overflow in the first place. This means, however, that neither a Plan A nor a Plan B will work as a technical quick fix. As personalized medicine evolve, priority setting problems and public controversies will continue to erupt; if we are right, they can only be resolved by rethinking and remaking science, technology and society.

References

Aggarwal, A., T. Fojo, C. Chamberlain, C. Davis, and R. Sullivan. 2017. Do patient access schemes for high-cost cancer drugs deliver value to society? – Lessons from the NHS Cancer Drugs Fund. *Annals of Oncology* 28 (8): 1738–1750.

Arnstein, S.R. 1969. A ladder of citizen participation. *Journal of the American Institute of Planners* 35 (4): 216–224.

Brekke, O.A., and T. Sirnes. 2011. Biosociality, biocitizenship and the new regime of hope and despair: Interpreting 'Portraits of Hope' and the 'Mehmet Case'. *New Genetics and Society* 30 (4): 347–374.

Callon, M. 1998. An essay on framing and overflowing: Economic externalities revisited by sociology. *The Sociological Review* 46 (1_suppl): 244–269.

Chen, E.Y., S.K. Joshi, A. Tran, and V. Prasad. 2019. Estimation of study time reduction using surrogate end points rather than overall survival in oncology clinical trials. *JAMA Internal Medicine* 179 (5): 642–647.

Collins, H., and T. Pinch. 1993. *The Golem: What everybody should know about science*. Cambridge: Cambridge University Press.

Davis, C., H. Naci, E. Gurpinar, E. Poplavska, A. Pinto, and A. Aggarwal. 2017. Availability of evidence of benefits on overall survival and quality of life of cancer drugs approved by European Medicines Agency: Retrospective Cohort study of drug approvals 2009-13. *BMJ* 359: j4530.

Duffy, M.J., and J. Crown. 2008. A personalized approach to cancer treatment: How biomarkers can help. *Clinical Chemistry* 54 (11): 1770–1779.

Fleck, L.M. 2010. Personalized medicine's ragged edge. *The Hastings Center Report* 40 (5): 16–18.

Fojo, T., and C. Grady. 2009. How much is life worth: Cetuximab, non–small cell lung cancer, and the $440 billion question. *Journal of the National Cancer Institute* 101 (15): 1044–1048.

Funtowicz, S.O., and J.R. Ravetz. 1985. Three types of risk assessment: A methodological analysis. In *Environmental impact assessment, technology assessment, and risk analysis*, ed. V.T. Covello, J.L. Mumpower, P.J.M. Stallen, and V.R.R. Uppuluri, 831–848. Berlin: Springer.

Gross, C.P., and A.R. Gluck. 2018. Soaring cost of cancer treatment: Moving beyond sticker shock. *Journal of Clinical Oncology* 36 (4): 305–308.

Hofmann, B., K. Bond, and L. Sandman. 2018. Evaluating facts and facting evaluations: On the fact-value relationship in HTA. *Journal of Evaluation in Clinical Practice* 24 (5): 957–965.

Jasanoff, S. 2004. *States of knowledge: The co-production of science and the social order*. London/New York: Routledge.

Jasanoff, S., and S.-H. Kim. 2009. Containing the atom: Sociotechnical imaginaries and nuclear power in the United States and South Korea. *Minerva* 47 (2): 119.

Kieslich, K., J.B. Bump, O.F. Norheim, S. Tantivess, and P. Littlejohns. 2016. Accounting for technical, ethical, and political factors in priority setting. *Health Systems & Reform* 2 (1): 51–60.

Latour, B. 1993. *We have never been modern*. Cambridge: Harvard University Press.

Meld. St. 34 (2015–2016). 2016. *Verdier i Pasientens Helsetjeneste* [Values in the patient's health care]. Oslo: Helse og omsorgsdepartementet.

Moreira, T. 2011. Health care rationing in an age of uncertainty: A conceptual model. *Social Science & Medicine* 72 (8): 1333–1341.

Moscow, J.A., T. Fojo, and R.L. Schilsky. 2018. The evidence framework for precision cancer medicine. *Nature Reviews. Clinical Oncology* 15 (3): 183.

Naci, H., C. Davis, J. Savović, J.P.T. Higgins, J.A.C. Sterne, B. Gyawali, X. Romo-Sandoval, N. Handley, and C.M. Booth. 2019. Design characteristics, risk of bias, and reporting of randomised controlled trials supporting approvals of cancer drugs by European Medicines Agency, 2014-16: Cross sectional analysis. *BMJ* 366: l5221.

NOU 2014:12. 2014. Åpent Og Rettferdig – Prioriteringer i Helsetjenesten [Open and fair – Priority setting in the health service]. *Norges Offentlige Utredninger*. Oslo: Helse- og sosialdepartementet.

Prasad, V. 2018. Tisagenlecleucel – The first approved CAR-T-cell therapy: Implications for payers and policy makers. *Nature Reviews. Clinical Oncology* 15 (1): 11–12.

Rawls, J. 1999. *A theory of justice*, rev. ed. Cambridge: Harvard University Press.

Saluja, R., V.S. Arciero, S. Cheng, E. McDonald, W.W.L. Wong, M.C. Cheung, and K.K.W. Chan. 2018. Examining trends in cost and clinical benefit of novel anticancer drugs over time. *Journal of Oncology Practice* 14 (5): e280–e294.

Sardar, Z. 2015. *Introducing philosophy of science: A graphic guide*. London: Icon Books Ltd.

Schork, N.J. 2015. Personalized medicine: Time for one-person trials. *Nature* 520 (7549): 609–611.

Stenmarck, M.S., C. Engen, and R. Strand. 2021. Reframing cancer: Challenging the discourse on cancer and cancer drugs – A Norwegian perspective. *BMC Medical Ethics* 22: 126.

Stirling, A. 2008. 'Opening up' and 'closing down' power, participation, and pluralism in the social appraisal of technology. *Science, Technology & Human Values* 33 (2): 262–294.

Strand, R. 2017. Expensive cancer drugs as a post-normal problem. In *Cancer biomarkers: Ethics, economics and society*, ed. A. Blanchard and R. Strand 2nd, 129–143. Kokstad: Megaloceros.

Sullivan, R., J. Peppercorn, K. Sikora, J. Zalcberg, N.J. Meropol, E. Amir, D. Khayat, et al. 2011. Delivering affordable cancer care in high-income countries. *The Lancet Oncology* 12 (10): 933–980.

Tranvåg, E.J. 2019. Confidential drug prices undermine trust in the system. *Journal of the Norwegian Medical Association* 139 (9). https://doi.org/10.4045/tidsskr.19.0284.

Tranvåg, E.J., O.F. Norheim, and T. Ottersen. 2018. Clinical decision making in cancer care: A review of current and future roles of patient age. *BMC Cancer* 18 (1): 546.

Tranvåg, E.J., Ø. Haaland, B. Robberstad, and O.F. Norheim. Submitted. *Balancing cost-effectiveness and fair distribution of health outcomes in drug coverage decisions: retrospective cohort study*.

Tranvåg, E.J., R. Strand, T. Ottersen, and O.F. Norheim. 2021. Precision medicine and the principle of equal treatment: A conjoint analysis. *BMC Medical Ethics 22*: 55.

Wilson, P.M., A.M. Booth, A. Eastwood, and I.S. Watt. 2008. Deconstructing media coverage of Trastuzumab (Herceptin): An analysis of national newspaper coverage. *Journal of the Royal Society of Medicine* 101 (3): 125–132.

Filled with Desire, Perceive Molecules

Roger Strand and Caroline Engen

Prologue: The Desire to Help

The Tao that can be named is not the Real Tao. This is one of the many translations of the opening sentence of *Tao Te Ching*, the masterpiece of ancient Chinese philosophy. The real world is richer than what can be expressed by human minds. The Universe is too vast; a single human being is too subtle to be fully put into words. In the Taoist tradition, as for mystics in the West and the East, the path to experience the richness of the Tao goes around and away from words: Wordless practice, silent meditation, the extinction of the ego. "Empty of desire, perceive mystery", *Tao Te Ching* reads. "Filled with desire, perceive manifestations."

Mystery is not the subject of modern science. As Niels Bohr said, the objective of science is to say what can be said about the world, nothing less, nothing more. Some scientists may express the desire to "know the mind of God"; by and large they get disappointed. Science describes the manifestations of the world, that is, how the world manifests *before* us, *for* us and by means of us and our cognitive, physical and emotional abilities.

Medicine was never embarrassed to admit the desire of its science: to help ill people by understanding disease. This noble desire has been crowned with success; it is the desire that made Richard Nixon declare war on cancer and René Descartes (1637) dream that "… we might free ourselves from countless diseases of body and

R. Strand (✉)
Centre for Cancer Biomarkers, Centre for the Study of the Sciences and the Humanities, University of Bergen, Bergen, Norway
e-mail: roger.strand@uib.no

C. Engen
Centre for Cancer Biomarkers CCBIO, Department of Clinical Medicine, University of Bergen, Bergen, Norway
e-mail: Caroline.Engen@uib.no

A. Bremer, R. Strand (eds.), *Precision Oncology and Cancer Biomarkers*, Human Perspectives in Health Sciences and Technology 5,
https://doi.org/10.1007/978-3-030-92612-0_15

of mind, and perhaps even from the infirmity of old age, if we knew enough about their causes and about all the remedies that Nature has provided for us." In the centuries following Descartes, this dream was pursued with ever more sophisticated concepts and languages and ever sharper extensions of the human senses. Medicine got to know the organs; the tissues; the cells and finally the molecules. If we control the molecules, can we correct the body and control the disease? Can we eradicate disease? From the physiology of the nineteenth century and the beginnings of molecular medicine in the twentieth, the slogan of early twenty-first century is that of precision medicine, of tailoring nanometre technologies to the molecular make-up of every individual patient.

Mary Shelley had her well-intended but ill-fated Doctor Frankenstein advise the readers "[…] never to allow passion or a transitory desire to disturb his tranquillity. I do not think that the pursuit of knowledge is an exception to this rule." (1993, chap. 4) On the whole this young woman's advice was ignored and dismissed as ignorant and unscientific. To the extent they were tolerated in academia, the practices of reflexivity and extinction of the ego were relegated to the soft sciences: philosophy, social anthropology, nursing science and the like. In science, *real* science, not the sort that Sir Rutherford once dismissed as stamp collecting, the higher the precision, the stronger the passion, tending towards the total imperative. In medicine, the imperative was to help: We have to help the patients, we have to act, we must never give up. Disease is intolerable, death is defeat.

This chapter tells a story about a science filled with a desire that enables it to perceive molecules. Its protagonists are the cancer scientists. They are courageous and persistent, as admirable as the heroes of the Greek tragedies, in their pursuit of heroic deeds with sharp tools and precise names. But what happens when the discrepancy between the Tao and its name shows itself? Is it too early in the story yet for the heroes to meet their downfall that will evoke fear and pity in us who are their spectators? Perhaps the story takes a turn to reveal other heroes, those who patiently ingest the molecules of desire and allow their bodies to be named a surgical and molecular battleground – heroes with whom we can empathize as they thrive and suffer while molecular soldiers fight for remission and the Tao, from the depths of its dark valleys, may decide otherwise.

Acute Myeloid Leukaemia

Heaven and Earth are not kind.
The ten thousand things are straw dogs to them. (Lao-Tzu: Tao Te Ching, Chapter 5)

Occasionally individual stories are shifted by abrupt, incomprehensible and devastating events, like the sudden and unexpected presentation of a life-threatening condition. Accounts of this are a source of great terror, and among the events most dreaded is that of being diagnosed with cancer (Vrinten et al. 2014). The perception of cancer as a source of horror is composite and involves interpretation of many dimensions of the disease, including the nature of cancer as a "stealthy, indestructible, indiscriminate killer", the toxicity and atrocities of cancer therapies, as well as death (Vrinten et al. 2016; Agustina et al. 2018; Murphy et al. 2018). As Peyton Rous articulated in his Nobel Prize Lecture in 1966, "The Challenge to Man of the Neoplastic Cell", "Tumors destroy man in an unique and appalling way, as flesh of his own flesh, which has somehow been rendered proliferative, rampant, predatory, and ungovernable" (Rous 1967).

Acute myeloid leukaemia – AML – a rare and aggressive haematological malignancy, exemplifies the terror of cancer. Annually, the disease is assigned as cause of some 150,000 deaths world-wide (GBD 2015 Mortality and Causes of Death Collaborators 2016). Originating and expanding from myeloid progenitor cells of the hematopoietic system, the disease usually presents itself by rapidly progressing symptoms, like fatigue, weakness, dizziness, shortness of breath and fever (Estey and Dohner 2006; Dohner et al. 2015; Short et al. 2018). It is not uncommon that the symptoms of AML initially are interpreted as a common viral infection, and in such cases suspicion of a more serious condition may arise abruptly and unexpectedly as the doctor is presented with blood sample results, often demonstrating aberrant blood cell counts. At the time of diagnosis approximately half of patients are in relatively good condition (Juliusson et al. 2009), but if left untreated, the condition typically advances quickly, resulting in bone marrow failure and ultimately death, often within weeks from initial presentation of signs and symptoms (Oran and Weisdorf 2012). The outcome in AML can, however, be improved by therapeutic intervention. The aggressiveness of the condition warrants rapid clarification of treatment goals and initiation of therapy, often commenced only a day or two after diagnosis. Current treatment options include chemotherapy-based regimens, hematopoietic stem cell transplantation, and targeted treatment. These regimens are severe and in many cases with little objective hope of success, while introducing their own risks of suffering and death from adverse effects. Accordingly, more lenient disease stabilizing treatment plans and supportive care are still real options (Dohner et al. 2017).

The diagnosis of AML, a lethal condition requiring potentially lethal treatment, thus, represents a true shock and horror story. It is not unreasonable to perceive the disease as a malicious enemy that attacks suddenly and without provocation. As such, AML is a medical emergency in which the patient's world is set in rapid and whirling motion.

Innocently thrown into this horror, who could be more worthy of help than the AML patient? Confronted with the naked and intense suffering of AML, futility becomes unbearable and physicians' and nurses' desire to help become an imperative: We have to help them. We *must* help them. The moral force of this imperative has led to extensive research on AML and decades of trying almost any therapy. And whereas Lao-Tzu and Mary Shelley warned against excessive desire, the strength of the desire to help AML patients could itself be seen as a sign of a human civilization that is willing and able to go at almost any length to protect and care for its frailest members, a part of humanity driven by compassion in the midst of a world also driven by darker and violent desires. Indeed, despite cancers of the blood being relatively rare, the discipline of haemato-oncology has been at the forefront of cancer research and oncological practice from the early 1950s. The first randomized comparative clinical trial was performed in patients treated for leukaemia (Frei et al. 1958), and the treatment of cancers of the blood has been at the forefront of oncological therapeutic development. Chemo-therapeutics, combinational regimes (Chabner and Roberts 2005), adaptive cell- and immune-therapy (Singh and McGuirk 2016) as well as gene-therapy (Rosenbaum 2017) were all of them first explored in leukaemia.

Yet, the AML diagnosis remains a death sentence to most who receive it. While treatment outcomes have improved, the overall five-year survival is still in the order of 25%. Some patients are regarded as cured, in the sense that there is remission and no relapse is observed until death from other causes. In the case of relapsed AML, however, prognosis is so poor that a case of long survival was officially deemed a miracle by the Vatican Church and used as evidence in the canonization of the Canadian "Mother of Universal Charity", Marie Marguerite d'Youville. In the case of AML, the helpers can feel the same urgency as was expressed by President Nixon when he declared war on cancer in 1971:

> … The time has come in America when the same kind of concentrated effort that split the atom and took man to the moon should be turned toward conquering this dread disease. Let us make a total national commitment to achieve this goal. (Nixon 1971, 53)

Urgency, desperation and also the sense of scandal that is implied in Nixon's statement. How can it be that Science split the atom and took man to the moon, and yet fail to find the cure for a trivial blood disease? Doctors must help these innocent suffering patients, but fundamentally and perhaps more importantly, Man has to tame and conquer this malicious expression of brute Nature, and it has to be conquered with Rutherford's Science, that is, by finding its precise causes and contravening them. "Human knowledge and human power come to the same thing, for where the cause is not known, the effect cannot be produced," Francis Bacon ([1620] 1994, 43) said.

The history of bone marrow transplantation is perhaps the most striking display of how much was considered to be at stake in Nixon's war, and how the desire to help was blended into persistence and commitment to show that what ought to work, indeed could work. To stay with Francis Bacon (1620), what Man can and ought to do, is to dominate and "penetrate the more secret and remote parts of Nature". Edward Donnell Thomas performed the first experiments with bone marrow transplants from donors to patients in 1957 and was awarded the Nobel Prize for his accomplishments in 1990. All the patients in the initial experiment died. A review in 1977 showed that among the first 100 patients to receive this treatment, three quarters died the first year, most of them because of the complications following the intervention. Yet, in the end persistence was crowned with success, or at least this is how the official story goes. Currently, only around 20% of AML patients subjected to allogeneic stem cell transplantation die from adverse effects, in part due to refinement of procedure and support and in part due to a better selection of the patients (Styczynski et al. 2020). These refinements are part of the explanation why overall survival for AML has somewhat improved.

A similar story can be told for the molecular level. Novel compounds have been developed and put into use. Yet, long-term survival rates in AML remain poor (Talati and Sweet 2018), for a variety of reasons: The cancer frequently adapts to the drug and develops drug resistance; or the cancer goes into remission but after a while, there is relapse; or the toxicity of the drugs impairs the patient or even causes death (Yeung and Radich 2017). Molecular studies of AML suggest ever new compounds and treatment regimens to be tested, and sometimes outcomes are improved for some subgroup of patients (Talati and Sweet 2018). On the whole, however, AML remains a horror to the patients and their carers and a scandal for Science.

Filled with Desire, Find the Molecule

Trying to control the world?
I see you won't succeed.
The world is a spiritual vessel
and cannot be controlled.
Those who control, fail.
Those who grasp, lose. (Lao-Tzu: Tao Te Ching, Chapter 29)

For the Taoist, the failure to tame and conquer AML comes as little surprise. Humans do not control the world. The desire for a cure does not imply the possibility of a cure, neither in logical or practical terms. It is a fundamentally modern, European and perhaps secular conception that desire implies existence, grounded in the belief that human imagination and ingenuity is omnipotent and limitless. The ancient Greeks called such beliefs *hybris* and explained the problem in the myth of Icarus.

And yet, sometimes Science appears to deliver the Silver Bullet. Chronic myeloid leukaemia – CML –, the less severe brother of AML, became one of the prominent success stories of molecular targeted therapy. Molecules known as tyrosine kinase inhibitors became game changers in the history of CML. With the drug *imatinib*, known under its brand name Gleevec, CML patients on the whole no longer die from their leukaemia and show the same overall survival rates as the general population (Deininger et al. 2005). For CML, imatinib became the silver bullet but also one of the first proofs-in-principle that the Art of oncology can become Science in Rutherford's sense. According to current scientific understanding, CML is nothing more than the result of a single mutation in blood cells. The mutation causes the cells to produce a protein, BCR-ABL fusion protein (Deininger et al. 2005), that is the phenotypic cause of the disease. Imatinib contravenes in the damage caused by BCR-ABL, and thus the disease is conquered.

Imatinib is not the only silver bullet in cancer medicine. The invention of immune checkpoint inhibitors to stop cancer cells defending against the immune system is a similar victory of molecular ingenuity (Demaria et al. 2019). New drugs such as *ipilimumab, nivolumab* and *pembrolizumab* do not only prolong life but also seem to induce durable remissions in a subset of patients with advanced malignant melanoma (skin cancer) who otherwise would have had very poor prognosis (Herrscher and Robert 2020). Likewise, the antibodies *pertuzumab* and *traztuzumab* dramatically improved outcomes for women with HER2-positive breast cancer in the scientifically most satisfactory way: They are monoclonal antibodies that bind to the HER2 receptor and thereby interfere in signalling pathways that are involved in the growth and division of cancer cells (Slamon et al. 2001). Not only do these drugs work; for many cancer researchers they are proof that they indeed are on the way to dominate and penetrate Nature in the innermost parts, as Sir Francis Bacon so picturesquely formulated it.

The desire to help and to cure is visible not only in the volume and ferocity of cancer research but also in the theoretical structure of scientific knowledge. The prominent causal theories are those that are mechanistic and lend themselves to immediate translation into technological practice (such as Somatic Mutation Theory). Other attempts at theorizing, such as the much cited "Hallmarks of Cancer" are not even causal in the mechanistic sense but rather inventories of sites for practical interventions. Alternative types of theorization such as Tissue-Oriented Field Theory (Sonnenschein and Soto 2000) or approaches inspired by evolutionary biology might have more biological merit but in practical cancer research there is little patience with them. The first question in the cancer research seminar will invariably be: So how can this help improve clinical practice? Time is running out, the patients are dying and there is no place for philosophising.

The problem, however, is that the silver bullets have been so rare, and the cases where they work are also rare. What works for skin cancer does not work for colon or prostate cancer. What works for CML, does not work for AML. There are numerous attempts at finding *the* mutation that causes AML, and they find different answers. The situation resembles that of rationalism after Descartes: The rationalists all agreed that what is self-evident, must be true. The problem is that they all disagreed on what is self-evident. With AML, some mutations are common, but not ubiquitous; there are really many of them; and none of them seem to work as a site for a molecular silver bullet (Dohner et al. 2017).

AML Is a Name

The Tao that can be named is not the real Tao.
Names can name no lasting name.

Nameless: the origin of heaven and earth
Naming: the mother of ten thousand things. (Lao-Tzu: Tao Te Ching, Chapter 1)

The desire to cure with molecules is rationalist in nature; it is an instance of what has been called the Cartesian Dream (Schei and Strand 2015). Such desires can be fulfilled if, and only if, reality and knowledge can be brought into the appropriate level of correspondence. To employ a cartographic metaphor: either the terrain has to be as simple as the map, or, alternatively, the terrain has to be changed so as to become as simple as the map, for instance by the use of bulldozers, lobotomy or other complexity-reducing technologies. As in the case of cancer; by cutting, burning and poisoning.

Sometimes, in the history of cancer, the terrain has emerged, seemingly, simple enough for the man-made map. That is, simple enough for scientists and physicians to fulfil their goal of improving outcome through precise and rational molecular approaches, for instance in the case of CML. The close relationship between clinical manifestations, morphological characteristics, the BCR-ABL fusion protein and response to targeted therapy led to the acceptance of a linear causal narrative of CML, a story in which the translocation is the ultimate cause of the disease. This story gradually grew so strong and compelling as to shape not only how CML was to be perceived, but also how cancer in general was to be explored and described, and how development of cancer therapy was to be pursued. More often, however, the intricacy of the gradually materialising landscape, of cancer at large, and individual tumours at small, has proven much more challenging to both map and navigate. Indeed, 20 years past the great success of imatinib and CML, this individual story by far remains the best example of precision oncology and its potential.

AML is one of the many diseases for which the ambition of precision oncology has struggled to become fulfilled. With time, and by force of evolving technologies, knowledge, and practices, the magnification of the AML landscape has gradually increased by changing the lenses through which the disease has been characterised and understood, gradually shifting from a clinical and macro-anatomical characterisation, to a focus on cells and morphology, and ultimately to one with emphasis on portrayal of molecular features and mechanisms.

In the case of CML, molecular characterisation led to therapeutic progress. The molecular characterisation of AML resulted instead in disintegration of the disease category. While a single genetic aberrancy characterises the clinical and morphological phenotype of CML, AML comprises multiple chromosomal rearrangements and more than 30 individual genes have been shown to be repeatedly mutated. In most cases more than one genetic variant is identified, and recurring mutational patterns suggest that several mutated gene-products may work together in leukemogenesis (Cancer Genome Atlas Research Network 2013; Metzeler et al. 2016; Tyner et al. 2018). The name of AML is currently understood to refer to a collective, a heterogenous collection of various acute blood cancers, grouped together by similar cytomorphological and clinical characteristics and the same type of causal story. According to this story, AML develops as the result of the manifestation and gradual dominance of a novel aberrant cell population. This cell population is still assumed to descend from a simple principle: It is thought to have resulted from a single haematopoietic stem or progenitor cell which has accumulated the sufficient set (and sequence) of somatic mutations to have become a cancer cell (with properties such as differentiation block, autonomous proliferation and immortality). Several observations, however, suggest that even this narrative is overly simplistic. Indeed, tumour evolution is a characteristic of AML disease trajectories. Across time, the cytogenetic characteristics and molecular patterns of recurrently mutated genes in AML show increasing complexity as the disease progresses. Furthermore, postulated causal factors, such as certain cytogenetic aberrancies or particular single gene variants, are sometimes lost through individual AML disease courses (Garson et al. 1989; Kern et al. 2002; Renneville et al. 2008; Ding et al. 2012; Welch et al. 2012; Hirsch et al. 2016; Dovey et al. 2017). Further, metaphase karyotyping, inferred cell population size by variant allele fraction patterns, single cell sequencing analysis as well as engraftment studies have demonstrated that individual AML samples frequently comprise numerous genotypically diverse cell populations (Welch et al. 2012; Bochtler et al. 2013; Cancer Genome Atlas Research Network 2013; Klco et al. 2014; Paguirigan et al. 2015; Vick et al. 2015; Shlush et al. 2017; Wang et al. 2017; Baron et al. 2018; Potter et al. 2018; van Galen et al. 2019). Moreover, treatment with targeted therapy is frequently followed by rapid emergence of alternate cell populations, characterised by mutations in unrelated genes (McMahon et al. 2019; Zhang et al. 2019). Several observations further suggest that gene variants may translate into phenotypic variation as a function of differentiation and that mutations and subsequent gene products may confer variable qualities dependent on individual context and connectivity (Sato et al. 2011; Yang et al. 2014; Karjalainen et al. 2017; Sung et al. 2019).

We have listed these molecular complexities to show that a simple, rationalist and reductionist account of AML is not merely challenged by complex phenomena at higher organisational levels, such as the life-world of the patients, his or her family, society and so on. For all diseases, also for those that are successfully treated with molecular silver bullets, the complexities of the life-world exist. CML patients survive with their Gleevec but the challenges of their quality of life are not at all trivial. Still, Gleevec works in line with its intention and well enough for it to be called a silver bullet and for the patients to be sufficiently well described by the name of CML. AML, however, is different. As mentioned above, AML emerged as a name for a condition discovered in the clinic and characterised in terms of its tissues and cells. As the Tao Te Ching reads, naming is the mother of the ten thousand things, and the naming of AML delineated the clinical, anatomical and cytological thing called by that name. By force of the desire to dominate and penetrate the innermost part of AML, however, the naming continued with ever more names, parts, mutations and aberrations until it became clear, at least for the prepared mind, that on the molecular level there is no one well-defined thing to be called AML. Rather, AML refers to a collective of conditions that are best understood and described as evolving processes of leukaemia where stability and constancy are only to be found at the clinical and anatomical level. At the cellular and molecular level, the Tao that can be named is not the real Tao. What is to be found, is heterogenous, dynamic and relational flux where casual contributions can be traced from several levels of biological organization. Ultimately, AML may best be understood as a disease of systems and cells rather than one of genes and molecules. Or at least, that is how AML may be best understood by our protagonists so far, the medical researchers.

Once again, it seems, that the story that has been told with medical researchers and health personnel as the protagonists, as the active subjects that observe and act upon their passive, suffering objects that suitably are called *patients*. How is AML to be understood for those who receive the diagnosis? What are their desires and how do they shape the manifestations of AML and its natural history?

We have to recall once again that naming is the mother of the ten thousand things. We now know that untreated AML usually presents itself by symptoms such as fatigue, weakness, dizziness, shortness of breath and fever, which quickly escalates until death arrives. In the absence of diagnoses and doctors, the misfortune that struck could be anything, a corona virus for that matter. In 1976, the mummy of the Egyptian pharaoh Ramses II was subjected to scientific investigation and lesions characteristic of tuberculosis were discovered. Bruno Latour famously asked if Ramses II died from tuberculosis, and indeed what it could mean that he died from something that only was named much after his death. Unfortunately it seemed that the lesions had been caused on the mummy itself by a fungal infection; however, the philosophical puzzle remained. From the Taoist perspective, Ramses II and all other predecessors died but we do not exhaust the truth of their death by giving it an anachronistic label. Indeed, we may expect that the vast majority of people whom we now would regard as AML victims, had no ideas of AML and its horrors at all. They did not perceive them. They fell ill and died.

In a modern welfare state, however, part of the destiny of becoming afflicted with AML is to acquire the diagnosis and with it, the knowledge of how AML manifests itself. The category of AML patients is now real by the process that Ian Hacking explained with his doctrine of dynamic nominalism: People are called by a name, and the naming changes them. They now know that they have a horrible disease and that they are likely to die very soon.

Not too much is scientifically known about how it is to be an AML patient; most of the research projects have served to fulfil the desire to help by technological means. We wrote above that the "aggressiveness of the condition warrants rapid clarification of treatment goals and initiation of therapy, often commenced only a day or two after diagnosis." While this is generally true, it does not mean that all patients will receive therapy. Quite a few of them are quite old and with comorbidities. We shall return that point later. For now, however, let us focus on the relatively young AML patient, meaning, in his early sixties or younger, who believed that he was quite healthy and now is subjected to the shock of the diagnosis and the extreme urgency of action. Complex decisions are to be made in a state of shock, devastation and confusion. Patients frequently describe a feeling of being overwhelmed and struggling to process information and make informed decisions (LeBlanc et al. 2017). The decision-making process is characterized by a lack of shared interpretation of the situation where patients tend to grossly overestimate their chances of cure and one-year survival and underestimate the risk of dying from the treatment (Sekeres et al. 2004).

The disease as well as the treatment result in physical deterioration and loss of bodily strength and function. With a compromised body, it is a struggle to maintain social functions and meaningful activities. Many patients experience their identities, relationships and worlds as threatened. It is not uncommon that patients suffering from AML experience psychological symptoms in line with those of anxiety disorders and depression (Tomaszewski et al. 2016; Deckert et al. 2018). More than 50 years ago, Kübler-Ross (1969) interviewed terminally ill individuals and named the stages of grief: denial, anger, bargaining, depression and, towards the end, acceptance. To this, we may add the sense of guilt that some cancer patients experience when they fail to mobilise the energy to "fight the disease" (Crawford et al. 2020).

To summarize, the course of AML is a matter of blood cells and bone marrows, of treatment choices and responses and of bodies that decay and die. More than that, however, it is a matter of travelling in landscapes of strong experiences and emotions, with shock, horror, fear and hope in the fore and middle ground. In our interpretation, then, our story has two sets of protagonists who meet each other in strong emotions and desires, under the name of AML. The patients, just being thrown off the cliff of apparent good health, are now suspended in desperate fears of instant suffering and death and equally desperate hopes that life can go on as before. The doctors and nurses want to help and must help. The first-hand solution for both is medical treatment. Perhaps it works, in the sense that the disease goes into remission or even is cured. Perhaps it does not work. In either case, it *might* work, and in this way the treatment already does two types of other work. It sustains the hope that the patient desperately needs and it releases the unbearable frustration into action for the doctor or nurse. They are doing something, they are doing the best they can, they are helping. A sense of meaning is produced. Hard work and expensive treatments confirm the dignity of a civilisation that spares no cost to try to protect its frailest. In an ironic twist of the plot, some of that cost has to be born by the patients themselves, if not financially in modern welfare states, by the suffering caused by adverse effects which even may be fatal. Still, regarding the modern project of medical science as heroic, these costs are also meaningful. Without the sacrifice of dozens and hundreds of patients who died because they received bone marrow transplantation, the technique would not have been developed into its current sophisticated form by which it cures thousands of patients.

More than 150 years ago, Claude Bernard, one of the fathers of modern physiology, explained his concept of modern medicine: "By normal activity of its organic units, life exhibits a state of health; by abnormal manifestation of the same units, diseases are characterised; and finally through the organic environment modified by means of certain toxic or medicinal substances, therapeutics enables us to act on the organic units" (Bernard [1865] 1957, 65). For Bernard, this is what medicine *is*. It is the application of toxic substances in order to reinstate chemical equilibrium or homeostasis in the organic body. What we can see so clearly in the case of AML, where that homeostasis only rarely is to be achieved, is the poverty of Bernard's concept. Medicine is its own Tao that cannot be named, and part of it is that real people meet with the real fears, hopes and despair and produce a sense of meaning together. Medical research plays an ever stronger role in those meetings and negotiations, in part because it actually, sometimes, delivers improvements in treatment à la Bernard, that act directly at the molecular level. This is good in itself and it sparks hope and fuels a positive process for the doctor-patient dyad. In part, however, research and above all clinical trials have their own value because they are future-oriented with a vision of progress and because they involve action. The more patients are inscribed into trials, the more is done for them: Not only are they receiving standard treatment but the extraordinary, with the latest and most exclusive promising drug, is being made available to them. In this way clinical trials are important sites of symbolic interaction and the creation of meaning.

Is all this good then? That question cannot be answered in the general and out of context. Where some see a human interaction that gives hope, others judge it as false hope created by false promises. Patients are known to systematically overestimate the benefit of the treatments that they accept to take and that the doctors so desperately need to give them. And behind the hospital scene there are the pharmaceutical companies and their shareholders who are creating huge profits for themselves on drugs that have modest clinical benefit in the usual sense and cause serious adverse effects.

Empty of Desire, Perceive Mystery

> Suddenly Master Lai grew ill. Gasping and wheezing, he lay at the point of death. His wife and children gathered around in a circle and began to cry. Master Li, who had come to ask how he was, said: "Shoo! Get back! Don't disturb the process of change!"
>
> Then he leaned against the doorway and talked to Master Lai. "How marvellous the Creator is! What is he going to make out of you next? Where is he going to send you? Will he make you into a rat's liver? Will he make you into a bug's arm?". (Zhuangzi: The Great and Venerable Teacher 2003)

Doctor Frankenstein, having been created by an English mind, came to recognise temperance as a major virtue and a yardstick by which passions and desires should be tested. Desire can be excessive at the expense of virtue and the good life. One interpretation of our story about AML medicine and research is that a particular desire to help, conditioned and constrained by the Cartesian and Baconian dreams of dominating, penetrating and controlling Nature, led, if not to excess, into a peculiar state of affairs where biomedical success has been scarce but where doctors, patients and researchers are dependent upon biomedical research to meet their emotional needs. To the extent such an interpretation can be said to be plausible, the ethical issues are multiple. From a deontological perspective, the entire enterprise, consisting of the research and innovation value chain from the research departments of pharmaceutical companies down to the individual patients, could be seen as immersed in dishonesty about the real potential of treatments. From a utilitarian view, the public expenditures on expensive cancer medicines would be seen as unjust and unfair. Finally, at the individual level the question remains if and when hope is to be maintained, and for what. Elisabeth Kübler-Ross, in her groundbreaking work on the stages of grief during terminal illness, described how the final phase is that of acceptance where the patient's "circle of interest diminishes" (101). She noted:

> There are a few patients who fight to the end, who struggle and keep a hope that makes it almost impossible to reach this stage of acceptance. They are the ones who will say one day, "I just cannot make it anymore," the day they stop fighting, the fight is over. In other words, the harder they struggle to avoid the inevitable death, the more they try to deny it, the more difficult it will be for them to reach this final stage of acceptance with peace and dignity. The family and staff may consider these patients tough and strong, they may encourage the fight for life to the end, and they may implicitly communicate that accepting one's end is regarded as a cowardly giving up, as a deceit or, worse yet, a rejection of the family. (102)

From a secularised Occidental perspective the acceptance that Kübler-Ross describes could be seen as a sign of the passage from being to nothingness, of, in her words, a diminishing circle of interest. Also, from what we might call spiritual Oriental perspectives such as Taoist or Zen Buddhist thought, acceptance would be a matter of emptying desires and interests. Still, it would not be seen as something void of significance. Rather than resignation it could be a passage of liberty from the self into transcendence and mystery. This is how to interpret Master Li's intervention. He tries to prevent the relatives from trivialising this significant moment at the end of Master Lai's life.

Secular or spiritual, Western healthcare is also sensitive towards the dignity of peaceful death. Hospices for cancer patients have long traditions and supportive and palliative care play important roles in the stories that we did not tell earlier in this chapter. Indeed, we have so far portrayed the AML patient as a mostly healthy person thrown into shock but then we disregarded that most AML patients are very old. One rule of thumb is that it is meaningful to treat AML if the patient is less than 80 years old and otherwise healthy. That excludes the majority of AML patients, who die after a short course of disease and in the presence of palliative care. The ethical issues mentioned at the beginning of this section can accordingly be reframed as the dilemma of finding the appropriate cut-off for offering treatment.

The verses of Tao Te Ching might suggest sharp dichotomies between being full or empty of desire; between manifestations and mystery; and between resistance and acceptance. Sharp dichotomies are, however, neither logically necessary nor a historically correct interpretation of these philosophical sources. In the ancient Chinese thought to which the Tao Te Ching belongs, opposites are themselves recognised to be names, that is, imperfect renderings of the real world. Nothing is merely a simple question of black and white, as indicated by the typical symbol associated with yin and yang (Fig. 1):

Fig. 1 A modern, simplified *tajitu*, symbolising the relationship between yin and yang

In this representation of the yin and the yang, there is always something black in the white and vice versa. This is not to say that all there is, are shades of grey. Rather, it is a message of ambiguity and complexity. It is possible to grieve and also suspend grief. One can experience tragedy and also acceptance.

The shades of grey, or rather differences by degrees, are also part of that complexity. In the received view of science it is often presented as the stance of disinterestedness and objectivity. Our story presents the science of AML as extremely passionate and driven by desires. We do not tell the story in this way to argue that it rather should become disinterested. Science cannot be disinterested. A wholly disinterested stance is mystical and not oriented towards words or action. Rather, as a matter of degrees, the science of cancer and of AML might benefit from relaxing just a bit from its urge to help, being slightly less medical and slightly more biological. Rather than spending all energy on "What molecules can help the patients?" one could ask the biological question "What is the function of cancer?" and perhaps learn a lot. And, as is well known in the sciences of nurses and other health professionals, one could learn a lot from the patients and their illness, if the illness is seen as something more than an enemy to be conquered or a deficiency to be removed. There is nothing new in this type of tactic; indeed, the history of physics and chemistry shows more than often that a temporary retreat from practical urgencies can give results that ultimately become highly applicable and useful.

Ultimately, however, these considerations will have to be made from within the practices that we have described. In Zhuangzi's story above, Master Li shooed the relatives away, apparently without hesitation, as if he knew the situation to the fullest, including the relatives' intentions. In that sense Master Li seemed to show quite strong opinions and desires himself. Of course we do not know what happened afterwards. Perhaps they threw him out. The stance true to the practice of telling such stories is neither that of the passionate scientist or the immutable mystic. Rather, we who present a Taoist perspective on AML position ourselves as the village fools. Our stories are tolerated, perhaps, and we may get to tell them to the end until we are told to leave. With some luck, they inspired some new desires, some new curiosities to explore the yin and yang, not only the yang, of AML.

References

Agustina, E., R.H. Dodd, J. Waller, and C. Vrinten. 2018. Understanding middle-aged and older adults' first associations with the word "cancer": A mixed methods study in England. *Psychooncology* 27 (1): 309–315.

Bacon, F. [1620] 1994. Aphorisms concerning the interpretation of nature: Book 1–77. The new organon: Or true directions concerning the interpretation of nature. In *Novum Organum*, ed. and trans. P. Urbach and J. Gibson. Chicago/La Salle: Open Court.

Baron, F., M. Stevens-Kroef, M. Kicinski, G. Meloni, P. Muus, J.P. Marie, C.J.M. Halkes, et al. 2018. Cytogenetic clonal heterogeneity is not an independent prognosis factor in 15-60-year-old AML patients: Results on 1291 patients included in the EORTC/GIMEMA AML-10 and AML-12 trials. *Annals of Hematology* 97 (10): 1785–1795.

Bernard, C. [1865] 1957. *Introduction to experimental medicine*. New York: Dover Publications.

Bochtler, T., F. Stolzel, C.E. Heilig, C. Kunz, B. Mohr, A. Jauch, J.W. Janssen, et al. 2013. Clonal heterogeneity as detected by metaphase karyotyping is an indicator of poor prognosis in acute myeloid leukemia. *Journal of Clinical Oncology* 31 (31): 3898–3905.

Cancer Genome Atlas Research Network. 2013. Genomic and epigenomic landscapes of adult de novo acute myeloid leukemia. *The New England Journal of Medicine* 368 (22): 2059–2074.

Chabner, B.A., and T.G. Roberts Jr. 2005. Timeline: Chemotherapy and the war on cancer. *Nature Reviews. Cancer* 5 (1): 65–72.

Crawford, R., K. Sully, R. Conroy, C. Johnson, L. Doward, T. Bell, V. Welch, F. Peloquin, and A. Gater. 2020. Patient-centered insights on treatment decision making and living with acute myeloid leukemia and other hematologic cancers. *Patient* 13 (1): 83–102.

Deckert, A.L., G. Gheihman, R. Nissim, C. Chung, A.D. Schimmer, C. Zimmermann, and G. Rodin. 2018. The importance of meaningful activity in people living with acute myeloid leukemia. *Leukemia Research* 67: 86–91.

Deininger, M., E. Buchdunger, and B.J. Druker. 2005. The development of imatinib as a therapeutic agent for chronic myeloid leukemia. *Blood* 105 (7): 2640–2653.

Demaria, O., S. Cornen, M. Daeron, Y. Morel, R. Medzhitov, and E. Vivier. 2019. Harnessing innate immunity in cancer therapy. *Nature* 574 (7776): 45–56.

Déscartes, R. 1637. *Le discours de la méthode pour bien conduire sa raison et chercher la vérité dans les sciences.* Translated by Jonathan Bennett as *Discourse on the method of rightly conducting one's reason and of seeking truth in the sciences* 2007. Published online at https://www.earlymoderntexts.com/assets/pdfs/descartes1637.pdf.

Ding, L., T.J. Ley, D.E. Larson, C.A. Miller, D.C. Koboldt, J.S. Welch, J.K. Ritchey, et al. 2012. Clonal evolution in relapsed acute myeloid leukaemia revealed by whole-genome sequencing. *Nature* 481 (7382): 506–510.

Dohner, H., D.J. Weisdorf, and C.D. Bloomfield. 2015. Acute myeloid leukemia. *The New England Journal of Medicine* 373 (12): 1136–1152.

Dohner, H., E. Estey, D. Grimwade, S. Amadori, F.R. Appelbaum, T. Buchner, H. Dombret, et al. 2017. Diagnosis and management of AML in adults: 2017 ELN recommendations from an international expert panel. *Blood* 129 (4): 424–447.

Dovey, O.M., J.L. Cooper, A. Mupo, C.S. Grove, C. Lynn, N. Conte, R.M. Andrews, et al. 2017. Molecular synergy underlies the co-occurrence patterns and phenotype of NPM1-mutant acute myeloid leukemia. *Blood* 130 (17): 1911–1922.

Estey, E., and H. Dohner. 2006. Acute myeloid leukaemia. *Lancet* 368 (9550): 1894–1907.

Frei, E., 3rd, J.F. Holland, M.A. Schneiderman, D. Pinkel, G. Selkirk, E.J. Freireich, R.T. Silver, G.L. Gold, and W. Regelson. 1958. A comparative study of two regimens of combination chemotherapy in acute leukemia. *Blood* 13 (12): 1126–1148.

Garson, O.M., A. Hagemeijer, M. Sakurai, B.R. Reeves, G.J. Swansbury, G.J. Williams, G. Alimena, et al. 1989. Cytogenetic studies of 103 patients with acute myelogenous leukemia in relapse. *Cancer Genetics and Cytogenetics* 40 (2): 187–202.

GBD 2015 Mortality and Causes of Death Collaborators. 2016. Global, regional, and national life expectancy, all-cause mortality, and cause-specific mortality for 249 causes of death, 1980-2015: A systematic analysis for the Global Burden of Disease Study 2015. *Lancet* 388 (10053): 1459–1544.

Herrscher, H., and C. Robert. 2020. Immune checkpoint inhibitors in melanoma in the metastatic, neoadjuvant, and adjuvant setting. *Current Opinion in Oncology* 32 (2): 106–113.

Hirsch, P., Y. Zhang, R. Tang, V. Joulin, H. Boutroux, E. Pronier, H. Moatti, et al. 2016. Genetic hierarchy and temporal variegation in the clonal history of acute myeloid leukaemia. *Nature Communications* 7: 12475.

Juliusson, G., P. Antunovic, A. Derolf, S. Lehmann, L. Mollgard, D. Stockelberg, U. Tidefelt, A. Wahlin, and M. Hoglund. 2009. Age and acute myeloid leukemia: Real world data on decision to treat and outcomes from the Swedish Acute Leukemia Registry. *Blood* 113 (18): 4179–4187.

Karjalainen, R., T. Pemovska, M. Popa, M. Liu, K.K. Javarappa, M.M. Majumder, B. Yadav, et al. 2017. JAK1/2 and BCL2 inhibitors synergize to counteract bone marrow stromal cell-induced protection of AML. *Blood* 130 (6): 789–802.

Kern, W., T. Haferlach, S. Schnittger, W.D. Ludwig, W. Hiddemann, and C. Schoch. 2002. Karyotype instability between diagnosis and relapse in 117 patients with acute myeloid leukemia: Implications for resistance against therapy. *Leukemia* 16 (10): 2084–2091.

Klco, J.M., D.H. Spencer, C.A. Miller, M. Griffith, T.L. Lamprecht, M. O'Laughlin, C. Fronick, et al. 2014. Functional heterogeneity of genetically defined subclones in acute myeloid leukemia. *Cancer Cell* 25 (3): 379–392.

Kübler-Ross, E. 1969. *On death and dying*. New York: Scribner.

LeBlanc, T.W., L.J. Fish, C.T. Bloom, A. El-Jawahri, D.M. Davis, S.C. Locke, K.E. Steinhauser, and K.I. Pollak. 2017. Patient experiences of acute myeloid leukemia: A qualitative study about diagnosis, illness understanding, and treatment decision-making. *Psychooncology* 26 (12): 2063–2068.

McMahon, C.M., T. Ferng, J. Canaani, E.S. Wang, J.J. Morrissette, D.J. Eastburn, M. Pellegrino, et al. 2019. Clonal selection with Ras pathway activation mediates secondary clinical resistance to selective FLT3 inhibition in acute myeloid leukemia. *Cancer Discovery* 9 (8): 1050–1063.

Metzeler, K.H., T. Herold, M. Rothenberg-Thurley, S. Amler, M.C. Sauerland, D. Gorlich, S. Schneider, et al. 2016. Spectrum and prognostic relevance of driver gene mutations in acute myeloid leukemia. *Blood* 128 (5): 686–698.

Murphy, P.J., L.A.V. Marlow, J. Waller, and C. Vrinten. 2018. What is it about a cancer diagnosis that would worry people? A population-based survey of adults in England. *BMC Cancer* 18 (1): 86.

Nixon, R. 1971. Annual message to the congress on the state of the union. In *Public papers of the Presidents of the United States*. Washington, DC: Office of the Federal Register, National Archives and Records Administration.

Oran, B., and D.J. Weisdorf. 2012. Survival for older patients with acute myeloid leukemia: A population-based study. *Haematologica* 97 (12): 1916–1924.

Paguirigan, A.L., J. Smith, S. Meshinchi, M. Carroll, C. Maley, and J.P. Radich. 2015. Single-cell genotyping demonstrates complex clonal diversity in acute myeloid leukemia. *Science Translational Medicine* 7 (281): 281re282.

Potter, N., F. Miraki-Moud, L. Ermini, I. Titley, G. Vijayaraghavan, E. Papaemmanuil, P. Campbell, J. Gribben, D. Taussig, and M. Greaves. 2018. Single cell analysis of clonal architecture in acute myeloid leukaemia. *Leukemia* 33 (5): 1113–1123.

Renneville, A., C. Roumier, V. Biggio, O. Nibourel, N. Boissel, P. Fenaux, and C. Preudhomme. 2008. Cooperating gene mutations in acute myeloid leukemia: A review of the literature. *Leukemia* 22 (5): 915–931.

Rosenbaum, L. 2017. Tragedy, perseverance, and chance – The story of CAR-T therapy. *The New England Journal of Medicine* 377 (14): 1313–1315.

Rous, P. 1967. The challenge to man of the neoplastic cell. *Cancer Research* 27 (11): 1919–1924.

Sato, T., X. Yang, S. Knapper, P. White, B.D. Smith, S. Galkin, D. Small, A. Burnett, and M. Levis. 2011. FLT3 ligand impedes the efficacy of FLT3 inhibitors in vitro and in vivo. *Blood* 117 (12): 3286–3293.

Schei, E., and E. Strand. 2015. Love life or fear death? Cartesian dreams and awakenings. In *Science, philosophy and sustainability: The end of the Cartesian dream*, ed. A.G. Pereira and S. Funtowicz, 45–58. London/New York: Routledge.

Sekeres, M.A., R.M. Stone, D. Zahrieh, D. Neuberg, V. Morrison, D.J. De Angelo, I. Galinsky, and S.J. Lee. 2004. Decision-making and quality of life in older adults with acute myeloid leukemia or advanced myelodysplastic syndrome. *Leukemia* 18 (4): 809–816.

Shelley, M. 1993. *Frankenstein; or, The Modern Prometheus*. Published online by the Project Gutenberg https://www.gutenberg.org/files/84/84-h/84-h.htm.

Shlush, L.I., A. Mitchell, L. Heisler, S. Abelson, S.W.K. Ng, A. Trotman-Grant, J.J.F. Medeiros, et al. 2017. Tracing the origins of relapse in acute myeloid leukaemia to stem cells. *Nature* 547 (7661): 104–108.

Short, N.J., M.E. Rytting, and J.E. Cortes. 2018. Acute myeloid leukaemia. *Lancet* 392 (10147): 593–606.

Singh, A.K., and J.P. McGuirk. 2016. Allogeneic stem cell transplantation: A historical and scientific overview. *Cancer Research* 76 (22): 6445–6451.

Slamon, D.J., B. Leyland-Jones, S. Shak, H. Fuchs, V. Paton, A. Bajamonde, T. Fleming, et al. 2001. Use of chemotherapy plus a monoclonal antibody against HER2 for metastatic breast cancer that overexpresses HER2. *The New England Journal of Medicine* 344 (11): 783–792.

Sonnenschein, C., and A.M. Soto. 2000. Somatic mutation theory of carcinogenesis: Why it should be dropped and replaced. *Molecular Carcinogenesis* 29 (4): 205–211.

Styczynski, J., G. Tridello, L. Koster, S. Iacobelli, A. van Biezen, S. van der Werf, M. Mikulska, et al. 2020. Death after hematopoietic stem cell transplantation: Changes over calendar year time, infections and associated factors. *Bone Marrow Transplantation* 55 (1): 126–136.

Sung, P.J., M. Sugita, H. Koblish, A.E. Perl, and M. Carroll. 2019. Hematopoietic cytokines mediate resistance to targeted therapy in FLT3-ITD acute myeloid leukemia. *Blood Advances* 3 (7): 1061–1072.

Talati, C., and K. Sweet. 2018. Recently approved therapies in acute myeloid leukemia: A complex treatment landscape. *Leukemia Research* 73: 58–66.

Tomaszewski, E.L., C.E. Fickley, L. Maddux, R. Krupnick, E. Bahceci, J. Paty, and F. van Nooten. 2016. The patient perspective on living with acute myeloid leukemia. *Oncology and Therapy* 4 (2): 225–238.

Tyner, J.W., C.E. Tognon, D. Bottomly, B. Wilmot, S.E. Kurtz, S.L. Savage, N. Long, et al. 2018. Functional genomic landscape of acute myeloid leukaemia. *Nature* 562 (7728): 526–531.

Lao-Tzu. 600 BC/2017. *Tao Te Ching*. Trans. Stephen Addiss and Stanley Lombardo. Boston/London: Shambhala.

van Galen, P., V. Hovestadt, M.H. Wadsworth Ii, T.K. Hughes, G.K. Griffin, S. Battaglia, J.A. Verga, et al. 2019. Single-cell RNA-Seq reveals AML hierarchies relevant to disease progression and immunity. *Cell* 176 (6): 1265–1281.

Vick, B., M. Rothenberg, N. Sandhofer, M. Carlet, C. Finkenzeller, C. Krupka, M. Grunert, et al. 2015. An advanced preclinical mouse model for acute myeloid leukemia using patients' cells of various genetic subgroups and in vivo bioluminescence imaging. *PLoS One* 10 (3): e0120925.

Vrinten, C., C.H. van Jaarsveld, J. Waller, C. von Wagner, and J. Wardle. 2014. The structure and demographic correlates of cancer fear. *BMC Cancer* 14: 597.

Vrinten, C., L.M. McGregor, M. Heinrich, C. von Wagner, J. Waller, J. Wardle, and G.B. Black. 2016. What do people fear about cancer? A systematic review and meta-synthesis of cancer fears in the general population. *Psychooncology* 26 (8): 1070–1079.

Wang, K., M. Sanchez-Martin, X. Wang, K.M. Knapp, R. Koche, L. Vu, M.K. Nahas, et al. 2017. Patient-derived xenotransplants can recapitulate the genetic driver landscape of acute leukemias. *Leukemia* 31 (1): 151–158.

Welch, J.S., T.J. Ley, D.C. Link, C.A. Miller, D.E. Larson, D.C. Koboldt, L.D. Wartman, et al. 2012. The origin and evolution of mutations in acute myeloid leukemia. *Cell* 150 (2): 264–278.

Yang, X., A. Sexauer, and M. Levis. 2014. Bone marrow stroma-mediated resistance to FLT3 inhibitors in FLT3-ITD AML is mediated by persistent activation of extracellular regulated kinase. *British Journal of Haematology* 164 (1): 61–72.

Yeung, C.C.S., and J. Radich. 2017. Predicting chemotherapy resistance in AML. *Current Hematologic Malignancy Reports* 12 (6): 530–536.

Zhang, H., S. Savage, A.R. Schultz, D. Bottomly, L. White, E. Segerdell, B. Wilmot, et al. 2019. Clinical resistance to crenolanib in acute myeloid leukemia due to diverse molecular mechanisms. *Nature Communications* 10 (1): 244.

Zhuangzi. 2003. *Basic writings*. Trans. Burton Watson. New York: Columbia University Press.

Conclusions: The Biomarkers That Could Be Born

Roger Strand and Anne Bremer

> *La crisi consiste appunto nel fatto che il vecchio muore e il nuovo non può nascere: in questo interregno si verificano i fenomeni morbosi più svariati.*
>
> *The crisis consists precisely in the fact that the old is dying and the new cannot be born: in this interregnum a great variety of morbid symptoms occur.* (Gramsci, 1930, Prison Notebooks)

The Biomarkers That Cannot Be Born: The Unsustainable Political Economy of Cancer Research

Having read more than a dozen chapters about cancer biomarkers and precision oncology, the readers of this book are likely to have made up their own minds about the issues at stake and the matters of concern. We expect and hope that the readers' thoughts differ from ours, to be presented in this final little chapter, and we furthermore hope that such differences can be a point of departure for new engagements with the fascinating topic of the future of cancer research and cancer medicine. We have called this part "Conclusions" but only to signify the finalisation of this book project and not at all as the final word in the debates into which it enters.

Biomedical literature *within* cancer biomarker research serves the function of organising the field, delineating and structuring what is known and highlighting promising avenues for further scientific work. This book is *about* cancer biomarker research as culture and practice, and precision oncology as a sociotechnical imaginary that inhabits and surrounds that culture and practice. As such it serves a quite different purpose than that of normal science: our task has been to open up the black

R. Strand (✉) · A. Bremer
Centre for Cancer Biomarkers, Centre for the Study of the Sciences and the Humanities, University of Bergen, Bergen, Norway
e-mail: roger.strand@uib.no; anne.bremer@uib.no

A. Bremer, R. Strand (eds.), *Precision Oncology and Cancer Biomarkers*, Human Perspectives in Health Sciences and Technology 5,
https://doi.org/10.1007/978-3-030-92612-0_16

boxes of the field and display complexities, tensions and paradoxes. This task does not help to streamline biomedical research or make it more efficient. As we argued in the Preface and Introduction, however, such an exercise may expose unsustainable realities and imaginaries and subject them to a broader debate that, with luck and with time, might make them better aligned with ethical, social and political values and in this sense become more sustainable and *responsible,* leading to *responsible research and innovation.*

Indeed, our first conclusion is that several of our chapters have shown various aspects of the *unsustainable political economy of cancer research.* There are very strong arguments in favour of the vision of personalised cancer medicine, in the sense of a medicine that successfully provides the right drug for the right person at the right time and the right dose to maximise clinical benefit and minimise harmful side-effects. Biomarkers play a key role in that vision. Yet, as discussed in the chapter by Bremer et al., very few biomarkers identified in the laboratory actually make it to the clinic and to the market. In this sense, some biomarkers cannot be born for biological reasons. Biomarkers are a means to account for and manage biological complexity but within limits. The story about the anti-HER2 treatment of trastuzumab indeed shows how neat and benign the complexity had to be in order for the drug and its accompanying biomarker to become such a success. The FLT3 story presented by Engen gives a quite different example in which possible biomarkers dissolve themselves in an almost fractal-like type of biological complexity.

Still, biology alone cannot fully explain why cancer biomarkers struggle to come into clinical use. Indeed, almost all chapters in this book contribute to casting light upon the *political economy of cancer research* and how the business models of cancer drugs get in the way of biomarker deployment. If we think of blockbuster drugs and “one-size-fits-all” as the old type of medicine that we would like to retire, and personalised medicine as the new to be born, the allusion to Antonio Gramsci’s famous statement (quoted above) can serve to suggest that innovation in the cancer field to some extent is in a state of crisis. The old “un-personalised medicine” is no longer fit for purpose, but still the political economy remains situated in the old by which revenues are created by economies of scale, implying that the pharmaceutical industry either has to keep selling big volumes of the same drug or increase the prices pr volume. This has led to an escalation of prices as well as total costs that indeed has created “a great variety of morbid symptoms”. These symptoms are seen in the clinic, when clinicians are not allowed to use the “right drug” because it cannot be afforded; in the innovative ways to get patients enrolled in trials and in this way receive novel treatments (see the chapter by Hillersdal & Svendsen); in the priority-setting institutions in countries with public health services (see chapters by Cairns, Fleck, Kang and Tranvåg & Strand); in the media (see Stenmarck & Nilsen), and in scientific practices where striving for hyper-precision is conducive to publication bias (see the chapter by Lie Lotsberg & D’mello Peters).

The response within cancer research as in any field of biomedicine has largely been to try to alleviate and overcome these morbid symptoms by incremental progress. This book was written in 2020–2021, after half a century that at least to us contemporaries appeared as marked by neoliberal ideologies and policies, and even more so after the end of the Cold War and the apparent victory of free-market capitalism at the entry of the 1990s. In this sense, it was not a period in which academic researchers were encouraged to challenge capitalism and the profit maximisation of big pharma. Rather, they were encouraged to play by the rules and engage in so-called triple helix interactions with private enterprises and governmental support that ultimately played into and contributed to support the political economy that was the old that arguably was dying. An additional driver for the *status quo* was the governmentality (*sensu* Michel Foucault) that disciplined citizens into accepting the political and economic structures as given and instead criticising and improving *themselves* and their own work effort to alleviate the morbid symptoms. And in the orchestration of the sciences, medicine – perhaps partly due to its orientation towards problem-solving at the individual level and partly due to its history - was definitely not the science with the strongest presence of structural critique and political protest. To give an example, the chapter by Tranvåg and Strand argues that the morbid symptoms in priority-setting institutions cannot be solved by incremental improvements but call for more radical change. This type of structural critique, however, is almost absent in the relevant academic fields of medical ethics and sociology of medicine.

So the first issue at stake is whether a biomarker-based personalised cancer medicine can be born at all, and the first matter of concern is the political economy of medical research and innovation that impedes it.

If we continue to liken this situation with Gramsci's notion of the crisis, it is both similar and different. What Gramsci had in mind was the type of social and political crisis in which the ruling class continues to stay in power by sheer dominance, but where the ruling ideology is no longer credible or legitimate among the subjects of that ruling class. As a thinker in the Marxist tradition, Gramsci conceptualised the crisis in material terms, as power differentials in physical force and in capital, but notably also in terms of power over cultural institutions. Developing Marxist thought and preparing the ground for Michel Foucault's concepts of power as enacted through discourse, Gramsci maintained that the bourgeoisie could oppress the working class by means of cultural hegemony.

Almost a century after Gramsci's reflections during imprisonment, and 30 years after the end of the Cold War, Marxist critique has become so rare, even in academe, that it almost feels idiosyncratic to point out the power differential between big, rich, globalised pharma on one hand and increasingly disempowered nation states on the other, to the extent that big pharma in some cases dictate states to keep their drug prices secret from the public. We shall refrain from suggesting how these power differentials can be overcome or even believing that there could be a quick fix of the pharmaceutical sector of capitalist societies.

Still, as scholars we are entitled to imagine what personalised cancer medicine might look like *if* its old political economy finally dies and gives rise to the new. Knowing that the link between market cost and product development cost of medical innovations is spurious at best, it makes perfect sense to ask: What if the price problem was solved, and the drugs and biomarkers actually became affordable? What if our friends in cancer research were provided with the resources they needed, and the outputs and outcomes of their research remained in the public domain, available to all? What if the toolbox of drugs and biomarkers became so big and well developed that by the time its tools went off license, there was no further need for new and therefore astronomically expensive drugs?

In this post-capitalist dream scenario, one could definitely imagine new clinical benefits produced and old harmful side effects being prevented. One could imagine more QALYs – quality-adjusted life years – for the patients, QALYs that at present are unavailable. And one could imagine not only more utility (for which QALY is a much used currency) but also other values, such as increased autonomy. Patients would perhaps be able to choose therapies better tailored to their own person, not only to maximise the standardised and abstract measures of overall survival, QALYs or "progression-free survival", but also taking into account their own capabilities, dispositions, needs and desires. One could certainly expect some good prospects there, but still few miracles and silver bullets, because cancer is a system disease that often is complex, heterogenous and adaptive even within one and the same patient, as explained in the chapters by Bremer et al., Engen, Gissum and Strand & Engen. If our relationship to cancer is war, the scenario would not necessarily be that of a quick surrender on the part of the disease.

Precision Oncology: Cartesian Dream or Nightmare?

And then again, to politicians from Richard Nixon to Barack Obama, to Nobel laureates such as Peyton Rous, and to citizens and patient organisations and even the communications department of the editors' own university, cancer must be fought by warfare and surrender must be demanded, with the greatest ferocity and persistence and at any financial cost and also, metaphorically and literally, accepting loss of civilians in the form of collateral harm in patients and beyond (see chapters by Dillekås and Strand & Engen). It is in this context that the imaginary of precision oncology emerged as the promise of making cancer research into exact science. If exact science could create rockets, dynamite, mustard gas and even the nuclear bomb, why shouldn't it be able to kill something as lowly as abnormal proliferating cells?

As explained by Engen (chapter "Introduction to the Imaginary of Precision Oncology"), the sociotechnical imaginary of precision oncology has come to influence and to some extent replace the earlier and less extreme imaginary of personalised medicine. The second issue at stake that we wish to highlight in this final chapter, is the role of the imaginary of precision oncology in cancer research. The main matter of concern is of the normative type: What does that imaginary entail, and would its realisation be a desirable development?

The chapter by Strand and Chu performed a formal and foundational examination of that claim and arrived at the general, abstract conclusion that in order for cancer medicine to become exact, the object of study has to change. Specifically, in order for exact science to work, its subject matter has to be tamed so as to not exert certain forms of complexity and indeterminacy; it has to "cross the Styx" and lose some of its features of being fully alive, as materially and semiotically open and evolving systems. The type of self-discipline and self-monitoring that we above connected to governmentality in neoliberal societies would be but a mere beginning of that journey towards Hades. For humans to transform themselves into proper subjects of exact science, they would have to engage in *biological* self-monitoring and discipline of unheard dimensions. This would not be a venture of a tinkering, personalised, patient-centred medicine where therapeutic strategies would be negotiated against individual needs and desires. *All* variables, inputs as well as outputs, would have to be precisely measurable and strictly defined. Worse, the type of knowledge produced would likely escape human sense-making faculties, leaving therapeutic trade-offs and other value-laden decisions to algorithms or artificial intelligences. We would all become cyborgs.

Fortunately, one might say, the rigorous vision of precision oncology as an exact science seems unlikely to be realised on philosophical grounds. Still, the imaginary in its less precise manifestations gives direction to actual cancer research and practice, as illustrated in several of the chapters (Gissum, Bremer et al., Engen, Strand & Engen, Hillersdal and Svendsen). In an ever more integrated and hybrid practice of cancer research-cancer treatment, scientific goals are pursued with patients also being the means and where measurable outcomes such as progression-free survival take on more meaning than just a scientific variable. By stating this observation, we do not intend to claim that the development goes against the will of the patients or that they are not being listened to. Rather, from the co-production perspective one can witness how scientific knowledge and medical technology advances together with the fears, hopes and desires of patients, scientists and health workers, giving ever more prominence to the delay of death and creating a culture in which the good cancer patients are those who subject themselves to clinical trials and do everything to fight, resist and refuse to surrender to their horrific enemy. This cultural framing effectively implies that cancer patients with incurable disease are bound to end up as losers, and their end a failure. It also implies a cultural *and* scientific encouragement to try every treatment and innovation to enact the proper fighting spirit and induce hope without necessarily asking if the treatment results in less suffering rather than more. In this way the Cartesian dream of perfect knowledge and total control, including over death, risks becoming a nightmare even as the science progresses.

From Precision to Personalized Oncology: Biomarkers for Dignity?

Returning to the Gramscian trope, it is accordingly possible to view the imaginary of precision oncology as a morbid symptom of another complicated generational shift that Descartes himself managed to undertake within his own lifetime, at least if we are to believe his own words: The shift from trying to conquer death to accepting it and living one's life in that acceptance (Strand 2021).

While reflecting on this alternative rendering of the crisis in the light of the present anthology on cancer biomarkers and the imaginary of precision oncology, we cannot avoid taking into account the chapter that was not part of the text but in which we have lived during this book project, namely the 2020–2021 COVID-19 epidemic.

Comments and interpretations around COVID-19 should not be made lightly, and at least not until the most acute phase of the epidemic is over. Nevertheless, it has been impossible for us not to use it as an additional lens to understanding the cultural and political relationship to illness, disease and death in our society. By "our society" we should be careful to point out that both editors, Bremer and Strand, live and work in Norway. We are aware that the trajectories of COVID-19 at the time of writing, March 2021, both in terms of disease and societal measures against it, vastly differ on different continents and even neighbouring countries.

In the case of Norway, however, what has been striking is the strict and disciplined compliance with which its population has followed the governmental measures to control the spread of the Sars-CoV-2 virus, and the strong degree of public support in the governmental policies. Without presenting empirical material as evidence to that claim, we have also lived the latter 13 months within a public sphere characterised by journalists', politicians' and citizens' expressions of fear of the disease and an intense if not obsessive focus on the statistics of infection rates, COVID-19 case numbers and numbers of COVID-19 related deaths. At the same time, due to the fortunate state of the Norwegian healthcare system and our affluent welfare state, even by the worst-case scenarios of Norwegian public health authorities the epidemic would not have given rise to any large increase in mortality in, say, a two- or three-year span even without vaccines and without strict measures. Unlike certain other parts of the world, the mortality of COVID-19 in Norway and Northern Europe is almost exclusively a phenomenon among the 70+ age group. Even with an uncontrolled spread of the disease, the health loss would have been much less than the health gain and the increase in life expectancy the latter 20–30 years.

So in terms of "rational" priority-setting, maximising the quality-adjusted lifetime of the population, the COVID-19 measures in our country are likely to be the most expensive QALYs in history, given not only the economic costs but also the collateral health loss from long-term unemployment, isolation, disrupted education et cetera.

We do not tell this story in order to criticise Norwegian authorities or citizens. If that were our aim, first of all we would have had to present rigorous evidence, and secondly it would not have been the right book for it. Rather, our aim is to use these observations to increase the scope of interpretation of what is at stake in the crisis of which precision oncology can be seen as morbid symptom.

One interpretation is that precision oncology and the Norwegian fear of COVID-19 are both symptoms of a culture characterised by the denial of the inevitability of death. In Norway, this cultural interpretation is above all due to the work of the philosopher Arne Johan Vetlesen (2009). If we were to paint with a broad brush, we could engage in the narrative of how the modern and secularised European subject has taken the shape of an atomised individual ego who finds incommensurately little value in everything but his (or even her) own persistent existence: Me, me, me. For this subject, death is not only unbearable: it is unacceptable. It should be prevented and delayed at any cost. The technological fantasies of immortality by deep-freezing the body or uploading the mind into the digital cloud belong to this realm; the insistence that all social activity must be controlled in order to control the virus, and the desire to transform medicine into exact science could be interpreted along the same lines. And from the co-production perspective, it would be easy to make the argument that modern Western biomedicine is closely co-produced with this subject. In terms of sustainability, such a trajectory seems to have its final destination in ecological collapse. Life on this planet cannot simply be humans.

Still, this is obviously not the only interpretation. As the book moves towards its end, we would like to propose another line of thought that nurtures more hope about a sustainable and responsible trajectory for modern society, modern medicine, cancer research and even down to cancer biomarkers.

If a single tumour can be heterogenous, then certainly a society and a culture is, and a single human being is capable of holding a rich variety of thoughts, attitudes and emotions. Perhaps Vetlesen's diagnosis about the denial of death is true – to some extent. This does not preclude that individuals and collectives still know perfectly well of their own mortality and frailty. In some of the chapters of this book patients appeared, with their fears, hopes and their suffering. Is it so certain that they have bought into the rationality of the QALYs, or does the prospect of a new drug represent something else – perhaps a distraction, a sense of meaning, a comforting thought that one is being taken care of, or the delay of the fear of the cruelty of the final phase of the disease?

At the same time, with a side glance to the Norwegian COVID-19 debates, the imagery that was most often invoked, was that of the Italian hospitals Spring 2020. It was an imagery not of the sheer number of deaths – they were still vastly outnumbered by the annual number of children who die from malaria – but of chaos, of desperately tired doctors who had lost control in the hospital, of patients lying unattended in the corridor as they perished. It was disordered, undignified death.

Studies that take a critical view on biomedicine (e.g. philosophy, sociology or nursing science) sometimes describe a tension between, on one hand, the biomedical focus on disease and hard endpoints and the utilitarian focus on the QALYs, and on the other, "softer" human values such as integrity, dignity and of sense of meaning and coherence.

We see no reason why personalised cancer medicine could not incorporate these human values. We wrote above that biomarkers could be developed to support a personalised medicine designed to tailor therapies according to the patient's own capabilities, dispositions, needs and desires. More generally, they could be developed to tailor therapies aiming not only to extend life and alleviate disease symptoms but to *reduce suffering*, which involves much more than pain and a reduced health state. To give just one example, one of our clinician friends in the CCBIO, the Centre for Cancer Biomarkers that is the source environment for this book, was asked what he thought is the most important for his (very ill) patients. His immediate reply was "a swift exit", that is, to know that if I do not get better, at least the course of death will not be unnecessary cruel. The division of labour as of today is to leave that concern to palliative care; one could very well imagine tailored therapy that is designed for the benefit of the entire course of disease from the beginning.

Such ideas may well not be easy to operationalise and implement. Indeed, we promised that this book will not make cancer biomarker research more streamlined or efficient in the short run. But we have not written this book for its thoughts to be implemented into practice in 2022, or even 2030. We have written it for 2100 and 2200. And while the old is dying, the new, in order to be born, has to be conceived and gestated in order to nourish and develop into existence.

References

Gramsci, A. [1930] 1992. *Prison Notebooks*. New York: Columbia University Press.

Strand, R. 2021. The impact of a fantasy. In *Personalized Medicine in the Making. Philosophical Perspectives from Biology to Healthcare*, ed. C. Beneduce and M. Bertolaso. Dordrecht: Springer.

Vetlesen, A.J. 2009. *A Philosophy of Pain*. London: Reaktion Books.

Index

A. Bremer, R. Strand (eds.), *Precision Oncology and Cancer Biomarkers*, Human Perspectives in Health Sciences and Technology 5,
https://doi.org/10.1007/978-3-030-92612-0

Printed in the United States
by Baker & Taylor Publisher Services